INNOVATION AND APPLICATION OF ENGINEERING TECHNOLOGY

PROCEEDINGS OF THE 1ST INTERNATIONAL SYMPOSIUM ON ENGINEERING TECHNOLOGY AND APPLICATION (ISETA 2017), MONTREAL, CANADA, 25–28 MAY 2017

Innovation and Application of Engineering Technology

Editors

Avon Pekano Oscar, Wenxue Chen & Shuren Wang

Concordia University, Québec, Canada
Henan Polytechnic University, Jiaozuo Shi, Henan Sheng, China

CRC Press
Taylor & Francis Group
Boca Raton London New York Leiden

CRC Press is an imprint of the
Taylor & Francis Group, an **informa** business
A BALKEMA BOOK

CRC Press/Balkema is an imprint of the Taylor & Francis Group, an informa business

Typeset by V Publishing Solutions Pvt Ltd., Chennai, India

Published by: CRC Press/Balkema
Schipholweg 107C, 2316 XC Leiden, The Netherlands
e-mail: Pub.NL@taylorandfrancis.com
www.crcpress.com – www.taylorandfrancis.com

ISBN: 978-1-138-05057-0 (Hbk)
ISBN: 978-1-315-16659-9 (eBook)

Innovation and Application of Engineering Technology – Oscar, Chen & Wang (Eds)
© 2017 Taylor & Francis Group, London, ISBN 978-1-138-05057-0

Table of contents

Innovation and Application of Engineering Technology – Oscar, Chen & Wang (Eds)
© 2017 Taylor & Francis Group, London, ISBN 978-1-138-05057-0

Preface

This publication holds the Proceedings of the 1st International Symposium of Engineering Technology and Application (ISETA2017) held from 25–28 May, 2017, in Montreal, Canada. ISETA 2017 was organized by the CU and IJSS.

The main theme of the Symposium was innovation and application of engineering technology. The papers in this publication are selected according to the sessions in which they were presented:

- ✓ Civil and Environmental Engineering
- ✓ Rock Mechanics and Mining Engineering
- ✓ Mechanical Engineering
- ✓ Geophysics
- ✓ Risk and Occupational Health and Management

The Symposium was comprised of Plenary Sessions, Parallel Sessions, Field Visits, a Technical Exhibition and Pre- and Post-Symposium Tours. An exhibition of products, equipment and software was held during the Symposium.

We would like to express our sincere gratitude to CU, IJRLHUSD, HPU and IJSS for the privilege given to us for organizing this Symposium. Thanks are due to all the members of the ISETA2017 Conference Advisory Committee for their guidance through the various stages of development to the successful completion of the Symposium. We also express our deep appreciation to the paper reviewers and co-reviewers. Their untiring efforts are gratefully acknowledged. The Symposium proceedings have been published by CRC Press/Balkema. Their efficient and high-quality work is appreciated.

Innovation and Application of Engineering Technology – Oscar, Chen & Wang (Eds)
© 2017 Taylor & Francis Group, London, ISBN 978-1-138-05057-0

Organizing and advisory committee

A.P. Oscar, *Canada*
W.X. Chen, *Canada*
S.R. Wang, *China*
Apurna Ghosh, *Australia*
X.Q. He, *China*
X.L. Liu, *China*
D.D. Tripathi, *India*
Min An, *UK*
M.J. Liu, *China*
Eunhye Kim, *USA*
M.G. Abdula, *Canada*
Malisa Wnika, *Canada*
Lisa Joseph, *Canada*
Q.G. Cao, *China*
Rapik Saat, *USA*
Jezan Md Diah, *Malaysia*
Raja Rizwan Hussain, *Saudi Arabia*
Y.W. Liu, *China*
M.R. Saharan, *India*
Frederick D. Hansen, *USA*
Ahmad Qasaimeh, *USA*
Radian G. Belu, *USA*
L.M. Marwax, *France*

Innovation and Application of Engineering Technology – Oscar, Chen & Wang (Eds)
© 2017 Taylor & Francis Group, London, ISBN 978-1-138-05057-0

Sponsors

Concordia University, International Joint Research Laboratory of Henan Province for Underground Space Development, Henan Polytechnic University and IJSS.

Innovation and Application of Engineering Technology – Oscar, Chen & Wang (Eds)
© 2017 Taylor & Francis Group, London, ISBN 978-1-138-05057-0

Practical model of 3D free-surface flow

Suhai Kang
Key Laboratory of Engineering Sediment of Communications Industry, Tianjin Research Institute for Water Transport Engineering of Ministry of Communications and Transportation, Tianjin, China

ABSTRACT: In this study, the 3D hydrodynamic numerical model is established on the basis of the Reynolds equation and the turbulent flow equation. We use hydrodynamic pressure with discrete model equations in combination with finite volume and finite difference. The hierarchical grid system is suitable and efficient with a planar unstructured grid. We examine the developed 3D model with the help of different examples of analytic solution and measured data, and inspect the capability of the model to handle complex boundaries and the practicability of the flow-field simulation program in combination with practical engineering. Both the examples and practical engineering reveal the accuracy of model calculation results.

1 INTRODUCTION

In the field of water transportation engineering, most of the hydrodynamic studies are conducted on the flow of fluid with free surface, such as the flow of water in rivers, lakes, and seas. It is necessary to build models for prediction research to solve engineering problems and predict the impacts on the local hydrodynamic force resulting from port construction, channel excavation, breakwater construction, jetty construction, and so on. Among other models, the numerical model is an important technological means, which is widely applied thanks to its advantages of low cost and short cycle, as it neither occupies the test site nor produces scale effect (Zhang Huaqing 1998, Jing Zhongqiang 1989, Wang Pingyi 2004 and Ying Qiang et al. 2004). In the past, the field of water transportation engineering has been focused using only one- and two-dimensional numerical models because of the limitations in computer capacity and calculation format. These numerical models could not achieve the distribution of horizontal velocity along the water depth direction or simulate the distribution of flow velocity in the vertical direction. In particular, two-dimensional models could not meet engineering requirements in local flow problems (such as flow field around spur dike and pier antiwater area). In order to overcome these drawbacks, a 3D numerical model is adopted for research (Shen Bo 1997, Zhao Shiqiang 1989, Rong Xuewen 2003, Kang Suhai 2011, Wang Jianjun 2007, Tan Weiyan 1997 and Jiang Changbo 1999).

The difficulties in solving the aforementioned problems using a 3D flow simulation numerical model mainly lie in the fact that it has no state equation due to the incompressibility of water; consequently, it is necessary to make a hydrostatic hypothesis under certain circumstances, otherwise it would be incredibly time-consuming to find the iteration solution to the pressure term. The amount of calculation was unacceptable for the general engineering application in the past due to the calculation format and computer performance.

Early stage 3D models adopted the layered 3D structure on the basis of hydrostatic hypothesis, that is, the 3D shallow water model, where the pressure term is replaced by the hydrostatic solution formula to avoid finding the iteration solution, reduce the overall amount of calculation required for the model significantly, and thus enable the application of the 3D shallow water model in salinity intrusion, thermal discharge, and other engineering fields (Zhou Guanlun et al. 2004, Cheng Changhua et al. 2001, Wang Fujun 2004, Zhang Baolin et al. 1994, Zhou Yilin et al. 2004 and 2001). However, the formation of the hydrodynamic pressure model equation shall be taken into consideration to find the solution in the study of certain engineering problems, such as the short-wave-driven water flow problems, high-density-gradient stratified flow, and water flow problems in terrain with great changes, for which the hydrostatic hypothesis is no longer reasonable as it is not gradually varied flow.

The study on the 3D numerical model applying hydrodynamic pressure has made some achievements. Casulli established a nonhydrostatic 3D model by using a semi-implicit control volume method; Mahadevan and others developed the mesoscale marine hydrodynamic calculation model of hydrodynamic pressure by using the finite-difference method; Jankowski established a nonhydrostatic 3D numerical model by adding a nonhydrostatic pressure term in the finite-element model; Koçyigit and others developed a nonhydrostatic model via the vertical depth using the discrete-ordinates method by applying methods similar to those used by Casulli and Mahadevan (Cui Zhanfeng 2006 and Peng Jing et al. 2002). All these nonhydrostatic models adopt the operator-splitting technique, that is, split the momentum equation into several differential operators according to their respective physical processes in requesting solutions; the original momentum equation is split into several simple equations, each containing only the relevant items in the physical process. In this case, the original momentum equation is taken as a combination of simple equations, one of which considers the influence of nonhydrostatics. Later, some scholars like Li and Fleming adopted ordinate system in the vertical direction, split the convective and diffusive terms from the pressure terms using the projection method, and established a 3D nonhydrostatic numerical model using the convective and diffusive terms via the MacCormack implicit algorithm. The model format is provided with accurate second-order time and space, which is applicable to 3D wave calculation and able to do perform well in modeling even high-order waves. Young applied a high-precision difference discrete to the pressure gradient term using the transformation theory of coordinate and established a vertical 2D nonhydrostatic numerical model of free-surface flow, which produces good results in the calculation of freak wave. Under the coordinate system, the whole computational domain is evenly distributed between 0 and –1; numerical discretization and computation may be performed within a fixed cassette area; the computational domain may be divided into the same number of layers in the vertical direction, making it convenient for numerical discretization.

In line with the development of 3D hydrodynamic numerical simulation techniques, requirements are raised for the post-treatment of 3D computational results, especially the analogue simulation of the 3D flow field. At present, research on the numerical simulation and visualization of 3D flow field worldwide basically focus on displaying 3D flow field data with sectional flow field, mainly due to the limitation that 2D display devices cannot display 3D scene immediately. Although the 3D flow field may be revealed through such method, the disadvantage thereof lies in the indirect display approach. Therefore, it seems pretty urgent to work out a solution to 3D flow-field simulation.

2 ESTABLISHMENT OF A 3D NUMERICAL MODEL

The 3D numerical model established by taking the Reynolds equation as the basic control equation reserves the vertical velocity component without performing vertical integration; it is able to reflect the characteristics of 3D flow and closes the model equation by the turbulent flow equation:

$$\frac{\partial u}{\partial x}+\frac{\partial v}{\partial y}+\frac{\partial w}{\partial z}=0 \tag{1}$$

$$\frac{\partial u}{\partial t}+u\frac{\partial u}{\partial x}+v\frac{\partial u}{\partial y}+w\frac{\partial u}{\partial z}=-\frac{1}{\rho}\frac{\partial p}{\partial x}+\nu^h\left(\frac{\partial^2 u}{\partial x^2}+\frac{\partial^2 u}{\partial y^2}\right)+\frac{\partial}{\partial z}\left(\nu^v\frac{\partial u}{\partial z}\right)+f_c v \tag{2}$$

$$\frac{\partial v}{\partial t}+u\frac{\partial v}{\partial x}+v\frac{\partial v}{\partial y}+w\frac{\partial v}{\partial z}=-\frac{1}{\rho}\frac{\partial p}{\partial y}+\nu^h\left(\frac{\partial^2 v}{\partial x^2}+\frac{\partial^2 v}{\partial y^2}\right)+\frac{\partial}{\partial z}\left(\nu^v\frac{\partial v}{\partial z}\right)-f_c u \tag{3}$$

$$\frac{\partial w}{\partial t}+u\frac{\partial w}{\partial x}+v\frac{\partial w}{\partial y}+w\frac{\partial w}{\partial z}=-\frac{1}{\rho}\frac{\partial p}{\partial z}+\nu^h\left(\frac{\partial^2 w}{\partial x^2}+\frac{\partial^2 w}{\partial y^2}\right)+\frac{\partial}{\partial z}\left(\nu^v\frac{\partial w}{\partial z}\right)-g \tag{4}$$

$$\frac{Dk}{Dt}-\nabla\left[\frac{\nu_t}{\sigma_k}\nabla k\right]=c_\mu\frac{k^2}{\varepsilon}G-\varepsilon \tag{5}$$

$$\frac{D\varepsilon}{Dt}-\nabla\left[\frac{\nu_t}{\sigma_\varepsilon}\nabla\varepsilon\right]=c_1\frac{\varepsilon}{k}G-c_2\frac{\varepsilon^2}{k} \tag{6}$$

In the formulas above, u, v, and w represent the components of velocity vector along the three coordinate axes x, y and z, respectively; p is the pressure; $f_c = 2\omega\sin\varphi$ is the coefficient of the Coriolis force (dimension s-1), in which φ is the dimension and ω is the rotational angular velocity of the earth; g is the acceleration due to gravity; ν^h and ν^v are horizontal and vertical viscosity coefficients, respectively; ρ is density; k is turbulent kinetic energy; ε is the dissipation rate of the turbulent kinetic energy; and G is the production term of turbulent kinetic energy, which can be expressed as: $G=\left(\partial u_i/\partial x_j+\partial u_j/\partial x_i\right)\left(\partial u_i/\partial x_j\right)$.

3 EQUATION DISCRETIZATION

Under normal conditions, the calculation under hydrodynamic pressure is carried out in two steps: first, hydrostatic calculation, that is, calculation of temporary variables $\tilde{u},\tilde{v},\tilde{w}$, and water level $\tilde{\eta}$, ignoring the implicit nonhydrostatic term in the momentum equation; second, nonhydrostatic corrective calculation, that is, calculation of the nonhydrostatic term and correction of the temporary variables $\tilde{u},\tilde{v},\tilde{w}$, and water level $\tilde{\eta}$ so as to obtain the final results.

3.1 *Hydrostatic calculation*

We perform flux calculation under the local coordinates of various interfaces of tri-prism finite control volumes during discretization in the first step by making use of the rotation invariance of equation; all physical quantities have the same discrete version. We apply semi-implicit finite differential discrete to the water-level gradient term in the horizontal momentum equation, implicit method to vertical viscosity term and explicit construction to other terms, such as the convective term, the Coriolis force term, and the horizontal viscosity term, ignoring the implicit nonhydrostatic term. We then get:

$$\begin{aligned}\tilde{u}^{n+1}_{m^j_{i,k}} &= F(u^n_{m^j_{i,k}})-g\frac{\Delta t}{\Delta\xi_{ms^j_i}}\left[\theta\left(\eta^{n+1}_{ms^{j(2)}_i}-\eta^{n+1}_{ms^{j(1)}_i}\right)+(1-\theta)\left(\eta^{n}_{ms^{j(2)}_i}-\eta^{n}_{ms^{j(1)}_i}\right)\right]\\ &+\frac{\Delta t}{(h^n_{ms^j_i}\Delta\sigma^n_{mz_i})^2}\left[\nu^v_{ml^j_{i,k-1/2}}\left(\tilde{u}^{n+1}_{m^j_{i,k-1}}-\tilde{u}^{n+1}_{m^j_{i,k}}\right)-\nu^v_{ml^j_{i,k+1/2}}\left(\tilde{u}^{n+1}_{m^j_{i,k}}-\tilde{u}^{n+1}_{m^j_{i,k+1}}\right)\right]\end{aligned} \tag{7}$$

where $\tilde{u}$ is the normal flow velocity of the central interface of the vertical section of the prism, with j = 1,2,3 representing the three vertical sections; $\Delta\sigma^n_{mz_i}=h^n_{mz_i}/N_l$ represents the vertical grid spacing; $\Delta\xi_{ms^j_i}$ is the projection on the normal vertical line of the distance between the right and left centroid points on the section, where the defined point of flow velocity to be solved is; n is the current moment, with n + 1 and Δt representing the next moment to be solved and the time step, respectively; F is the explicit differential operator, including the explicit discrete of the Coriolis force term, the convective term, and the horizontal viscosity term; the value range of θ factor is 0~1; and ν^v is given by the turbulent flow model equation at moment n.

The matrix form discrete equations are obtained after reorganization by taking the free surface boundary conditions and the bottom boundary conditions into consideration:

$$A_{ij}^{n}\tilde{U}_{ij}^{n+1} = G_{ij}^{n} - \theta\frac{g\Delta t}{\Delta\xi_{ms_i^j}}\left[\eta_{ms_i^j(2)}^{n+1} - \eta_{ms_i^j(1)}^{n+1}\right] \tag{8}$$

The matrix forms of A_{ij}^{n} and G_{ij}^{n} are as follows:

$$\tilde{U}_{ij}^{n+1} = \left(\tilde{u}_{m_{i,1}^j}^{n+1}, \tilde{u}_{m_{i,2}^j}^{n+1}, \ldots\ldots, \tilde{u}_{m_{i,N_l-1}^j}^{n+1}, \tilde{u}_{m_{i,N_l}^j}^{n+1}\right)^T$$

$$A_{ij}^{n} = \begin{bmatrix} (1 + a_{ml_{i,3/2}^j} + \frac{\gamma_B^{n+1}}{(h_{ms_i^j}^n)^2\Delta\sigma_{mz_i}^n}) & -a_{ml_{i,3/2}^j} & 0 \\ -a_{ml_{i,1/2}^j} & +(1 + a_{ml_{i,1/2}^j} + a_{ml_{i,3/2}^j}) & -a_{ml_{i,3/2}^j} \\ & \cdots & \\ 0 & -a_{ml_{i,N_l-1/2}^j} & (1 + a_{ml_{i,N_l-1/2}^j} + \frac{\gamma_T^{n+1}}{(h_{ms_i^j}^n)^2\Delta\sigma_{mz_i}^n}) \end{bmatrix}$$

$$G_{ij}^{n} = \begin{bmatrix} F(u_{m_{i,1}^j}^n) - (1-\theta)\frac{g\Delta t}{\Delta\xi_{ms_i^j}}\left(\eta_{ms_i^j(2)}^n - \eta_{ms_i^j(1)}^n\right) - \theta\frac{g\Delta t}{\Delta\xi_{ms_i^j}}\left(\eta_{ms_i^j(2)}^{n+1} - \eta_{ms_i^j(1)}^{n+1}\right) \\ F(u_{m_{i,2}^j}^n) - (1-\theta)\frac{g\Delta t}{\Delta\xi_{ms_i^j}}\left(\eta_{ms_i^j(2)}^n - \eta_{ms_i^j(1)}^n\right) - \theta\frac{g\Delta t}{\Delta\xi_{ms_i^j}}\left(\eta_{ms_i^j(2)}^{n+1} - \eta_{ms_i^j(1)}^{n+1}\right) \\ \vdots \\ F(u_{m_{i,N_l}^j}^n) - (1-\theta)\frac{g\Delta t}{\Delta\xi_{ms_i^j}}\left(\eta_{m_{i,N_l}^j(2)}^n - \eta_{m_{i,N_l}^j(1)}^n\right) - \theta\frac{g\Delta t}{\Delta\xi_{ms_i^j}}\left(\eta_{m_{i,N_l}^j(2)}^{n+1} - \eta_{m_{i,N_l}^j(1)}^{n+1}\right) + \frac{\gamma_T^{n+1}}{(h_{ms_i^j}^n)^2\Delta\sigma_{mz_i}^n}W_\xi^{j\,n+1} \end{bmatrix}$$

In order to obtain water level η^{n+1}, we perform coordinate transformation to the water-level equation and obtain the water level equation under the local coordinate; we also perform finite-volume semi-implicit discrete, and finally we obtain the following equation expressed in matrix form:

$$\begin{aligned} S_{mz_i}\eta_{mz_i}^{n+1} - g\theta^2\Delta t^2\sum_{j=1}^{3}\left(\frac{\lambda_{ms_i^j}\Delta\sigma_{ms_i^j}^n}{\Delta\xi_{ms_i^j}}A_{ij}^{n-1}\left(\eta_{ms_i^j(2)}^{n+1} - \eta_{ms_i^j(1)}^{n+1}\right)\right) = \\ S_{mz_i}\eta_{mz_i}^{n} - (1-\theta)\Delta t\sum_{j=1}^{3}\left(\lambda_{ms_i^j}\Delta\sigma_{ms_i^j}^n\sum_{k=1}^{N_l}\tilde{u}_{m_{i,k}^j}^n\right) - \theta\Delta t\sum_{j=1}^{3}\left(\lambda_{ms_i^j}\Delta\sigma_{ms_i^j}^n A_{ij}^{n-1}G_{ij}^n\right) \end{aligned} \tag{9}$$

By solving equations (9) to get the water-level value at moment n + 1 and substituting this value at moment n+1 into equation (8), we can obtain the intermediate normal flow velocity $\tilde{u}$ and then the flow velocity components u and v along the x and y axes, respectively, by using equation (7) through rotational inverse transformation.

With regard to the vertical momentum equation (4), we ignore the nonhydrostatic term and discretize it using the finite differential method and express it in matrix form as follows:

$$B_{ij}^{n}\tilde{W}_{ij}^{n+1} = Q_{ij}^{n} \tag{10}$$

where $\tilde{W}_{ij}^{n+1} = \left(\tilde{w}_{m_{i,1}^j}^{n+1}, \tilde{w}_{m_{i,2}^j}^{n+1}, \ldots\ldots, \tilde{w}_{m_{i,N_l-1}^j}^{n+1}, \tilde{w}_{m_{i,N_l}^j}^{n+1}\right)^T$

$$
B_{ij}^{n}=\begin{bmatrix}
\left(1+\dfrac{\Delta t}{(h_{mz_i}^{n}\Delta\sigma_{mz_i}^{n})^{2}}\left(\nu^{v}_{ml^{j}_{i,1/2}}+\nu^{v}_{ml^{j}_{i,3/2}}\right)\right) & -\dfrac{\Delta t\,\nu^{v}_{ml^{j}_{i,3/2}}}{(h_{mz_i}^{n}\Delta\sigma_{mz_i}^{n})^{2}} & 0\\
\dfrac{\Delta t\,\nu^{v}_{ml^{j}_{i,3/2}}}{(h_{mz_i}^{n}\Delta\sigma_{mz_i}^{n})^{2}} & \left(1+\dfrac{\Delta t}{(h_{mz_i}^{n}\Delta\sigma_{mz_i}^{n})^{2}}\left(\nu^{v}_{ml^{j}_{i,3/2}}+\nu^{v}_{ml^{j}_{i,5/2}}\right)\right) & -\dfrac{\Delta t\,\nu^{v}_{ml^{j}_{i,3/2}}}{(h_{mz_i}^{n}\Delta\sigma_{mz_i}^{n})^{2}}\\
\cdots & \cdots & \cdots\\
0 & \left(1+\dfrac{\Delta t}{(h_{mz_i}^{n}\Delta\sigma_{mz_i}^{n})^{2}}\left(\nu^{v}_{ml^{j}_{i,N_l-1/2}}+\nu^{v}_{ml^{j}_{i,N_l+1/2}}\right)\right) & -\dfrac{\Delta t\,\nu^{v}_{ml^{j}_{i,N_l+1/2}}}{(h_{mz_i}^{n}\Delta\sigma_{mz_i}^{n})^{2}}
\end{bmatrix}
$$

$$
Q_{ij}^{n}=\left(F(\tilde{w}^{n}_{m^{j}_{i,1}}),\,F(\tilde{w}^{n}_{m^{j}_{i,2}}),\,\ldots,\,F(\tilde{w}^{n}_{m^{j}_{i,N_l}})\right)^{T}
$$

We can get the value of the vertical flow velocity component by solving equation (10).

3.2 *Nonhydrostatic corrective calculation*

Through equations (9), (8), and (10), we determine the intermediate variables of flow velocity and water level at moment n+1 without considering the hydrodynamic pressure; as the flow velocity at this point is not yet the final real flow velocity value, it will inevitably not satisfy the continuity equation. Therefore, it is necessary to correct the hydrodynamic pressure term to obtain the variables that satisfy the continuity equation and momentum equation. If q^* is the hydrostatic pressure and q' is the nonhydrostatic correction term, then the hydrodynamic pressure can be expressed as:

$$
q=q^{*}+q' \tag{11}
$$

The pressure correction diffusive equation expressed with q' may be derived from the momentum equation discretized through the differential method and the discrete version with the divergence of incompressible fluid as 0 $\left(\text{i.e., } \nabla\cdot\vec{U}=0\right)$; the equation is expressed as follows:

$$
\begin{aligned}
&\frac{1}{3}\sum_{j=1}^{3}\frac{\Delta S_j}{\Delta\xi_{ms_i^j}}\left(\sum_{jj=1}^{3}\frac{\Delta t}{\rho\Delta\xi_{ms_i^{jj}}}\left(q'^{n+1}_{m^{jj(2)}_{i,k}}-q'^{n+1}_{m_{i,k}}\right)-\sum_{jj=1}^{3}\frac{\Delta t}{\rho\Delta\xi_{ms^{j}_{ms_i^{j(2)}}}}\left(q'^{n+1}_{m^{jj(2)}_{ms_i^{j(2)},k}}-q'^{n+1}_{m^{j(2)}_{i,k}}\right)\right)(n_x^{ij}+n_y^{ij})\\
&-\frac{\Delta S_{mz_i}\Delta t}{2\rho\Delta\sigma_{mz_i}}\left(q'^{n+1}_{m_{i,k+2}}-3q'^{n+1}_{m_{i,k+1}}+3q'^{n+1}_{m_{i,k}}-q'^{n+1}_{m_{i,k-1}}\right)\\
&=\frac{1}{3}\sum_{j=1}^{3}\frac{\Delta S_j}{\rho\Delta\xi_{ms_i^j}}\left(\sum_{jj=1}^{3}\tilde{u}^{n+1}_{m^{jj}_{i,k}}-\sum_{jj=1}^{3}\tilde{u}^{n+1}_{m^{jj}_{ms_i^{j(2)},k}}\right)n_x^{ij}+\frac{1}{3}\sum_{j=1}^{3}\frac{\Delta S_j}{\rho\Delta\xi_{ms_i^j}}\left(\sum_{jj=1}^{3}\tilde{v}^{n+1}_{m^{jj}_{i,k}}-\sum_{jj=1}^{3}\tilde{v}^{n+1}_{m^{jj}_{ms_i^{j(2)},k}}\right)n_y^{ij}\\
&-\frac{\Delta S_{mz_i}\Delta t}{2\rho\Delta\sigma_{mz_i}}\left(\tilde{w}^{n+1}_{m_{i,k+1}}-2\tilde{w}^{n+1}_{m_{i,k}}+\tilde{w}^{n+1}_{m_{i,k-1}}\right)
\end{aligned}\tag{12}
$$

where j and jj represent the three element boundaries requiring self-circulation, with $u^{n+1}_{m^{j(2)}_{i,k}}$, $u^{n+1}_{m^{j(1)}_{i,k}}$, $v^{n+1}_{m^{j(2)}_{i,k}}$, $v^{n+1}_{m^{j(1)}_{i,k}}$, $w^{n+1}_{m_{i,k+1/2}}$, $w^{n+1}_{m_{i,k+1/2}}$, and $w^{n+1}_{m_{i,k-1/2}}$ representing the physical quantities at the centroid of the control volume, which may be obtained through defined adjacent physical quantity interpolation; n_x^{ij}, n_y^{ij}, and n_z^{ij} represent the direction cosines of the normal vector outside the three vertical interfaces of plane element mz_i corresponding to control volume element $m^j_{i,k}$ when j = 1, 2 and 3, and the upper and lower interfaces of control volume element $m^j_{i,k}$ when j = 4 and 5.

3.3 *Turbulent flow model equation discretization*

Discretization of the $k-\in$ equation adopts the semi-implicit first-order upwind finite volume; the discrete version conserves the mass balance and is characterized by monotonic conservation:

$$\begin{aligned}
S_i\Delta\sigma^{n+1}_{m_{i,k}}k^{n+1}_{m_{i,k}} &= S_i\Delta\sigma^{n}_{m_{i,k}}k^{n}_{m_{i,k}} - \Delta t\sum_{j=1}^{3}\lambda_{m^j_{i,k}}\Delta\sigma^n_{m_{i,k}}u^{n+\theta}_{m^j_{i,k}}k^n_{m^{j(2)}_{i,k}} \\
&- \Delta t S_i\left(w^{n+\theta}_{m^j_{i,k+1/2}}k^n_{m_{i,k+1}} - 2w^{n+\theta}_{m^j_{i,k-1/2}}k^n_{m_{i,k}} + w^{n+\theta}_{m^j_{i,k-3/2}}k^n_{m_{i,k-1}}\right) \\
&+ \Delta t\sum_{j=1}^{3}\lambda_{m^j_{i,k}}\Delta\sigma^n_{m_{i,k}}c_\mu\frac{(k^n_{m_{i,k}})^2}{\varepsilon^n_{m_{i,k}}}\left(\frac{k^n_{m^{j(2)}_{i,k}} - k^n_{m_{i,k}}}{\Delta\xi_{m^j_{i,k}}}\right) \\
&+ \Delta t S_i\left[c_\mu\frac{(k^n_{m_{i,k+1}})^2}{\varepsilon^n_{m^j_{i,k+1}}}\left(\frac{k^{n+1}_{m_{i,k+1}} - k^{n+1}_{m_{i,k}}}{\Delta\sigma^n_{m_{i,k}}}\right) - c_\mu\frac{(k^n_{m^j_{i,k-1}})^2}{\varepsilon^n_{m^j_{i,k-1}}}\left(\frac{k^{n+1}_{m_{i,k}} - k^{n+1}_{m_{i,k-1}}}{\Delta\sigma^n_{m_{i,k}}}\right)\right] \\
&+ \Delta t S_i\Delta\sigma^n_{m_{i,k}}\left[c_\mu\frac{(k^n_{m_{i,k}})^2}{\varepsilon^n_{m_{i,k}}}G^n_{m_{i,k}} - \varepsilon^n_{m_{i,k}}\frac{k^{n+1}_{m_{i,k}}}{k^n_{m_{i,k}}}\right]
\end{aligned} \quad (13)$$

$$\begin{aligned}
S_i\Delta\sigma^{n+1}_{m_{i,k}}\varepsilon^{n+1}_{m_{i,k}} &= S_i\Delta\sigma^{n}_{m_{i,k}}\varepsilon^{n}_{m_{i,k}} - \Delta t\sum_{j=1}^{3}\lambda_{m^j_{i,k}}\Delta\sigma^n_{m_{i,k}}u^{n+\theta}_{m^j_{i,k}}\varepsilon^n_{m^{j(2)}_{i,k}} \\
&- \Delta t S_i\left(w^{n+\theta}_{m^j_{i,k+1/2}}\varepsilon^n_{m_{i,k+1}} - 2w^{n+\theta}_{m^j_{i,k-1/2}}\varepsilon^n_{m_{i,k}} + w^{n+\theta}_{m^j_{i,k-3/2}}\varepsilon^n_{m_{i,k-1}}\right) \\
&+ \Delta t\sum_{j=1}^{3}\lambda_{m^j_{i,k}}\Delta\sigma^n_{m_{i,k}}\frac{c_\mu(k^n_{m_{i,k}})^2}{h^n_{m^j_{i,k}}\varepsilon^n_{m_{i,k}}}\left(\frac{\varepsilon^n_{m^{j(2)}_{i,k}} - \varepsilon^n_{m_{i,k}}}{\Delta\xi_{m^j_{i,k}}}\right) \\
&+ \Delta t S_i\left[\frac{c_\mu(k^n_{m_{i,k+1}})^2}{h^n_{m^j_{i,k}}\varepsilon^n_{m^j_{i,k+1}}}\left(\frac{\varepsilon^{n+1}_{m_{i,k+1}} - \varepsilon^{n+1}_{m_{i,k}}}{\Delta\sigma^n_{m_{i,k}}}\right) - \frac{c_\mu(k^n_{m^j_{i,k-1}})^2}{h^n_{m^j_{i,k}}\varepsilon^n_{m^j_{i,k-1}}}\left(\frac{\varepsilon^{n+1}_{m_{i,k}} - \varepsilon^{n+1}_{m_{i,k-1}}}{\Delta\sigma^n_{m_{i,k}}}\right)\right] \\
&+ \Delta t S_i\Delta\sigma^n_{m_{i,k}}\left[c_1\frac{\varepsilon^n_{m_{i,k}}}{k^n_{m_{i,k}}}G^n_{m_{i,k}} - c_2\frac{\varepsilon^n_{m_{i,k}}}{k^n_{m_{i,k}}}\varepsilon^{n+1}_{m_{i,k}}\right]
\end{aligned} \quad (14)$$

In the above equation, $k = 1, 2, \ldots, N_l$, $u^{n+\theta}$ is the normal flow velocity of the vertical interface at different calculation moments, and $w^{n+\theta}$ is the corresponding normal flow velocity of the top and bottom interface. All of the above are given values and therefore it is evident that the coefficient matrices of equations (13) and (14) are tridiagonal, and both equations may be immediately solved by taking the boundary conditions into consideration.

4 MODEL VERIFICATION

4.1 *Analytic solution verification*

Analog computation of the 3D linear standing wave with a theoretical and analytical solution is chosen for verification. Suppose there is a 3D closed pool with length and width of 10 m and depth of >10 m; let the initial velocity be zero and the initial water level be:

$$\eta(x, y) = A\cos(k_x x)\cos(k_y y) \quad (15)$$

In the equation above, amplitude A = 0.1 m and $A/h = 0.01$; k_x and k_y represent wave numbers in x and y, respectively; $k_x = \pi/L$, $k_y = \pi/W$, the total wave number $k = \sqrt{k_x^2 + k_y^2} = 0.44$; and the corresponding wave period is T = 3.01 s.

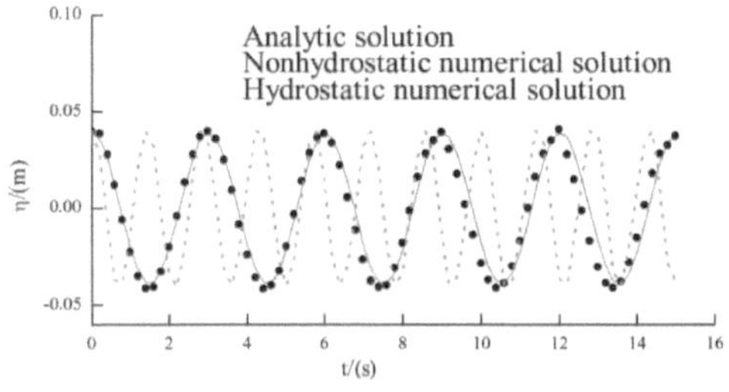

Figure 1. Comparison of the numerical solution analytic solution of the water level at (0.125, 0.125).

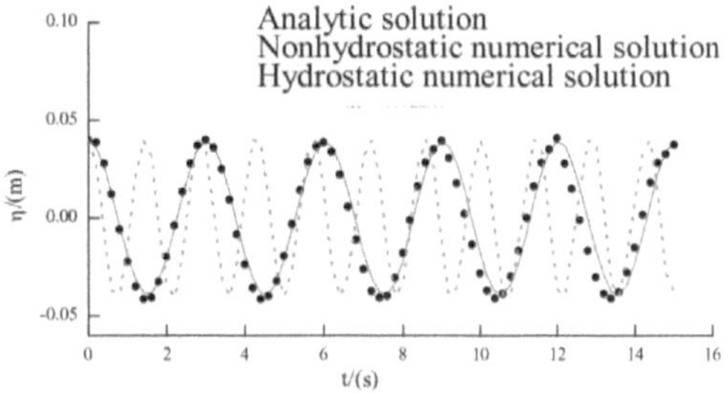

Figure 2a. Comparison of the nonhydrostatic pressure and the model numerical solution and the analytic solution of velocity *u*.

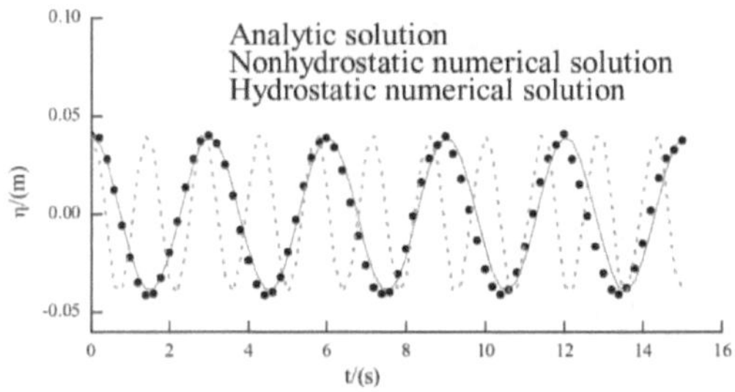

Figure 2b. Comparison of the nonhydrostatic pressure model.

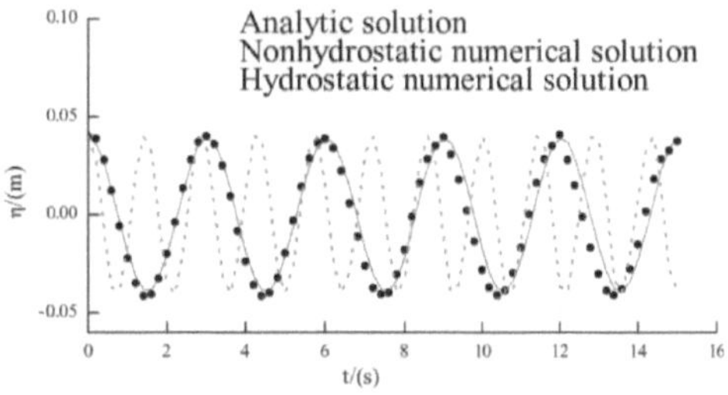

Figure 2c. Comparison of the nonhydrostatic numerical solution and the analytic solution of velocity *v* pressure model numerical solution and the analytic solution of velocity *W*.

In numerical calculation, the computational domain is discretized by adopting the triangle elements in a plane that is divided into 10 layers in the vertical direction without considering the viscosity. Figure 1 shows the comparison of hydrostatic and nonhydrostatic computational solutions and the analytic solution of the water level at point (0.125, 0.125). It is evident from the figure that the nonhydrostatic computational solution matches with the analytic solution well and the result given by the hydrostatic model is not satisfactory, which is mainly because the hydrostatic assumption is not valid, as the wavelength in this computational example is not a small value relative to the water depth. Figure 2 compares the numerical solutions and the analytic solutions of *u*, *v*, and *w* under hydrodynamic pressure at point (3.75, 1.125, 1.125); both solutions match well.

The hydrostatic and nonhydrostatic computational results of the established 3D numerical model are verified through the analytic solution of a 3D linear standing wave; considering the conformity of the nonhydrostatic computational results and the analytic solution, the model computation compares the hydrostatic and nonhydrostatic computational results in the meantime, further proving that the hydrodynamic pressure conforms to the theoretical solution better and that the hydrodynamic pressure under the complex water flow state is more reasonable.

4.2 *Experimental verification*

Three-dimensional water flow surrounding the spur dike is experimented via simulation with regard to the generalized model test of spur dike flume by adopting a 3D hydrodynamic numerical model of dynamic pressure. The spur dike test flume is 600 cm long, 200 cm wide, and 10 cm deep; the crest is 7.5 cm wide; the upstream slope is 1:1.5; and the downstream slope is 1:2.

First, verification is performed to the upstream and downstream section water levels of the spur dike, and the results are shown in Table 1. The difference between the computational

Table 1. Verification of water level of arc and linear spur dike with obtuse current angle (unit: m).

	1# cross section			2# cross section			3# cross section			4# cross section		
	Measured	Calculated	Difference	Measured	Calculated	Difference	Measured	Calculated	Difference	Measured	Calculated	Difference
Longitude 1	0.1664	0.1634	–0.0030	0.1708	0.1755	0.0047	0.1637	0.1639	0.0002	0.1504	0.1441	–0.0063
Longitude 2	0.1665	0.1633	–0.0032	0.1706	0.1747	0.0041	0.1654	0.1643	–0.0011	0.1503	0.1476	–0.0027
Longitude 3	0.1666	0.1633	–0.0033	0.1700	0.1738	0.0038	0.1649	0.1634	–0.0015	0.1624	0.1664	0.0040
Longitude 4	0.1673	0.1631	–0.0042	0.1701	0.1729	0.0028	0.1644	0.162	–0.0024	0.1621	0.1661	0.0040
Longitude 5	0.1673	0.163	–0.0043	0.1692	0.172	0.0028	0.1629	0.1607	–0.0022	0.1607	0.1656	0.0049
Longitude 6	0.1670	0.1629	–0.0041	0.1686	0.1714	0.0028	0.1620	0.1598	–0.0022	0.1622	0.166	0.0038
Longitude 7	0.1672	0.1628	–0.0044	0.1681	0.1706	0.0025	0.1616	0.1591	–0.0025	0.1637	0.1668	0.0031
Longitude 8	0.1659	0.1627	–0.0032	0.1667	0.1702	0.0035	0.1604	0.1589	–0.0015	0.1634	0.1673	0.0039
Longitude 9	0.1649	0.1625	–0.0024	0.1659	0.17	0.0041	0.1598	0.1588	–0.0010	0.1630	0.1675	0.0045
	5# cross section			6# cross section			7# cross section			8# cross section		
	Measured	Calculated	Difference	Measured	Calculated	Difference	Measured	Calculated	Difference	Measured	Calculated	Difference
Longitude 1	0.1492	0.1484	–0.0008	0.1551	0.1513	–0.0038	0.1531	0.1528	–0.0003	0.1605	0.1643	0.0038
Longitude 2	0.1499	0.1489	–0.0010	0.1543	0.1511	–0.0032	0.1534	0.1527	–0.0007	0.1606	0.1643	0.0037
Longitude 3	0.1510	0.1492	–0.0018	0.1541	0.1508	–0.0033	0.1533	0.1527	–0.0006	0.1602	0.1643	0.0041
Longitude 4	0.1516	0.1493	–0.0023	0.1548	0.1515	–0.0033	0.1543	0.1529	–0.0014	0.1610	0.1643	0.0033
Longitude 5	0.1553	0.153	–0.0023	0.1552	0.1524	–0.0028	0.1547	0.1532	–0.0015	0.1612	0.1643	0.0031
Longitude 6	0.1570	0.153	–0.0040	0.1557	0.1532	–0.0025	0.1547	0.1534	–0.0013	0.1611	0.1643	0.0032
Longitude 7	0.1582	0.1547	–0.0035	0.1568	0.1541	–0.0027	0.1551	0.1539	–0.0012	0.1611	0.1643	0.0032
Longitude 8	0.1581	0.1556	–0.0025	0.1563	0.1547	–0.0016	0.1543	0.1543	0.0000	0.1604	0.1643	0.0039
Longitude 9	0.1580	0.156	–0.0020	0.1565	0.1551	–0.0014	0.1540	0.1545	0.0005	0.1597	0.1643	0.0046

value and the measured value of the water level at each test point is within ±0.005 m, which means that the simulation of the changes between the water levels before and behind the spur dike via the 3D water flow model established in this study conforms to the measured results of the experiment.

Verification is performed to the upstream and downstream flow field of the spur dike. In order to visually represent the verification result, the calculated and measured values of the composed flow velocities u and v at a series of grid points on the horizontal section at 0.6H from the bottom of the generalized flume are compared, as shown in Fig. 3a. We take the calculated and measured values of the composed flow velocities w and v at a series of grid points on the same horizontal section for comparison, as shown in Fig. 3b. In the following figures, the coordinate unit is centimeter, the unit of flow velocity is centimeter/second, the black arrow denotes the measured value, and the red arrow denotes the calculated value from the numerical flume.

As we can see from Fig. 3a, the verification result and the measured value of the flow velocity match, including the value and direction of the flow velocity, with different flow directions at few points behind the dike, which is caused by the post-dike circulation swing. The verification of the flow velocity at the horizontal section reveals that the verification of components u and v on x and y axes is correct.

In the flume test, the flow velocity components v and w are small, relative to the flow velocity component u, and the values of v and w are close, so the composed flow velocity of w and v is chosen to draw a flow velocity vector diagram for comparison of graphical verification, with verification results as shown in Fig. 3b. It is important to note in particular that the value of the flow velocity component drawn in the figure in the x direction actually represents

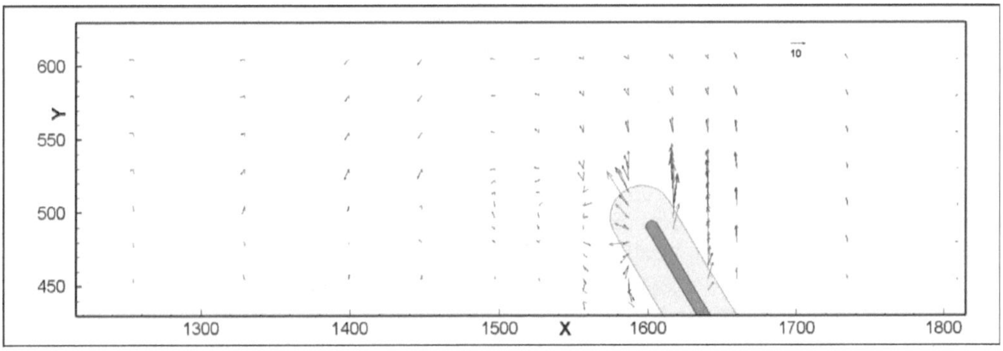

Figure 3a. Comparison of the calculated and measured values of the composed flow velocity u and v on the horizontal section at 0.6H from the bottom of the generalized flume.

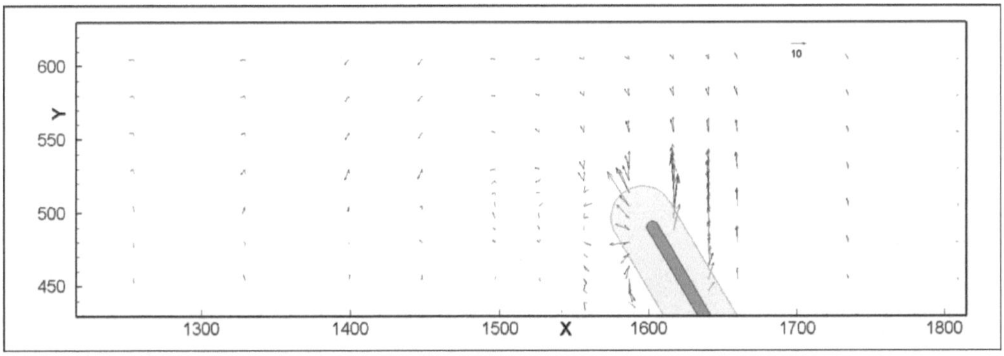

Figure 3b. Comparison of the calculated and measured values of the composed flow velocity w and v on the horizontal section at 0.6H from the bottom of the generalized flume.

the value of the component w; if the flow velocity points to the positive direction of x, the value of w is positive; otherwise, the value of w is negative. As the absolute values of w and v are relatively small and usually not in the main flow direction, it is very difficult to perform verification accurately. As we can see from the figure, the verification of the points with larger values of flow velocity components w and v in the vicinity of the spur dike via model computation is relatively accurate, while the individual points with a larger difference is caused by the complex flow state and the dramatic space change of the flow velocity around the spur dike; the tiny variation between the measured point and the model computation point may also result in the comparatively large deviation between the two values; therefore, individual larger differences will not affect the success of the overall verification results.

5 ENGINEERING APPLICATION

The confluence reach of Yangtze River and Tuojiang River is controlled by themselves. The flow state of the confluence area is comparatively complicated; in addition, the confluence area of the two rivers is located at the turning of the main stream of Yangtze River, making the flow state of this water area even more complicated (as shown in Figure 4). In order to study the channels at the confluence reaches of main streams and tributaries and the changes of the port navigation water flow conditions, it is very important to accurately simulate the 3D flow field of the confluence area, which is the precondition to reveal the mechanism of local scour and deposition of sediment.

The model computation range is as shown in Fig. 4. The model plane adopts triangular grid, whereas layered grid is adopted in the vertical direction. In order to guarantee the calculation accuracy around the spur dike, the mesh encryption method is applied in such area with minimum grid scale of 1.5 m, which ensures the ability to accurately reflect the neighboring spur dike, while the gird scale is enlarged by means of gradual transition in other area, with a maximum grid scale of 50 m. In the entire plane area, there are 11,600 grid nodes, 22,666 elements, and 10 vertical layers.

Because of space limitation, only the comparison result of the measured water level and the flow velocity during the medium flood period is given. As shown in Fig. 2, the verification result of the measured water level during the flood period below medium flow level is given, with the water level deviation at all the verification points smaller than 9 cm. The measured flow velocity along the river by two hydrological balls and the calculated results of the surface flow velocity from the 3D computation are compared on the basis of the measured flow velocity at the site, as shown in Fig. 5. The calculated results conform well to the distribution of the measured flow velocity along the river.

We achieved the simulated computational results in the flow field simulation. Figure 6a compares the vertical-section flow field in the direction perpendicular to the trend of the river, whereas Fig. 6b compares the horizontal section flow field at different heights.

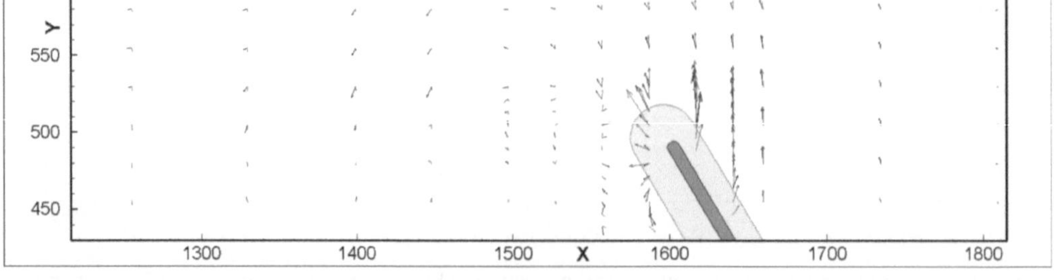

Figure 4. Confluence reach of Yangtze River and Tuojiang River as well as their model range.

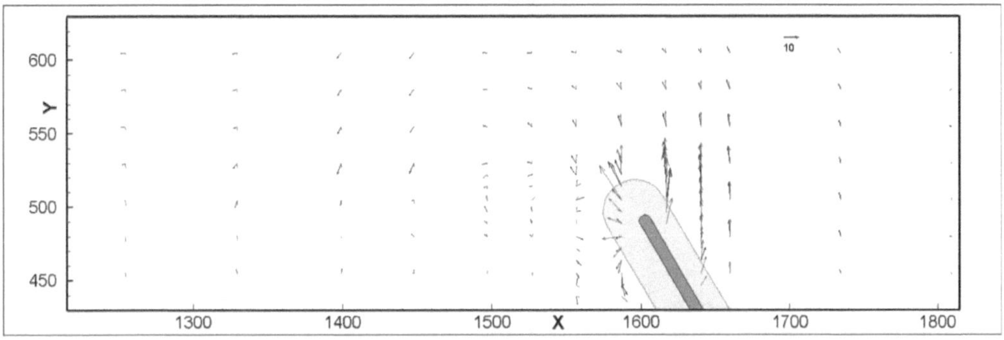

Figure 5a. Comparison of the measured value by 1# hydrological ball along the river and the calculated value of flow velocity during medium flood period.

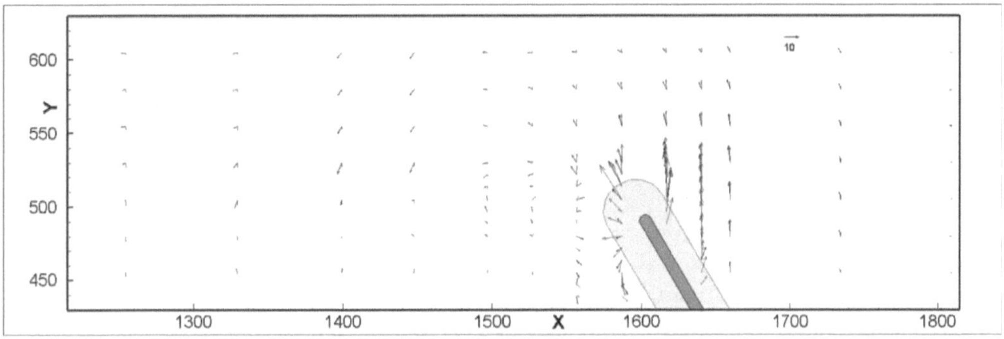

Figure 5b. Comparison of the measured value by 2# hydrological ball along the river and the calculated value of the flow velocity during medium flood period.

Table 2. Verification table of water level during medium flood period (flow rate of Yangtze River is 9840 m^3/s while that of Tuojiang River is 250 m^3/s).

Water gauge	Grid section	Distance (km)	Measured water level (m)	Calculated water level (m)	Calculation deviation (m)
5#L	115	2.7	228.134	228.141	0.007
6#R	103	2.0	228.212	228.270	0.058
7#L	123	3.1	228.082	228.110	0.028
8#R	112	2.5	228.154	228.160	0.006
9#L	144	3.7	228.038	228.120	0.082
10#R	130	3.0	227.888	227.950	0.062
11#L	161	4.3	228.023	228.100	0.077
12#R	159	3.6	227.878	227.885	0.007
13#L	173	4.8	227.881	227.860	–0.021
14#R	174	4.0	227.794	227.862	0.068
Tuojiang River1#	113	6.7	228.121	228.110	–0.011

Figure 7 shows the 3D flow field by means of 3D Laplace particle tracing using 3D imaging technology (red and blue).

As the flow rate of Yangtze River far outweighs that of Tuojiang River, the water flow at the confluence reach moves to the left bank at the Tuojiang River side, with two streams of

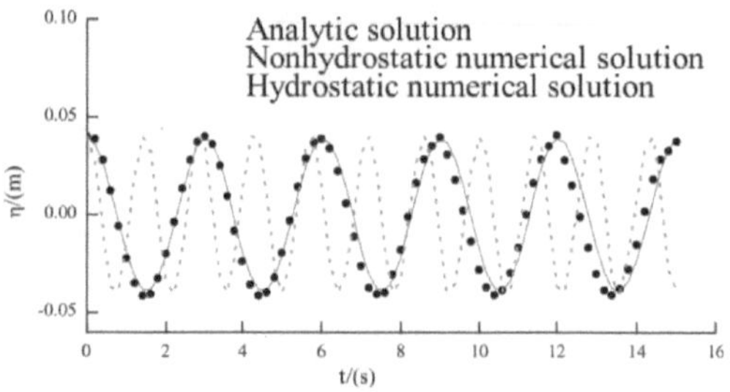

Figure 6a. Vertical section flow field at the confluence reach.

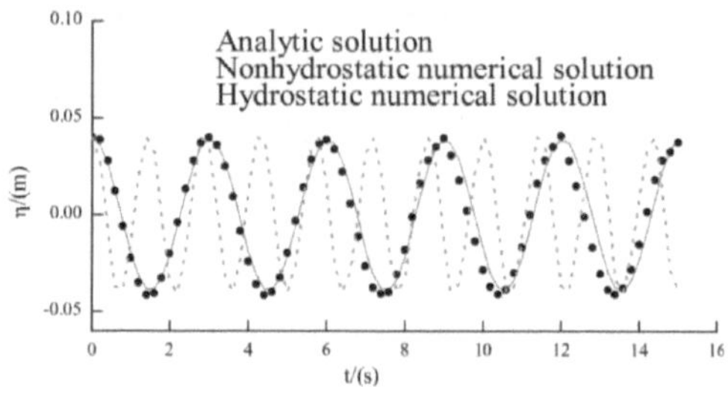

Figure 6b. Horizontal section flow field at the confluence reach.

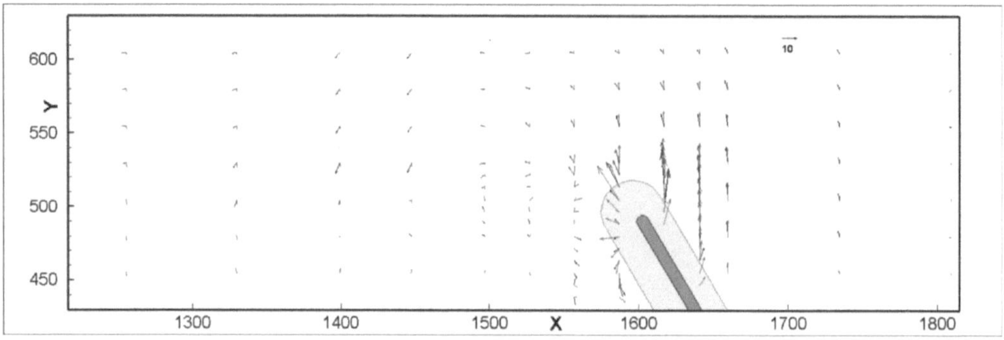

Figure 7. Three-dimensional flow field at the confluence reach by 3D imaging technology.

water moving forward while pressing against each other, forming the spiral flow toward the downstream reaching the turning of the river, which presents a vertical counterclockwise rotation. The existence of the spiral flow makes the sediment on the left bank and the bottom of Yangtze River prone to scour, which is why this water area may maintain greater water depth.

6 CONCLUSION

In this study, we established a 3D hydrodynamic numerical model based on hydrodynamic pressure, verified the model by adopting analytic solution and experimental data, and proved that the hydrodynamic processing approaches are better applicable to the numerical simulation of complex water flow. The confluence reach section of Yangtze River and Tuojiang River chosen for numerical simulation computation, which studies the 3D water flow characteristics of this section, instructed the researchers on local scour and deposition of sediment and verified that the 3D model can adapt to the simulation of 3D water flow under complex terrain. The fact that the 3D hydrodynamic numerical model based on the hydrodynamic pressure can solve actual engineering problems will powerfully promote the application of 3D water flow calculation in the actual engineering and advance the development of the 3D numerical model in the water transportation engineering field.

REFERENCES

Cheng Changhua; et al. Waterway Engineering [M]. *Beijing: China Communications Press*, 2001.
Cui Zhanfeng; Zhang Xiaofeng. Application of 3-D Turbulent Flow Model in Spur Dike [J]. *Engineering Journal of Wuhan University*, 2006(1): 15–20.

Jiang Changbo. 2-D Large-Eddy Numerical Simulation of Flow Field Around Spur Dike [J]. *Journal of Changsha Jiaotong University*, 1999, 15(3): 68–72.
Jing Zhongqing. *Numerical Solution to N-S Equation and Turbulent Flow Model* [M]. Nanjing: Hehai University Press, 1989.
Kang Suhai. Numerical Simulation and Visualization on 3-D Hydrodynamics [D]. *Ph. D. Dissertation of Dalian University of Technology*, 2011.
Peng Jing; et. al. 3-D Numerical Simulation of Scour and Deposition Around Spur Dikes [J]. *Journal of Sediment Research*, 2002(1): 25–29.
Rong Xuewen. Water Destruction Theory of Spur Dike and Plane 2-D Water Flow Numerical Simulation [D]. *Master Dissertation of Chongqing Jiaotong University*, 2003.
Shen Bo, Plane 2-D Numerical Model of Local Scour and Deposition of Spur Dike. *Journal of Xi'an Highway University* [J], 1997, 17(3): 31–36.
Tan Weiyan. Hydrodynamics of Shallow Water Computation—Application of Finite Volume Method [M]. *Beijing: Tsinghua University Press*, 1997.
Wang Fujun. Computational Fluid Dynamics Analysis—Principle and Applications of CFD Software [M]. *Beijing: Tsinghua University Press*, 2004.
Wang Jianjun. On Numerical Model of 3-D Water Flow Suspended Sediment by Finite Volume Method [D]. *Mater Dissertation of Tianjin University*, 2007.
Wang Pingyi. *Water Destruction Theory and Simulation Technology Regarding Channel Regulation Structures* [M], China Communications Press, 2004.
Wang Pingyi; Zhang Jisheng; Cheng Changhua. *Fuzzy Comprehensive Assessment of Security and Stability of Channel Regulation Structures* [J]. Journal of Hydrodynamics, 2004, 19(3): 383–388.
Ying Qiang; Jiao Zhibin. Spur Dike Hydraulics [M]. China Ocean Press, 2004.
Zhang Baolin; Yuan Guoxing. Parallel Finite Differential Method of Partial Differential Equation [M]. *Beijing: Science Press*, 1994.
Zhang Huaqing. *The Numerical Model of Flow and Sediment and Its Applications in Rivers and Estuaries* [D]. Ph. D. Dissertation of Hehai University, 1998.
Zhao Shiqiang. Scour Mechanism of Spur Dike and Local Scour Computation [J]. Journal of Chongqing Jiaotong University, 1989, 28(1): 13–21.
Zhou Guanlun; et al. Waterway Engineering Manual [M]. *China Communications Press*, 2004.
Zhou Yilin. Large-Eddy Numerical Simulation of 3-D Water Flow Movement Around Submerged Spur Dike [J]. *Journal of Yangtze River Scientific Research Institute*, 2001(5): 28–31.
Zhou Yilin; Dao Shang Zheng Yan. Research on Characteristics of 3-D Water Flow Movement Around Non-submerged Spur Dike [J]. *Journal of Hydraulic Engineering*, 2004(8), 53–46.

Innovation and Application of Engineering Technology – Oscar, Chen & Wang (Eds)
© 2017 Taylor & Francis Group, London, ISBN 978-1-138-05057-0

Evaluation of structural complexity and optimization of permeability coefficient based on fractal theory

Hui Xu, Yi Mu & Hao Qiu
Mine Safety Technology Branch of China Coal Research Institute, Beijing, China
State Key Laboratory of Coal Resource Efficient Mining and Clean Utilization (China Coal Research Institute), Beijing, China
The Geophysical Engineering Branch Company of Jincheng Anthracite Mining Group Technology Research Institute, Jincheng, China

ABSTRACT: On the basis of fractal theory, the capacity dimension in the fractal dimension is taken as the basic feature quantity. The grid-covering method was used to quantitatively evaluate the structural complexity of Donghuantuo mine. Using this method, a positive relationship was found between the structural complexity and the permeability coefficient using the fractal theory. The value of permeability coefficient is optimized to match the empirical formula to the real situation. Thus, the prediction accuracy of water inflow is improved.

1 INTRODUCTION

The complexity of the geological structure has a significant effect on coal mining conditions, water inrush, gas outburst, and so on. In a mine, water prevention, geological structure, and water inflow have a direct relationship (Mandelbort 1983). Because of the constraints of tectonic stress field, the distribution of faults and folds in the zone is widespread. The degree of structural development has a controlling impact on the permeability of the rock, which affects the accuracy of the prediction of water inflow. Therefore, it is very important to quantify the complexity of geological conditions and understand the permeability of rock in specific areas (mining, roadway, working face, etc.) in mathematical language.

It is difficult to quantitatively reflect the complexity of the geological structure of the mining area by using conventional mathematical statistical methods. This paper introduces fractal theory in capacity dimension, whose results are more accurate than those of conventional statistical methods of measurement. In recent years, some experts and scholars worldwide have shown that the fractal dimension in fractal theory can comprehensively reflect the fault extension length scale and distribution of faults in the evaluation of fault network complexity in geological blocks. The fractal dimension can be used to quantitatively evaluate the complexity of the fracture network compared with the conventional mathematical statistical method.

2 FRACTAL THEORY

2.1 *Fractal definition*

The word fractal was coined by B.B.Mandelbrot[1] in 1973 from the Latin word "Frangere". It means irregularly fragmented, meaning fractal geometry is a geometric shape with irregular geometry as the object of study. Because irregularity is common in nature, fractal geometry is also called geometrical description of nature. However, there is no concrete definition of the term fractal. Fractal is a general term for shapes and structures without self-similarity in length.

2.2 *Fractal determination method*

At present, there is no definite method for judging fractal and nonfractal. Some experts and scholars worldwide used methods such as (1) artificial determination; (2) correlation coefficient test method; (3) strengthening coefficient method; (4) fitting error method; (5) fractal dimension error method; and (6) overall fitting method (Liu Ying et al 2006 and Chen Jianan 1999).

2.3 *Definition of fractal dimension*

Fractal dimension is an important parameter to describe fractal, which represents the basic characteristics of the fractal. However, there are many kinds of definition and calculation method because of the difference of the focal plane. Fractal dimension is of many types such as similar dimension, Hausdorff dimension, capacity dimension, and box-counting dimension, with different applications. In this paper, only the capacity dimension is described.

According to Soviet mathematician Kolmogorov, capacity dimension is the use of the same size and shape of the ball or cube covering geometric objects. Let S be a ball-shaped object. If the minimum value of the diameter of the ball to cover S is N (ε), then the capacity of S dimension is:

$$Dc = \min_{\varepsilon \to 0} \frac{\log N(\varepsilon)}{\log(1/\varepsilon)}$$

3 EVALUATION OF STRUCTURAL COMPLEXITY

In this paper, we take Donghuantuo mine as an example to introduce the fractal theory to evaluate the structure complexity of mining area.

3.1 *Introduction to the coal mine*

Donghuantuo mine is located in Tangshan City, Hebei Province, Fengrun District, in the long, asymmetric Chezhoushan syncline. The synclinal axis is tilted toward the northwest direction, and the hinge is tilted to the southwest direction, extending in the shape of a "tongue" on the plane. The strata of synclinal wings vary greatly, the southeastern wing is gentle and the northwest limestone is steep. In the syncline, the fault structure is more developed and the strike direction is more consistent with the synclinal axis. The main aquifer of the mine has a quaternary bottom pebble aquifer, A~A_0 aquifer, $5^{\#}$ coal roof aquifer, $12_{-2}^{\#}$~$14_{-1}^{\#}$ aquifer, $14_{-1}^{\#}$~$K_3^{\#}$ aquifer, and Ordovician limestone aquifer, with each of them having a hydraulic connection (Xuhui 2015, Yin Shangxian et al 2015).

3.2 *Fractal calculation*

On the basis of the tectonic data in the mining plan of Donghuantuo mine, the fractal dimension (volume dimension) of each unit area is calculated by using the grid cover method in mine structure.

The capacity dimension of the fault grid is calculated as follows[4]: The grid is divided into four levels of dimensions 1000 m × 1000 m, 500 m × 500 m, 250 m × 250 m, and 125 m × 125 m, in order to form a square grid with an exponential increase in grid density (Figure 1). The number of grids containing fault traces in different units is counted in turn, and the specific method is referred to the fractal dimension tomography (Fig. 2).

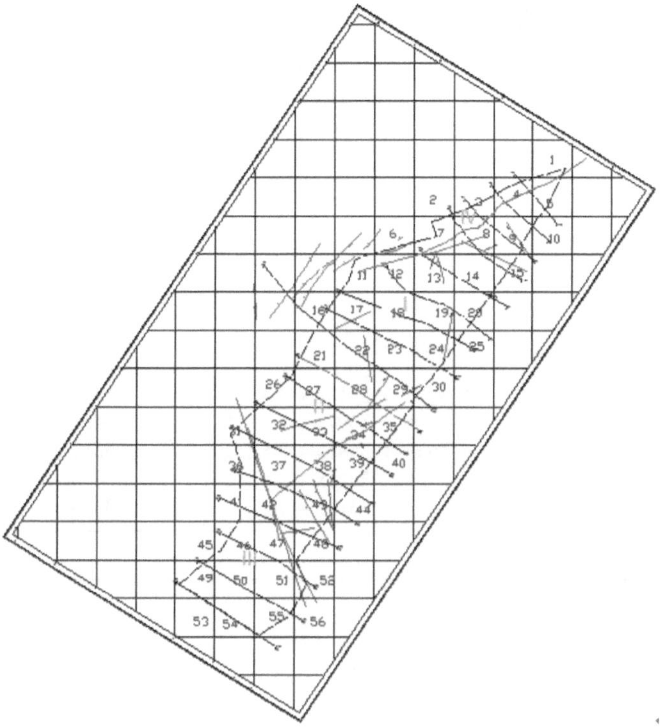

Figure 1. Unit partition diagram of the fractal evaluation of Donghuantuo mine.

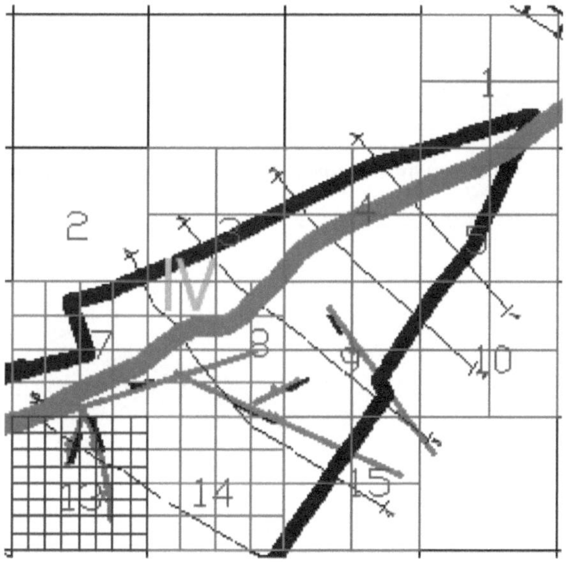

Figure 2. Fault grid partition diagram of fractal dimension.

3.3 *Calculation results and evaluation*

Table 1 shows the structural unit of the fault volume dimension data of Donghuantuo mine. The fractal dimension values represent the values of the center points of the partitioned units. As a result, the complexity of the fault network within the Donghuantuo mine is quantified.

On the basis of the double-logarithmic linear regression analysis, the capacity dimension of Donghuantuo mine is 0.3312 (Fig. 3), and most of the correlation coefficients of each unit are more than 0.95, indicating that the fault network is distributed in the Donghuantuo mine area. The correlation of unit data is good. The coordinate values and capacity dimension values of the center points of the partitioned units are input to the computer, and the contour map of the fault network capacity of Donghuantuo mine is drawn (Fig. 4).

Table 1. Dongtuo coal mine fault structure fractal dimension calculation results.

Numbering	Area	Regression equation $\lg N(r) = a + (-D) \log r$	Fractal dimension D	Correlation coefficient R^2
1	Whole mine	y = −0.3312x + 2.8732	0.3312	0.9973
2	I	y = −0.3497x + 2.4464	0.3497	0.9943
3	II	y = −0.3581x + 2.5226	0.3581	0.9872
4	III	y = −0.3243x + 1.5731	0.3243	0.9819
5	IV	y = −0.3314x + 2.219	0.3314	0.9988

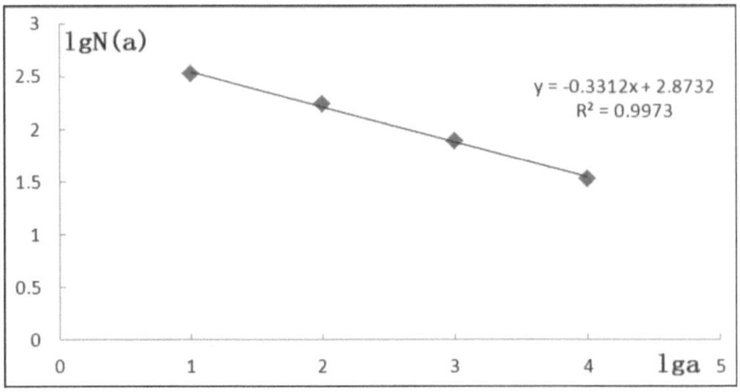

Figure 3. Statistical data of Donghuantuo mine region obtained by double-logarithmic fitting.

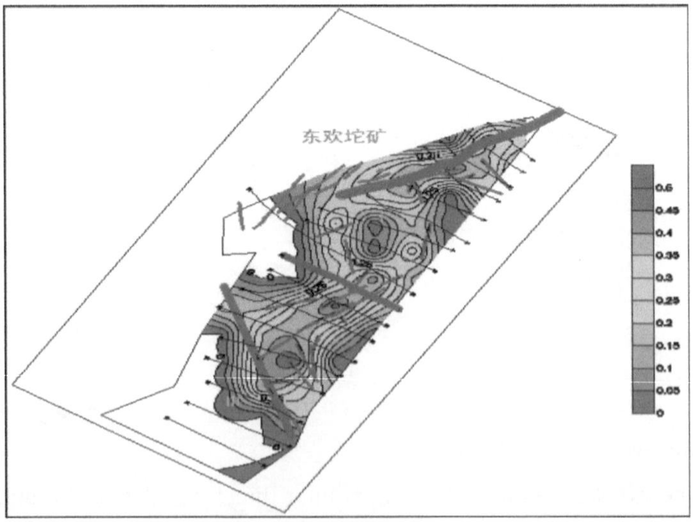

Figure 4. Donghuantuo mine capacity dimension plane diagram of the fault network.

The analysis shows that the capacity dimension value of area I is 0.3497. The regional boundary is east of the 16 section line to the southeast of the synclinal axis, which is a tectonic block section. The block section shows a slight change in strike and trending and a simple monoclinic structure with 58 faults, with a fault density of 1.38/km. As most faults are concentrated in the northeast region, I (capacity dimensions: 0.36) development is more complex.

The capacity dimension of the area α is 0.3581. The area boundary is a 16 section line to the southeastern wing of the F_{35} fault. The block section is characterized by a high-angle inclination or an oblique normal fault, which is more developed and the drop is larger. There are 17 faults (F22–F37) in the area, with 16 of them being normal, and the faults greater than 50 m are F_{22}, F_{27}, F_{31}, F_{32}, and F_{35}. The gap between faults F_{23}, F_{26}, and F_{30} is 30–49 m. Six faults have gaps of less than 30 m. As most faults are concentrated in the southeast region, α (capacity dimensions: 0.5) development is more complex.

The fractures in the above two block sections are more developed than those in other block sections, and the values of the capacity dimensions are significantly higher than those of the weak faults. The capacity dimensions value of the whole area is 0.3312, which shows that the structure is simple in the field, but the heterogeneity of the structure in the area is complex.

4 OPTIMIZATION AND APPLICATION OF PERMEABILITY COEFFICIENT

The degree of structural development has a controlling effect on the permeability of the rock, and are there any links between the two? With this question, the authors found the fractal dimension of the location of the pumping hole and the permeability coefficient measured by the hole and also found a positive correlation between the fault capacity dimension and the permeability coefficient (Figure 5). They used this relationship to optimize the permeability coefficient[4–5] and obtain the optimization formula.

Optimization formula:

$$K^* = 44.7653 - 102.0408\text{D}$$

K^*: The optimized value of permeability coefficient K.

This optimization formula is applied to the prediction of water inflow in the working face of Donghuantuo mine.

Method of prediction: “big well method”.

Calculation formula:

$$Q = 1.366K^* \frac{(2H - M^*)M^*}{\lg R_0 - \lg r_0}$$

The results are shown in Table 2.

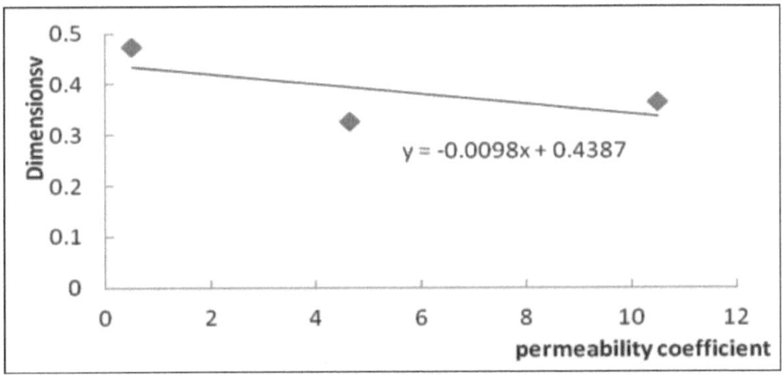

Figure 5. Relationship chart of dimension and permeability coefficient of fault capacity.

Table 2. Results of the prediction of water inflow.

		Value (m^3/min)	Error (%)
Measured value		2.3	
Predictive value	Conventional	3.4	47.83
	Optimization	2.35	2.2

5 CONCLUSION

In this study, fractal theory is used to evaluate the complexity of the coal mine structure. Taking Donghuantuo mine as an example, we described the complexity in the method for the evaluation of the coal mine structure and realized the quantitative evaluation of the structure complexity. The method provides a certain analysis method for mine water inrush and gas outburst during mining, which also has an important significance in guiding mine water prevention correctly.

We found a positive correlation between the fault capacity dimension and the permeability coefficient. This relationship is used to optimize the permeability coefficient, and the optimization formula is obtained. This optimization formula is applied to predict water inflow in the working face of Donghuantuo mine, and the result shows that the prediction error is improved significantly compared with the original prediction error.

REFERENCES

[1] Mandelbort, B.B. 1983. The Fractal Geometry of Nature[M] *New York, Freeman.*

[2] Liu Ying, Hu Min, Yu Guiying, Li Xiaobing & Liu Xiaolin. 2006. Fractal Theory and Its Application[J]. *Journal of Jiangxi Sciences* 24(2):205–209.

[3] Chen jianan. 1999. Definition and Measurement of Fractal Dimension[J]. *Journal of Electronic Science and Technology* 4:44–46.

[4] Xu Hui. 2015. Study of the Mine Water Inflow Prediction[D], North China institute of science and technology.

[5] Yin Shangxian, Zhang Xiangwei, Xu Hui, Liu Ming & Xu Bin. 2015. Optimization of permeability coefficient and aquifer thickness in large-well-method[J]. *Coal Geology & Exploration* 5:53–56.

Innovation and Application of Engineering Technology – Oscar, Chen & Wang (Eds)
© 2017 Taylor & Francis Group, London, ISBN 978-1-138-05057-0

Occupational adaptation-related safety, a literature review of China till 2014

Xiaofei Lin
School of Economics and Management, Beijing Jiaotong University, Beijing, China
Civil Engineering and Agriculture School, Anhui University of Technology, Ma'anshan, Anhui, China

Shouxin Song & Huaiyuan Zhai
School of Economics and Management, Beijing Jiaotong University, Beijing, China

ABSTRACT: This paper aims to summarize the literature characteristics on occupational adaptation that has recently taken place in China and suggests directions for future research. Occupational aptitude of drivers has been paid attention to since the late 1980s, and the number of related research papers presents a wave trend, which are increasing especially in recent years. There are two focuses concerning research contents: (1) drivers' adaptability and (2) analyzing occupational adaptivity from the viewpoint of safety. The chief research institutions are Beijing Jiaotong University and Anhui Sanlian Group. Their research objects were mainly drivers and power workers and focused on evaluation criteria. Two directions of future research are suggested: (1) the impact of organization on the individuals should be considered in the evaluation indicators and (2) more attention should be paid to the adaptive testing, learning, and training, which are the key to promoting adaptation ability.

1 INTRODUCTION

Research on security from the perspective of occupation adaptation is a popular way. Occupation adaptability concerns mainly about people; safety research will be more practical and universal by putting its bases on people. Second, study on occupation adaptability has a history of almost 80 years; its results can be adapted for safety research, especially in high-risk industry, special occupation adaptability with security as the goal. Safety occupation adaptation can be understood as special occupation adaptability.

The remainder of this paper is organized as follows. Section 2 describes the conceptual framework and difference of occupation adaptability, occupation suitability, position adaptability, and so on. Section 3 discusses the literature analysis of the occupational adaptation-related safety in China, which took the perspectives of time, contents, and research institutions. Section 4 concluded the paper by describing the characteristics of driver's occupation adaptability, occupation adaptability of other high-risk industries and special operators, and the safety occupation test method. Future research directions are summarized in Section 5.

2 CONCEPTUAL FRAMEWORK

Although the content of occupation adaptation is the same, there is no commonly used term for it. Some scholars called occupation adaptability which is zhiyeshiyingxing in Chinese (such as Shouxin Song and Sen Li of Beijing Jiaotong University), some others called occupation suitability which is zhiyeshihexing in Chinese (such as Kaili Xu and Baozhi Chen of Northeastern University and), and some researchers called position adaptability which is gangweishiyingxing in Chinese (such as Safety and Environmental Protection Research Institute of Ministry of metallurgical industry). Regarding safety (special) occupation adaptability, more scholars focused on the job or specific position, such as crane operators, motor vehicle driver, fireman, electrician, personnel operating boilers in dangerous plants, and power system operation personnel and some on a comprehensive perspective, such as safety occupation aptitude based on psychological resource model. At the same time, some scholars also call it zhiyeshixing in Chinese.

Chinese profound culture has rich text expression meanings; however, "occupation adaptability" is a translated term. For the rigor of research, clarification of the concept is very necessary.

2.1 *Concept-related occupation*

Ancient Chinese books have mainly two kinds of interpretation of the occupation: (1) office and "Four" of the industry. The office is the official position. "Four", namely "four people", denotes the division of China in ancient times according to the social stratum: scholar, agricultural, industrial, and commercial; "Fixed position or office four" of the industries of scholar, farmer, artisan, and merchant class. (2) We should do. Office selects a person who can fulfill the responsibilities. Overall, occupation is called duty in Chinese ancient books, which is zhiwu or zhizhang in Chinese (Xiao & Suo, 2011).

Modern Chinese dictionary explanted the vocation as: the individual in society are engaged as the main source of work (Dictionary of Language Research Institute of Chinese Academy of Social Sciences editorial office, 2005). Yanpei Huang put the occupation as "labor or labor in exchange for the needs of daily work" (the Chinese occupation education agency, 1985). From the perspective of terminology, occupation is involved in the social division of labor, with specialized knowledge and skills, to create material wealth and spiritual wealth for the society, obtain reasonable reward, as the material source of life, and meet the spiritual needs of the work. From the viewpoint of society, occupation is the social role, assuming certain duties and responsibilities for society and obtaining the corresponding reward; from the perspective of national economic activities by human resource, occupation refers to the special labor positions of different nature, different content, different forms, and different operation (Baidu Encyclopedia, 2013).

Usually we think that occupation is work, which involves long-term production and life habits of view, conventional ideas, and the idea of never ambiguous. However, in academic applications, it should be carefully distinguished.

2.2 *Concept-related adaptation*

2.2.1 *Difference between adaption and suitable for a job*

Adaptation was first used in the field of biology and widely followed in psychology, architecture, computer, and so on. The most typical explanation of adaptation is the evolution theory of biology, which gave the definition of "survival" as the fittest. Psychology explains adaptation from three aspects: physiological adaptation, psychological adaptation, and social adaptation (Yi, 2006).

At present, the concept of "dictionary of psychology" is also emerging. "Adaptation is a term from biology, used to represent the physical and behavioral changes can increase the chance of survival of organisms" (Zhu, 1989), while suit is agreeable or acceptable (Baidu Encyclopedia, 2013).

Suit emphasized the subject as passive and static; for example, one set of people suit for the role of a typist, meaning the person's quality has met the working physiological and psychological needs of a typist; adaptation emphasized the subject as active and dynamic; for example, a person adapted as a typist focusing on changing his/her own quality to adapt to another work process. Furthermore, "adaptation" is stronger than "suitable for" at a job.

2.2.2 *Relationship of adaptability (shixing), adaptation (shiyingxing), and adaptive capacity (shiyingnengli)*

There are three kinds of explanations for adaptation (shiyingxing): from the perspective of biology, it is the survival potential given certain genetic biological composition (Baidu Encyclopedia, 2013); from the individual perspective, it refers to individual changes in the system of social organization, group, or cultural and economic factors, with changing survival capacity, development, and goals (Wen, 2001); from the perspective of social psychology, it refers to the individual for the completion of certain social life adaptation process and the ability to form a model of psychological behavior correspondingly (Xu, 2000).

Adaptability (shixing), meaning satisfactory and consensual (Baidu Encyclopedia, 2013), emphasizes the psychological feeling and the experience of things; in the papers of occupation adaptation, adaptability (shixing) and adaptation (shiyingxing) are used interchangeably.

Adaptive ability (shiyingnengli) refers to subjective conditions to adapt to the process.

Adaptability has the same meaning to adaptation, and we should treat the differences between the three words in academic research.

2.2.3 *Adaptability and competency*

The representative definition of competency was put forward by Hay-McBer consulting company in 1973, according to which, competence refers to the individual features of staff for a job or task needs, including personal knowledge, skills, attitudes, and values. Then, Spenser (Spencer, 1993) gave a representative definition, that is, competence refers to the qualities, skills, self-image, social roles, attitudes, values, knowledge, and skills, which can be reliably measured and can put any distinction between high-performance employees and the general performance characteristics of an individual (Li, 2011).

In summary, the biggest difference between adaptation and competence lies in that adaptation is a measure of employees' basic ability to change their quality and behavior to complete an activity, and competency is based on individual inherent characteristics, which can reflect the employee performance.

2.3 *Vocational adaptation*

Occupational adaptation refers to whether objective ability and the physiological and psychological characteristics of workers adapt the occupation or job demands, which can match with it (Gao, 1996). It indicates people and occupation in social and economic activities in the process to achieve coordination and integration (Tian, 1999). It points out a person having psychological quality needed for the physiological characteristics of a job. It is formed and developed on the basis of the interaction of the congenital factors and acquired environment (Wu, 2002). Occupational adaptation in its essence is a kind of occupation ability; it is the product of the individual with a particular occupation environment interaction. It is the role of occupation, ability, attitude, values, and interpersonal relationship, which are evaluated and adjusted to achieve the harmonious process (Tan, 2012). Occupation adaptation is realized by the fusion process, which means the interaction between individuals and organizations (Bakke, 1953).

To sum up, occupation adaptation is a kind of occupation ability on physiology and psychology and is a dynamic balance process. The subject of this process is human, object is the occupation, and subsidiary is the environment.

2.4 *Safety vocational adaptation*

Safety occupational adaptation consists of two core parts: safety occupation aptitude and safety psychological adaptability. Safety occupation aptitude is the self-protection tendency of individual adjusting psychological state, controlling and dominating the mental resources to complete the task during occupation and safety activity. It is a physiological and psychological safety quality that refers to the human engaged in a certain work, the completion ability of the work needed, and the quality and ability used for self-protection and regulation (Li, 2009, 2010).

The safety occupational adaptation is the ability to work and take safety behavior by psychological and physiological quality with self-organization and regulation in the interaction process between people and occupation. Its characteristics are as follows: compared with position adaptation, occupational adaptation considered more influencing factors; it highlights the learning and control ability of personal safety, which are generally used for high-risk accidents and occupation disease population of industry.

3 PROCEDURE OF LITERATURE ANALYSIS

In accordance with the exact title occupational adaptability and theses strategies from the knowledge service platform database Wanfang Data, 127 dissertations were retrieved from 1989 to 2014. Among them, 94 were about students and teachers, seven were about traffic staff, six were about power workers, two were method researches, and the others were about farmers, police, and people with intellectual disabilities and so on. We analyzed the dissertations and concluded 15 related to safety occupational adaptability, according to Sen Li's viewpoint.

4 RESULTS AND PRACTICAL VALUES

4.1 *Safety vocational adaptation*

The earliest occupation adaptability study conducted in China was on driver occupation adaptability. The original Anhui Sanlian Accident Prevention Research Institute (now Anhui Sanlian Group) has been devoted to the study of occupation adaptability of road drivers. It has published "accident", "epidemiology of driving adaptation", and other works and completed the "Chinese road traffic accident prevention technology research and application", "the ontology-based transportation system of driver training and technology development and standardization", and so on. Driver testing and training examination equipment have been widely used in 234 cities of China's public security, transportation, health, and military institutions and colleges, which screened 6–8% drivers who are not suitable for driving.

Beijing Jiaotong University has dedicated a series of railway driver safety occupation aptitude from the 1990s. It identified eight indicators of railway driver occupation safety inspection (attention, allocation, and stability; transient memory; learning ability; speed prediction ability; personality; color light reaction time; visual acuity; and left and right handgrip) and formed a set of locomotive driver occupation adaptability inspection systems, which have been applied in railway systems and achieved good results.

The "Driver occupation adaptability testing center" developed by Southwest Jiao Tong University was applied to detect cargo & passenger drivers as well as taxi drivers. The national standard GB18463 "motor vehicle driver physical conditions and evaluation requirements" was promulgated by the State Quality Inspection Administration. Sichuan provincial standard DB51/T302 "motor vehicle driver mental condition evaluation code" was promulgated in 1999. Application of Jinan Apt Technology Trade Limited Company of "a military vehicle driver occupation adaptation detecting system" was approved in 2011.

At present, the road traffic driver index, occupation adaptability evaluation standard, testing center, and others have been widely used in China and produced appreciable results. Attention is also paid to index of crane and ship drivers. However, related standards have not been established and officially promoted.

4.2 *Occupation adaptability research on other high-risk industries and special operators*

Beijing Jiaotong University conducted research on the electric power enterprise personnel and identified six indicators of target: accuracy, mean velocity prediction time, attention, speed prediction coefficient of variation, coefficient of variation and complex reaction, and low concave face motion tracking time. These indicators are used for the selection, appointment, and elimination of electric power operators and the analysis of the important degree of these indexes as security factors. It is applied in power industries and shown to have achieved good results.

In 1997, the research institute of labor protection in Zhejiang Province investigated 27 psychological and physiological indicators related to dangerous occupation operators and divided them into four groups: personality traits, general mental ability, cognitive ability, and operation ability of movement. Furthermore, no relevant application effects have been reported. In the same year, the Ministry of Safety and Environmental Protection of Metallurgical Research Institute studied the evaluation system composed of high-temperature operating personnel, comprehensive evaluation of human behavior, the operating environment temperature, and labor physical intensity through the application of Fisher criterion function (i.e., by reducing the dimension to find the easiest classified projection line), established the comprehensive evaluation of a high-temperature operation model, and used clustering analysis to obtain evaluation standard. No systematic studies have been conducted for other occupations such as fireman, miners, and pressure vessel and boiler operators.

4.3 *Test method*

At present, research on the adaptive index of safety occupation has adopted basic methods such as literature analysis, questionnaire, and statistics (Lin, 2013); the measurement of indicators mainly depends on the instrument (Zhang, 2010 & Lin, 1996). Computer simulation is also used. The sources of these indexes root in Cattell 16 personality factors, Holland occupation evaluation theory, Yeatts and Fotts's theory of work adjustment, the "general occupation aptitude test battery" (GATB) method issued by United States Department of Labor, Eysenck EPQ—personality test scale, and Japan Railway driver safe driving adaptability measurement index (Gui, 2005), and so on. At present, developed countries have adopted a portable occupation adaptability evaluation system and realized the man–machine dialog (Chan, 2011 & Elatri, 2012).

5 FIGURES AND TABLES

5.1 *Figures*

Position referred to the guarding place of military police formerly, and now it refers to an office, not without obvious differences. Every organization requires one or more individual responsibilities. The office is in accordance with the provisions of the job and engages in to achieve a clear purpose of work behavior, which is composed of a set of positions with similar responsibilities (Baidu Encyclopedia, 2013).

The position is the basic part of an infrastructure; in particular, a job is undertaken by a person's work; office is a part of organization, which is important to a job and usually regards some tasks, duties, and responsibilities as a whole. For example, the production department in a machinery manufacturing enterprise of the operator in an office, consisting of many position employees, may be composed of drilling operators, laminated operators, screen printing operators, and so on (Baidu Encyclopedia, 2013), whereas the production department operators, maintenance workers, and laboratory assistants hold positions in the mechanical manufacturing industry.

The job is a kind of labor and focuses on verb semantic interpretation and general explanation; position, office, and occupation focus on semantic interpretation and professional interpretation.

The relationship between occupation, office, position, and job is shown in Figure 1.

The total 85 papers related to safety occupational adaptability show a trend changing with time. "Car driver occupation adaptation research" marked the beginning of the safety occupation adaptation research in China (Li, 1989). "Research on locomotive driver personality factors of occupation adaptation" marked the sustainable development stage of

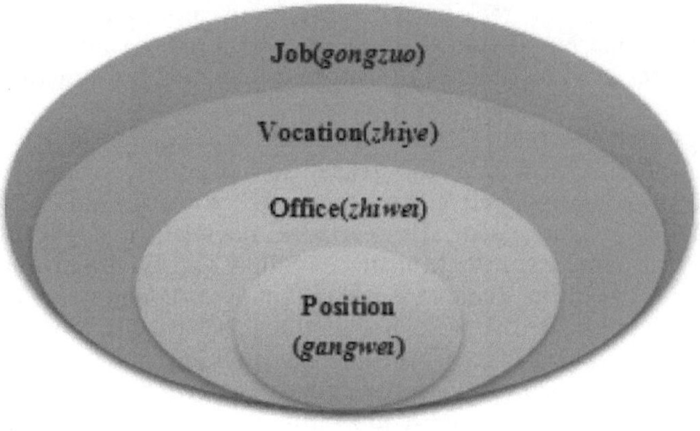

Figure 1. Relationship between occupation, position, and job.

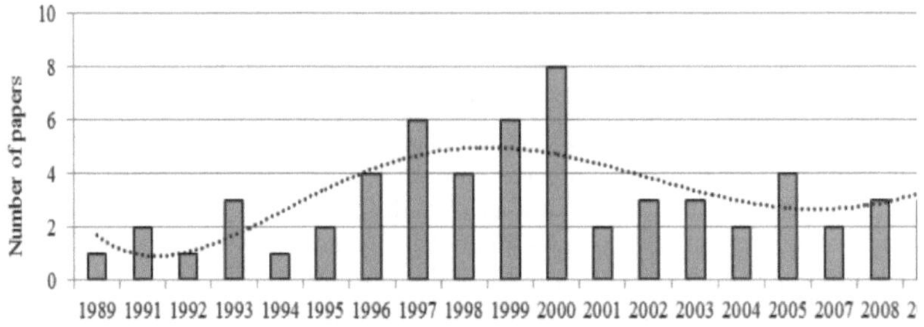

Figure 2. Number of papers on safety occupation adaptation.

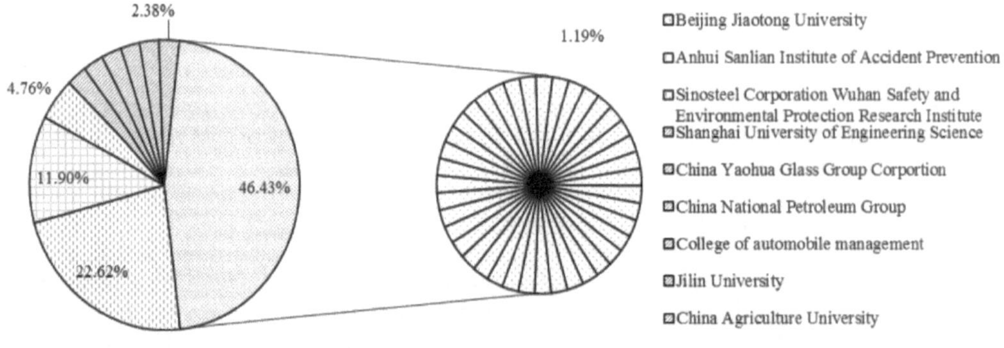

Figure 3. Distribution of research institutions about safety occupation adaptation.

the safety occupation adaptation research (Gui, 2005). The details are shown in Figure 2, which presents an obvious wave and increasing figures in the last 6 years before 2014.

The main institution of the authors is Beijing Jiaotong University, which accounted for 22.62%, followed by Anhui Sanlian Institute of Accident Prevention (11.9%), Sinosteel Corporation Wuhan Safety and Environmental Protection Research Institute (4.76%), and then Shanghai University of Engineering Science, China Yaohua Glass Group Corportion, China National Petroleum Group, College of automobile management, Jilin University, and China Agriculture University, accounting for 2.38% separately and 14.28% totally. The last part has 39 institutions, which accounted for 46.43% in all and 1.19% individually, and the institutions included North China Electric Power University, North China Institute of Science and Technology, Taiyuan University of Science and Technology, Huazhong University of Science and Technology, Capital University of Economics and Business, China Academy of Railway Science, Wuhan University of Science and Technology, and Shandong University of Science and Technology. The distribution details are shown in Figure 3.

5.2 *Tables*

Studies on occupational adaptability in China began in the late 1980s. The field was named occupational adaptation, which mainly focuses motor vehicle drivers, disabled people, car drivers, high-temperature workers, risk device operators, and so on. Its objective is accident prevention, using computer simulation. By the beginning of the 21st century, it was gradually applied to human resources management, first for enterprises and then for city residents, doctors, police, college students, teachers, and so on. The contents included index analysis, psychological load, personality factors, path of ascension, model construction, and scale. Most of them were qualitative. Some representative literature summaries are listed in Table 1.

Table 1. Safety occupational adaptation research abstracts in the last 5 years.

Reference	Sample	Method	Results
Xiaojing Liu, Shangquan Ma, 2014	Coal mine new workers	Experiment rest.	Design an experiment that is based on safety man-machine engineering laboratory. According to the relationship between temperament and safety in safety psychology, make a few tests through the combination of different machines. And on the basis of the scores, decide whether the person is suitable for the job.
Xinyi Zheng, Yanqun Yang, et al., 2014	Drivers	Review analysis, statistical analysis.	The factors influencing safe driving of drivers are risk attitude (0.84), attack (0.58), emotional (0.57), anxiety (0.49), and sensation seeking (0.37).
Yang Song, Lai Jin, et al., 2013	High-speed railway driver	Interview, investigation, and experimental test.	Test indicators were divided into five categories. An analysis of the test data on the adaptability was done. A discriminate function was given for the adaptability.
Haiyan Zhu, Huaping Ye, et al., 2013	Urban Rail Transit workers	Questionnaire survey, data processing	A general index of the key post occupation aptitude is established. The result shows that this system could help select candidates who are suitable for the corresponding positions and are expected to prevent possible human-factor accidents in urban rail transit operation after special training.
Ke Wang, 2012	Electric operators	Literature review, Delphi method, AHP, fuzzy evaluation method	Build the model of safety occupation aptitude evaluation index of electric power enterprise operators (2 grade one, 4 grade two, 22 grade three indexes) and fuzzy comprehensive evaluation model (five grades).
Xianlun Xu, 2011	Different occupation and age groups	Questionnaire survey, data processing software Epidata and SPSS	Preparation of occupational adaptation of personality assessment inventory identifies seven dimensions of personality structure occupation adaptability: innovative, emotional stability, extraversion, integrity, interpersonal relations, adhere to, and tradition.

(*Continued*)

Table 1. (*Continued*).

Reference	Sample	Method	Results
Fei Li, 2011	Operators in power enterprises	Literature, fuzzy mathematics method and fuzzy neural network (MATLAB)	Propose the concept of security capabilities: in special industries, it shows excellent performance and can maintain the stability of no individual characteristics and quality accident at work; establish the concept model of the safety occupation aptitude and evaluation model of fuzzy neural network (grade five).
Hongsheng Shi, Yongxiang Qiu, et al., 2011	Train attendants from Beijing to Lhasa	Questionnaire survey, statistical analysis	Screening to determine the respiratory rate, pulse, oxygen saturation of three indicators can reflect the relative sensitivity to hypoxia; symptoms associated with acute high-altitude reaction can be used as a physiological index of rapid detection of Tibet Plateau Adaptability of the train crew
Fei Li, Sen Li, et al., 2010	Ability model	Literature review	A definition of ability put the human resources management theory into the research on safety occupational aptitude, build safety competency model of special industry, and evaluate the model.
Feng Dong, 2010	Different occupation and age groups	Questionnaire survey, statistical analysis	Preparation of "the Chinese occupation adaptability personality vocabulary by 158 words", personality structure consists of 54 adjectives items, divided into eight dimensions, which is the study of social maturity, inquiry, nonrational, prudent, extroversion, integrity, perseverance, conservative.
Haiqi Zhang, 2010	Operator in Power Supply Bureau	Questionnaires, observation and Delphi method	Put forward the safety occupational adaptation evaluation index system on the basis of five cognitive abilities: attention, memory, thinking ability, observation, and imagination.

6 CONCLUSIONS

6.1 *The concept of safety occupational aptitude*

According to the definition provided by Professor Li Sen, safety occupation aptitude is a physiological, psychological safety quality of self-protection, self-regulating that human must have for adaptation to a safety job.

In a broad sense, safety occupational adaptation belongs to the category of ergonomics, and not man–machine engineering category in China. The following two directions are worth studying. First is the delineation of the physiological and psychological safety aptitude and in turn the differences between occupational adaptation standard and safety occupational adaptation standards. Second is whether the responses of individual in a group, that is, the interaction between the individual and the organization should be putting into our consideration.

So far, the index of driver occupation adaptability in China has emphasized individual ability to adapt to machine, without considering individual reactions to people around and social indicators. For example, a driver's work environment is special and he is not susceptible to other peer influence (such as herd mentality); however, his responsibility for passengers or his passion toward his work will affect the safety occupational adaptation. For example, CO leak detection personnel must work in pairs, as both will influence each other and directly affect the safety. Therefore, the social characteristics of individuals are worth studying.

6.2 *Future research of safety occupation aptitude*

Many scholars think that occupation adaptation can be divided into general occupational aptitude and special occupational aptitude. The general occupational aptitude directs at physiological and psychological characteristics engaged in general occupation, whereas special occupational aptitudes refer to special physiological and psychological quality engaged in particular occupation.

Safety occupation adaptation focuses on adaptation research, emphasizing the process; occupational aptitude focuses on the occupation tendency, emphasizing the results. Therefore, safety occupation adaptation should study more on "adaptation". Taking the very important factor of safety occupation adaptation of cognitive ability as an example, the safety occupation adaptation should study more on how to improve cognitive ability and achieve safety behavior by means of testing, learning, and training, not just evaluating individuals with safety features.

In accordance with the classification of occupation aptitude, safety occupational adaptation belongs to a special category of general occupation adaptability, mainly used for occupation guidance and employment counseling, while safety occupational adaptation is more related to safety physiology and psychological traits of high-risk industry. Safety occupational adaptation can be further divided into coal mine, chemical industry, construction, traffic fields, and so on. It can also be divided into man–machine, interpersonal, man–environment areas and so on. At present, studies on other directions except traffic, human–machine, and human–environment are still lacking. There is still great potential for study on safety occupation adaptation.

ACKNOWLEDGMENTS

This paper was supported by the National Natural Science Foundation of China (Grant No. B16 A300010).

REFERENCES

BoChe, Wen. 2001. *Psychological counseling Encyclopedia*. Zhejiang: Zhejiang science and Technology Publishing.

Chan, Alan H.S. 2011. Driving Aptitude Test (DAT): A new set of aptitude tests for occupational drivers. *International Multi Conference of Engineers and Computer Scientists* 1322–1326.

Chao, Wu. 2002. Occupation adaptation in human resources management. *Journal of Teachers in Dandong College* 24 (4):73–74.

Chuansi, Gao & Zhongshui, Zhu. 1996. Occupation adaptability test evaluation and application. *Industrial safety and dust control* (1):21–25.

Chunlin, Gui. 2005. *Research on locomotive driver personality factors of occupation adaptation*. Beijing: Beijing Jiaotong University Publishing.

Dictionary of Language Research Institute of Chinese Academy of Social Sciences editorial office. 2005. *Modern Chinese dictionary*. Beijing: the Commercial Publishing.

Dongfang, Feng. 2010. *Occupation adaptability personality characteristics of lexical analysis and adjective rating scale compilation*. Zhengzhou: Zhengzhou University Publishing.

Elatri, S. 2012. European legislation and occupational psychological evaluation of train drivers and other safety related personnel. *Rail Human Factors around the World: Impacts on and of People for Successful Rail Operations* 716–727.

Fei, Li & Sen, Li (eds). 2010. Construction and evaluation of capability to model the safety occupation aptitude. *China Safety Science Journal* 15:135–137.

Fei, Li. 2011. *Study on the model construction and evaluation of safety occupation Aptitude*. Beijing: Beijing Jiaotong University Publishing.

Feng, Xu. 2000. The analysis of social psychology training on the adaptability of education research and experiment. *Educational Research and Experiment* (06): 58–61.

Fengxiang, Xiao & Jing, Suo. 2011. Occupational education. *Journal of Tianjin University (Social Science Edition)* 13 (5): 435–440.

Haili, Zhang & Weining, Fang. 2011. Subway driver occupation adaptability test management system design. *Urban Mass Transit* (4):69–73.
Haiqi, Zhang. 2010. *Power Supply Bureau safety occupation aptitude evaluation system research.* Beijing: Beijing Jiaotong University Publishing.
Haiyan, Zhu & Huaping, Ye (eds). 2013. Test and evaluation system of occupation aptitude for urban rail transit key operation position. *Urban mass transit* 11:53–56.
JianShen, Liu & Xiaoyi, Wang (eds). 2013. Analysis on occupational adaptability of young employees in the production of oil and gas. *New West* 15:22–26.
Jinglei, Yao & Xiao, Lin, e. 2013. Altitude acclimatization of coach attendants serving Beijing-Tibet railway. *Journal of Environmental & Occupational Medicine* 30(8):604–607.
Ke, Wang. 2012. *Safety occupation aptitude fuzzy comprehensive evaluation model and evaluation study.* Beijing: Beijing Jiaotong University Publishing.
Lin, Chen & Chuansi, Gao. 1996. Computer in occupation adaptability test application. *Computer and Communications* 14 (4):63–65.
Ming, Tan & Hanqing, Fang (eds). 2012. Study on the review of our occupation adaptability. *Chinese occupation technology education* (18):34–39.
Sen, Li & Shouxin, Song & Yuting, Chen. 2009. The evaluation of safety occupation adaptability. *Productivity Research* 15:135–137.
Sen, Li & Shouxin, Song. 2010. The safety occupation adaptability index of power workers. *Productivity Research* (7):97–98.
Sen, Li. 2010. *The study of the safety occupation aptitude base on metal resource*. Beijing: Beijing Jiaotong University Publishing.
Shi, HongSheng & Qiu, Yongxiang (eds). 2011. Occupational adaptability index of passenger train crew in Plateau. *Railway Energy Saving & Environmental Protection & Occupational Safety and Health* 1 (1): 34–41.
The Chinese occupation education agency. 1985. *Huang Yanpei education.* Shanghai: Shanghai Education Publishing.
Xianlun, Xu. 2011. *Contemporary Chinese occupation adaptation of personality structure.* Zhengshou: Zhengzhou University Publishing.
Xiansheng, Li. 1989. *Car driver occupation adaptation research.* Jilin: Jilin University Publishing.
Xiaojing, Liu & Shangquan, Ma. 2014. The discussion of coal mine new workers' job adaptability test system. *Journal of North China Institute of Science and Technology* 11(8):25–33.
Xinyi, Zheng & Yanqun, Yang (eds). 2014. Analysis of safety driving adaptability factors in twenty years in China. *Chinese Journal of Ergonomics* 20(5):64–69.
Yang, Song & Lai, Jin (eds). 2013. Research on high speed railway drivers' safety adaptability and its detection technology. *China Safety Science Journal* 23(6):9–13.
Yanqiu, Tian. 1999. On the occupation adaptability. *The Ministry of agriculture of Beijing Land Reclamation Management Cadre Institute* (3):50–53.
Yi, Jin. 2006. *Study on the problems and Countermeasures of secondary vocational school graduates occupation adaptability of Shanghai*. Shanghai: Shanghai Normal University Publishing.
Zhixian, Zhu. 1989. *The dictionary of psychology*. Beijing: Beijing Normal University Publishing.
Zhongmin, Li. 2011. *Knowledge type human capital competency research.* Beijing: Science Publishing.

Innovation and Application of Engineering Technology – Oscar, Chen & Wang (Eds)
© 2017 Taylor & Francis Group, London, ISBN 978-1-138-05057-0

Application of field test on gas drainage by long-distance bedding directional borehole

Wei Qiao
State Key Laboratory of Gas Disaster Detecting, Preventing and Emergency Controlling, Chongqing, China
Chongqing Research Institute Co. Ltd., China Coal Technology and Engineering Group, Chongqing, China

ABSTRACT: The gas outburst control method is adopted to pre-evacuate the serious overburden of gas in the Huoerxinhe No. 3 coal seam by the long-distance bedding directional borehole. The results show that the gas content in the test point decreases from 13.81 to 7.81 m^3/t after 25 days of gas drainage by long-distance bedding directional borehole and the gas pre-pumping rate is about 43.30%. The driving speed of the main transport roadway is 1.48 times that of the traditional ordinary drilling drainage measures, and the tunneling efficiency indicators have not exceeded. This method confirmed that the long-distance bedding directional borehole can effectively reduce the amount of drilling holes constructed, increase the gas extraction rate compared to that of the traditional ordinary drilling hole, effectively eliminate the danger of protruding, and realize the safe and rapid excavation of the coal mine roadway.

1 INTRODUCTION

The Shanxi Huoerxinhe coalmine area is 71.39 km^2, and the main mining is No. 3 coal seam, which is a single thick coal seam. The average thickness is 5.65 m, and the average dip angle is 5°. According to the mine gas emission characteristics, the mining face gas mainly come from the No. 3 coal seam.

At present, the depth of the main roadway excavation area is about 536 m, and the coal seam gas content measured at 104 m away from the No. 14 contact lane is 11.51 m^3/t. It is found that when the roadway is excavated normally, the gas concentration of the backflow wind varies greatly, and the maintenance gas concentration is generally between 0.3% and 0.5%. The gas concentration of production classes generally varies between 0.4% and 0.8%. The average gas emission rate is about 6.2 m^3/min in the tunneling face. The gas emission of coal wall is about 3.33 m^3/min, and the gas emission rate is about 2.87 m^3/min. No gas drainage is carried out in the heading face of the mine, and only ventilation was adopted to solve the gas emission. However, with the gradual extension of the roadway, coal seam gas content increases and ventilation cannot completely solve the digging gas emission and occasionally gas close to the phenomenon of overrun. During the main transport roadway under normal tunneling, the return air gas concentration has a big change, and the average gas emission is about 6.2 m^3/min. It can be seen that it urgently needs to control and manage the main transport roadway driving face gas drainage.

Therefore, this study mainly adopts the technology of gas drainage with directional long boreholes to carry out the application in the tunneling face of the main transport roadway in Huoerxinhe No. 3 coal seam. During the test, the ZYWL-6000DS type crawler type full hydraulic drilling rig, independently developed by the China Coal Science and Industry Group Chongqing Research Institute Co. Ltd., is used to carry out the construction of long-borehole gas drainage.

2 LONG-DISTANCE BEDDING DIRECTIONAL BOREHOLE CONSTRUCTION AND PROGRAMING

The head of the main transport roadway constructs two main holes: a drilling hole in the roadway head with a height of 1.6 m and a hole to control the coal roadway. The drilling length is 400 m, with horizontal and vertical directions to open branches, and the hole spacing is 4 m. The drilling layout parameters are shown in Table 1 and Figure 1.

The main transport roadway driving face distance from the 14th contact lane is 104 m after the construction of the first group, with hole spacing of 4 m and 80 m after the construction of the second group, with hole spacing of 3 m. After the completion of the second group ordinary drilling and extraction test, the gas content in the coal seam is measured after excavation at 80 m (distance: 14 m, contact hole: 264 m, maximum gas content: 13.81 m^3/t), construct of directional bedding drilling and two main holes, opening the hole with eight branches, and controlling 5.3 m coal seam gas and the entire coal thickness with each branch length of 24–114 m.

Table 1. Drilling layout parameters.

Hole number	Aperture orientation	Target azimuth	Opening height (m)	Final hole diameter (mm)	Main hole length (m)	Branch hole length (m)
1#	0°	0°	1.6	94	400	144
2#	0°	0°	1.6	94	400	144
Total	1088 m					

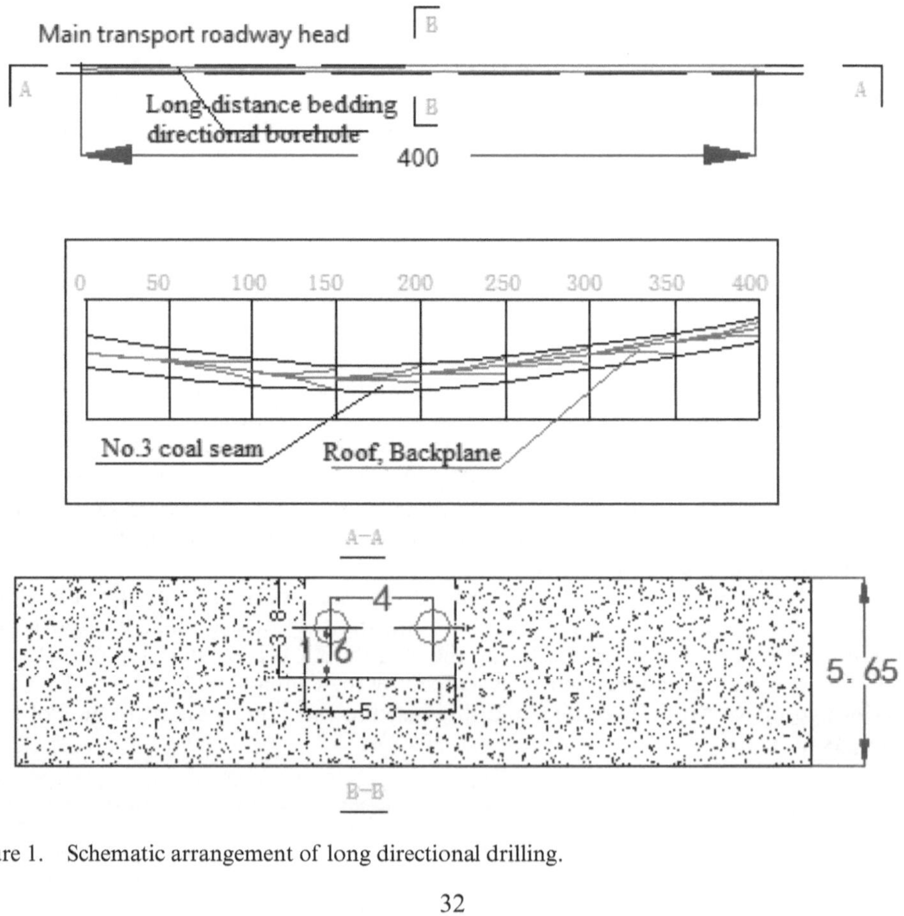

Figure 1. Schematic arrangement of long directional drilling.

3 LONG-DISTANCE BEDDING DIRECTIONAL BOREHOLE GAS DRAINAGE EFFECT

3.1 *Statistical analysis of the gas extraction capacity*

Directional long boreholes are constructed in front of the main transport roadway heading face. The observations were carried out from 29 February 2016 to 24 March 2016. The drilling gas concentration is above 71% after pumping 25 days, and the drilling single hole standard gas extraction rate is above 0.77 m^3/min. The statistical results are shown in Table 2.

As can be seen from Figures 2–3, the drilling gas extraction concentration of Huoerxinhe coalmine main transport roadway heading face is very high, the single-hole gas concentration is above 70%, and the single-hole pure gas extraction volume is 3.73 times that of the ordinary single-hole pure gas extraction, which successfully eliminates the gas emission anomaly and effectively eliminates the danger of coal and gas outburst. The negative pressure of gas drainage in the drilling section farther away from the location of hole drilling begins to decay. The drilling gas extraction volume is reduced, and the residual gas content is large. It is verified that the distribution of negative pressure along the long direction is gradually reduced.

Table 2. Long bedding directional borehole gas extraction test data.

		1# hole			2# hole		
No.	Date	Gas concentration (%)	Mixed flow (m^3/min)	Extraction volume (m^3/min)	Gas concentration (%)	Mixed flow (m^3/min)	Extraction volume (m^3/min)
1	29 February	89	1.52	1.35	90	1.47	1.32
2	3 March	90	1.27	1.14	88	1.34	1.18
3	7 March	85	1.44	1.22	81	1.51	1.22
4	10 March	80	1.11	0.89	83	1.06	0.88
5	13 March	78	1.10	0.86	79	1.05	0.83
6	19 March	75	1.08	0.81	76	1.04	0.79
7	24 March	72	1.07	0.77	71	1.20	0.85

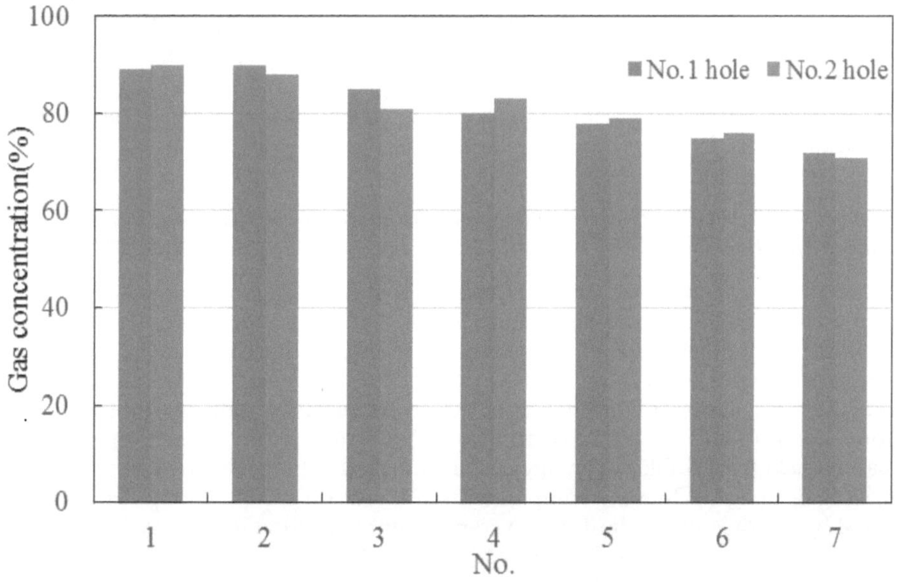

Figure 2. Long bedding directional drilling gas extraction concentration.

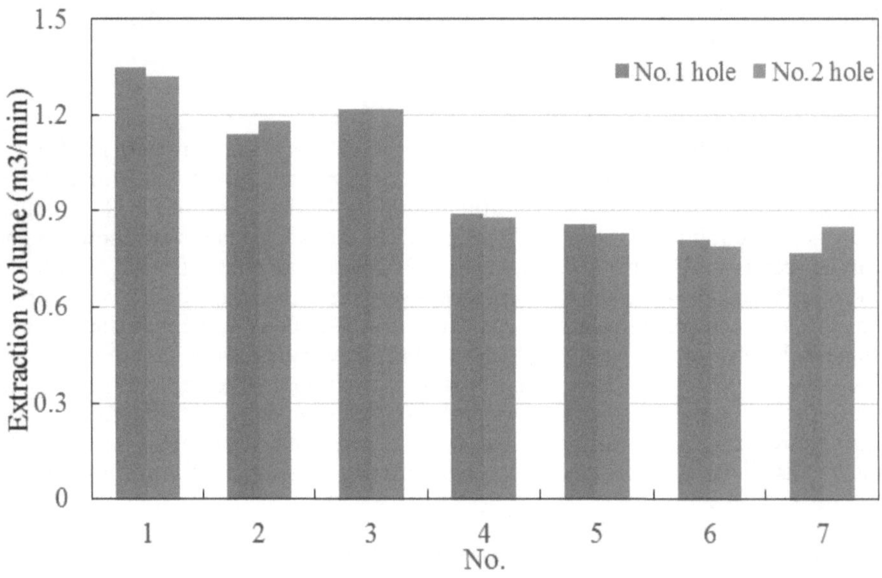

Figure 3. Pure gas drainage capacity with long bedding directional drilling.

3.2 *Analysis of gas content change*

The average thickness of the main transport roadway No. 3 coal seam is 5.52 m, the gas content of the original coal seam is 13.81 m^3/t, and the length of the drilled hole is 400 m. The coal seam gas reserves in the front area of the main roadway face controlled by the directional long drilling in this area is calculated to be about 16,500 tons, and the coal seam gas reserve is controlled to be 227,900 m^3.

During the main transportation roadway driving face driving process, the cumulative statistics of the residual gas content determination is eightfold. Combining with the downhole determination of the directional control region long boreholes in the driving working face of the main transportation roadway, the average gas drainage volume of the single-hole extraction is about 0.252 m^3/min.hm, the total extraction volume is about 98,700 m^3 for 25 days, and the residual gas content of coal seam is between 5.64 and 7.83 m^3/t (Table 3).

Table 3 shows that the long-distance bedding directional borehole measure significantly reduces the coal seam gas content in the heading face of the main transport roadway in the Huoerxinhe coalmine, and the residual gas content in the coal seam decreases gradually along the drilling direction.

3.3 *Long-distance bedding directional borehole drilling with long construction quantity analysis*

Taking the main transport roadway excavation 400 m for example, the ordinary drilling rig construction is carried out to drainage, then each 80 m digging needs to construct a cycle, and each cycle drilling footage is 200 m. The gas content drops to the target gas content of 8 m^3/t, which takes five cycles. According to the ordinary drilling construction quantities in January 2016, the comparison of comprehensive advance rate between the long-distance bedding directional borehole and the ordinary drilling is shown in Table 4.

As can be seen from Table 4, the integrated tunneling is the slowest when using the ordinary drilling drainage. While the comprehensive excavation speed of long-distance bedding directional borehole is the highest, the average monthly drilling footage is 133 m. And it is proved that the long-distance bedding directional borehole can effectively reduce the amount of drilling holes constructed.

Table 3. Determination of gas content during the directional control region long-distance boreholes driving.

Drilling section (m)	Original gas content (m^3/t)	Residual gas content (m^3/t)	Extraction rate (%)
45–55	13.81	5.81	57.93
95–105	13.81	5.64	59.16
145–155	13.81	7.32	46.99
195–205	13.81	7.83	43.30
245–255	13.81	7.61	44.90
295–305	13.81	7.59	45.04
345–380	13.81	7.81	43.45

Table 4. Comprehensive advance rate.

Classification	Ordinary drilling		Long-distance bedding directional borehole
	4 m hole spacing	3 m hole spacing	
Drilling work amount	1000 m	1000 m	1088 m
Extraction time	at least 95 days	at least 70 days	25 days
Drilling construction time	5 days	5 days	8 days
Normal advance speed	7 m/day	7 m/day	7 m/day
Comprehensive advance speed	76 m/month	90 m/month	133 m/month

4 SUMMARY

The gas content of the main transport roadway driving face decreases from 13.81 to 7.81 m^3/t after 25 days by the implementation of the ordinary borehole gas drainage, long-distance bedding directional borehole gas drainage test, and the gas pre-pumping rate of about 43.30%. With using the ordinary drilling drainage, the integrated tunneling is the slowest. While the comprehensive excavation speed of long-distance bedding directional borehole is the highest, the average monthly drilling footage is 133 m. Through the application of long-distance bedding directional borehole in the main transport roadway of Huoerxinhe coalmine, it is proved that this borehole can effectively reduce the amount of drilling holes constructed and increase the gas extraction rate compared with the traditional ordinary drilling holes, so as to achieve the purpose of effectively eliminating the outburst danger and finally reduce the construction quantity of drilling, thereby saving costs as well as achieving a safe and rapid excavation of the coal mine roadway.

ACKNOWLEDGMENTS

This study was financially supported by the State Key Research Development Program of China (Grant No. 2016YFC0801404) and Chongqing Technological Innovation Leading Talent Support Program (CSTCKJCXLJRC14).

REFERENCES

Du Z.J. & Liu Z.L. 2007. Technology of gas drainage efficiency about feather-veined 1000 m length-drillhole along seam. *Coal Safety and Environment* 34(1): 28–29.

Li S.W. 2013. Study on the negative pressure distribution at different boreholes depths in Xinan mine. *Coal Engineering* 5: 103–104.

Nie B.S., Hu S.T. & Liu M.J. 2012. Application of outburst elimination technology for pre-pumping coal seam gas of long boreholes along seam in Pansan coal mine. *Safety in Coal Mine* 43(12): 120–122.
Tong B. Zhao M.M. & Liu M.J. 2011. Application of outburst elimination technology with large diameter and long borehole drilled along seam to Panyi mine. *Coal Science and Technology* 39(9): 30–33.
Wang K., Zhao X.S. & Dai L.C. 2014. Research and application on negative pressure distribution law of drilling for consequent landslide gas extraction. *Safety in Coal Mine* 45(5): 1–4.
Wang Z.F., Tian F.C. & Zhao B. 2010. The testing of gas drainage efficiency about feather-veined 1000 m length-drillhole. *Journal of China Coal Society* 35(1): 76–79.
Zhou J.L., Dong H.K. & Li G.Q. 2013. Research on effect test method for regional gas pre-drainage measure against outburst by directional long boreholes along coal seam. *Mining Safety and Environment Protection* 40(6): 69–72.

Innovation and Application of Engineering Technology – Oscar, Chen & Wang (Eds)
© 2017 Taylor & Francis Group, London, ISBN 978-1-138-05057-0

Research on the ground pressure features of longwall top coal caving working face in extra-thick coal seam

Chuang Liu
School of Energy Science and Engineering, Henan Polytechnic University, Jiaozuo, China
Department of Mining and Materials Engineering, McGill University, Montreal, Canada

Huamin Li & Wen Wang
School of Energy Science and Engineering, Henan Polytechnic University, Jiaozuo, China

Dongjie Jiang
School of Energy Science and Engineering, Henan Polytechnic University, Jiaozuo, China
School of Safety Science and Engineering, Henan Institute of Engineering, Zhengzhou, China

Wenxue Chen
School of Civil Engineering, Henan Polytechnic University, Jiaozuo, Henan, China

ABSTRACT: On the basis of the raw data of shield pressure and overlying strata movement measured in the field, the ground pressure features and overlying strata movement were analyzed in 8202 extra-thick coal seam longwall top coal caving working face of Tongxin coal mine after its extraction. The research results show that during mining the interval of the lower sub-key strata caved for the first time is 120 m, and the periodic weighting interval ranges from 15 to 30 m, with an average of 25 m. Because of the large mining space of longwall top coal caving working face in extra-thick coal seam and the large rotary angle of the overlying strata, the lower sub-key strata cannot connect with each other to form a stable bearing structure; therefore, the shield pressure is a given load of roof rock, which results in a constant pressure of the shield.

Keywords: Extra-thick coal seam; top coal caving face; strata movement; ground pressure features; shield pressure

1 INTRODUCTION

With the development of China's coal mining technology, mine construction, and coal mining toward large-scale, intensive direction, extra-thick coal seam mining is an urgent need for high yield and efficiency of coal mining methods (Li H.M. et al., 2015; Yu B. et al., 2016). Compared with the traditional layered mining caving technology, longwall top coal caving

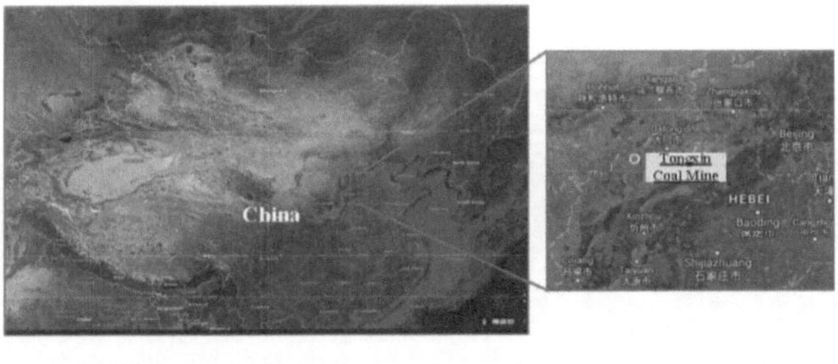

Figure 1. Location of Tongxin coal mine.

(LTCC) is a good method to exploit the extra-thick coal seam, with high yield, high efficiency, low cost of roadway excavation rate, low cost of raw coal, strong adaptability, and high safety (Huang Z.Z., 2013; Jia X.R. et al., 1998; Yan S.H., 2009). LTCC has been rapidly gaining popularity nowadays, nearly 30 years after its introduction in China.

Most research works in China are concentrated in the aspects of working surface equipment, ground pressure control, and the selection of mining method on the extra-thick coal seam. However, there are only few studies on the mutual influence of dual coal seam mining on extra-thick coal seam. Two main coal seams in Tongxin coal mine (Figure 1), have been mined, with the upper coal seam being the Jurassic and coal being stable. The distance between the two coal seam layers is 150 m. In the process of the lower coal seam mining, the overlying strata are affected by the mining action, the stability of surrounding rock is poor, the movement features of overlying rock is irregular, and the randomness is strong, which brings hidden danger to the safety production of mine. On the basis of this, taking Tongxin coal mine 8202 LTCC working face as an example, research on extra-thick coal seam under gob-caving roof strata movement and ground pressure features, in order to achieve the purpose of safe and efficient mining working face, provides a reference for the similar conditions of coal mining.

2 GENERAL SITUATION OF ENGINEERING

8202 LTCC working face is mining 3–5# carboniferous coal seam. The mining height of the 8202 working face ranges from 13.61 to 28.92 m, with an average value of 20.58 m. The 3–5# coal seam is stable and consists of 8–15 layers; among them, the 1.96 m gangue accounts for 9.6%. The Polodyakonov coefficient of 3–5# coal seam is 1–2, resulting in an easy collapse of the top coal. The 8202 LTCC working face uses comprehensive mechanized, low top coal caving, longwall retreating mining, multiround and hierarchical sequential caving method, and the natural collapse method for the overlying roof, with a 0.8 m cycle progress and coal caving step distance, until the working face advances to 30 m before the mining stop line. 8202 top coal caving face operates at the cutting height, with no caving top coal between the stop mining line and 30 m before. The layout of gate roads of 8202 top coal caving face is shown in Figure 2.

The thickness of the lower sub-key strata is 17.04 m, which consists of sandy mudstone, No. 1 coal seam, coarse sandstone, and fine sandstone. The immediate roof thickness is 3.87 m, for carbon mudstone and 2# coal seam. The immediate bottom is 1.98 m thick, grayish brown sandy mudstone, containing carbonaceous clip and kaolin coal line. The main floor is dark gray, thick–layered sandy mudstone.

The ZF15000/27.5/42 low top coal caving type shield is selected for the 8202 LTCC working face. The support height of this type shield is 2.75–4.20 m, the center-to-center distance between adjacent shields is 1.75 m, the yield pressure is 36.86 MPa (15000 KN), and the setting pressure is 31.4 MPa (12778 KN). In actual applications, because of the supply pipe pressure and other factors, the yield and setting pressure of the shield are 31.0 MPa (12615 KN) and 14.0 MPa (5700 KN), respectively. In order to adapt to the shields in the middle of the working face, the ZFG13000/27.5/42H low caving top coal type is selected as transition shield and ZTZ20000/30/42 as end position shield. The working face is arranged in a total of

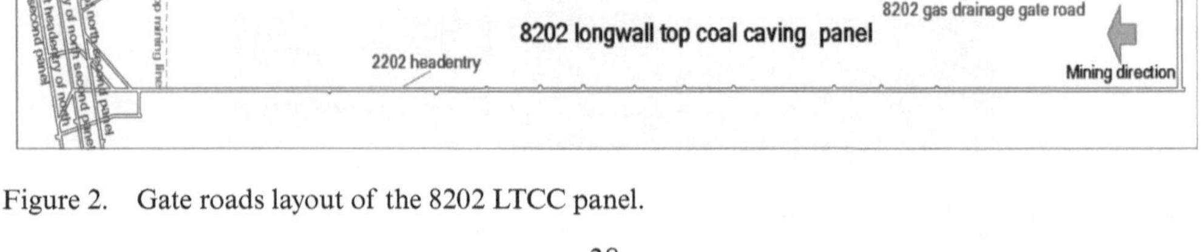

Figure 2. Gate roads layout of the 8202 LTCC panel.

108 middle shields and nine transition shields (four transition shields arranged in the upper working face, and five transition shields arranged in the lower working face). The upper part of the working face is arranged with one set of end shield, including two shields.

3 GROUND PRESSURE OBSERVATION

A total of 11 sets of KJ21 online ground pressure monitoring system recorder are installed in the middle working face shields of the 8202 LTCC working face, that is, one measuring point is arranged every 10 shields, and the monitoring points are arranged as shown in Figure 3. The monitoring system records data every 5 min, that is, real-time record of the changes of the shield pressure during its process of rising, moving, and falling. The monitoring process spanned from 13 August 2016 to 9 October 2016, during which 278 m of working face advances.

3.1 *Features of overlying strata periodic weighting*

The sum of the average of shield pressure at the end of a cycle P_t and its square deviation σ ($P_t + \sigma$) is taken as the main index to judge the cycle pressure of the overlying roof strata. According to the calculation results of the periodic weighting criterion, the middle period of the working face is higher than the two ends of the working face.

The periodic weighting criterion of 48# shield is maximum, that is, 30.71 MPa, and that of 88# shield is 25.73 MPa, which is only 70% of the yield working pressure and indicates that the overlying strata strength around the 88# shield is weak and the roof pressure appears to ease. The periodic weighting criterion is obtained after the calculation of the average of shield pressure at the end of a cycle P_t and its square deviation σ. 18#, 58#, and 98# shields in the upper, middle, and lower working faces, respectively, are selected to analyze the working pressure of each shield.

As shown in Figure 4, according to the raw data of 18#, 58#, and 98# shield pressures during the 25–278 m mining period, the overlying roof strata of the working face is periodic caving. The lower sub-key strata initial weighting distance is 100–140 m, with an average of 120 m; meanwhile, the periodic weighting distance is 15–30 m, with an average of 25 m. The periodic weighting characteristics is not obvious, and the shield pressure slightly increases during the periodic weighting time.

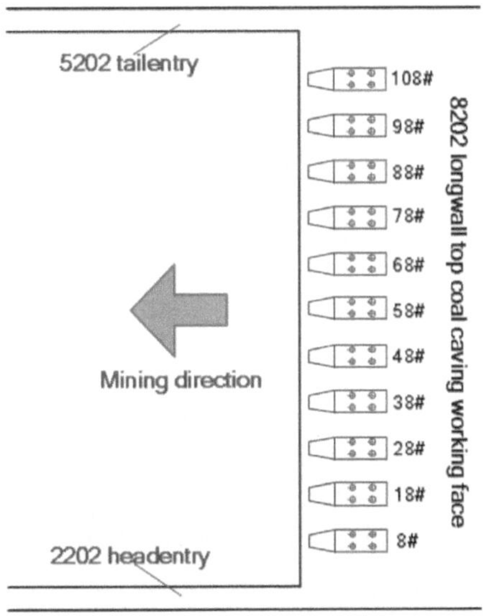

Figure 3. Layout of shield pressure monitoring station.

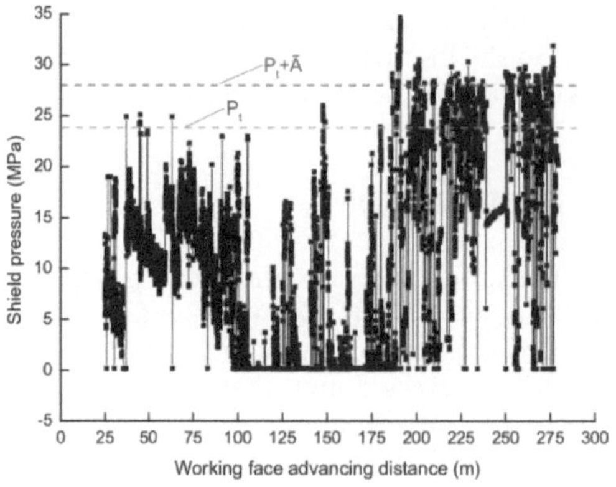

(a) 18# shield

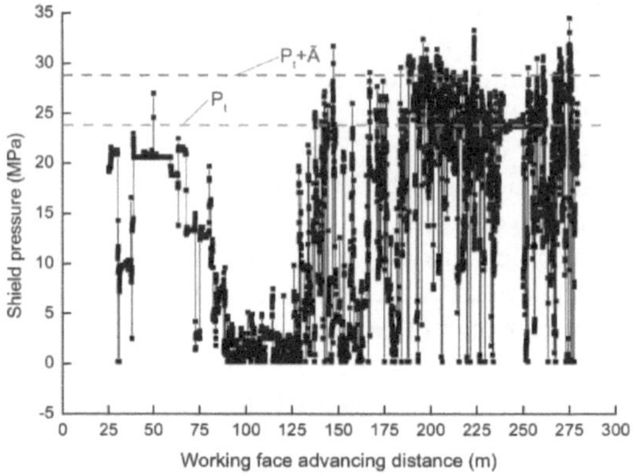

(b) 58# shield

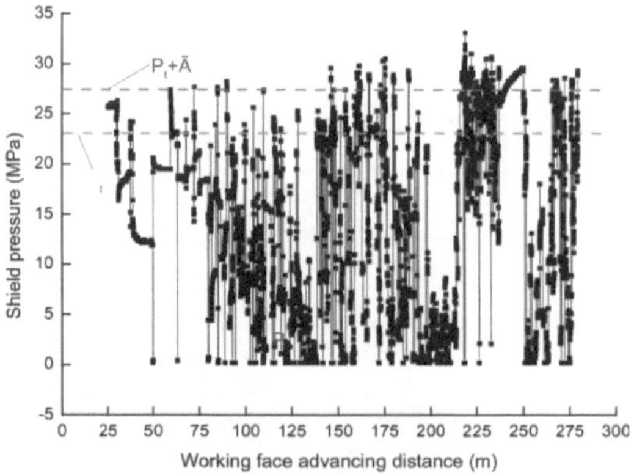

(c) 98# shield

Figure 4. Changes of shield pressure with advancing of the working face.

3.2 *Shield pressure features along the working face direction*

Figure 5 shows that the square deviation of the average of 38#–68# shield pressure at the end of a cycle is large, indicating that the ground pressure around the range appears more strongly.

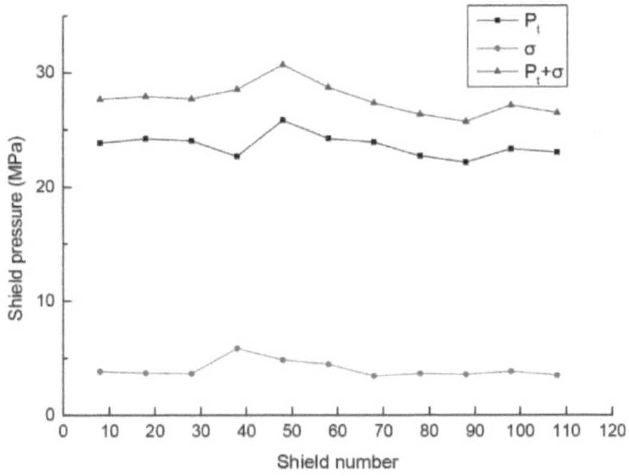

Figure 5. Curve of average of shield pressure at the end of a cycle and its square deviation.

The square deviation of the average of 38# shield pressure at the end of a cycle is the maximum at 5.87 MPa, and that of the 108# shield is the least, which is 3.47 MPa. In the whole working face, the square deviation of the average of 38# shield pressure at the end of a cycle presents a big trend in the middle and both ends, indicating that the ground pressure in the middle of the working face is higher than that at the two ends.

4 ANALYSIS OF SHIELD PRESSURE

From the shield pressure curve, it can be seen that the working pressure of shield in the mining period changes little (without considering the shield maintenance and falling stage) relative to the working face with large mining height. This is because the mining space of the top coal caving face is not consistent with the support space, as the transfer medium to apply the load directly to the shield is top coal rather than rock (Zhu S.S. et al, 1996; Gu Q.Z. et al, 1996). The density of coal is small, the compression is brittle and loose, the joint and crack is much more than that of rock, the fluidity is high, and the overall stability is low. After being discharged and recovered, the top coal has sufficient free space in the mined out area. Under the support pressure, the top coal body is compressed in the longitudinal direction and expands and breaks in the lateral direction (Kang T.H. et al, 2007). During the process of the working face mining, the energy is gradually released to the mined-out area, and the top coal is broken and crushed.

Because of the recovery of top coal in top coal caving face, the space needed to fill in the gob is increased exponentially, and only the higher full caving zone can maintain the balance of the whole stope rock mass. In the case of extra-thick immediate roof, the collapse height of the caving zone is increased; when the thickness of the immediate roof is not sufficient, the upper overlying strata with a certain thickness will be used as the rock block of the regular collapse zone to compensate for the lack of gob filling. Because of the large rotation angle of the lower sub-key strata, the horizontal projection length of the rock beam is shortened greatly. When the change of the contact position between the rock beams is not enough to compensate for the shortened displacement, the rock beam loses the lateral force of contact. The lower sub-key strata above this layer is equivalent to the fractured-zone strata of slicing mining. Although it can form a balanced structure of "Voussior Beam", because of the long formation time, the structure is formed far away from the mining face and hence it does not have a significant impact on the mining space (Li H.M. et al, 2014; Yu B. et al, 2014).

Therefore, under different geological conditions of top coal caving faces, the shield pressure is small and has little difference. Furthermore, there is no significant increase or decrease in pressure before and after the periodic weighting because the external load of the shield is a given load, which is the function of top coal and the caving zone that can collapse.

5 SUMMARY

1. The initial weighting distance of lower sub-key strata is 120 m, and the periodic weighting interval is 25 m in Tongxin 8202 LTCC working face.
2. The work pressure of shield in the middle section of the top coal caving face is higher than that at the two ends, and the ground pressure of middle shield is higher than that of the two ends.
3. For the extra-thick top coal caving face, the working pressure of shield is a given load, which changes relatively slightly during mining.

ACKNOWLEDGMENTS

The authors acknowledge the contributions of Associate Prof. GP Zhang, XH Zhang, and QL Zhang in the setup of the apparatus and the collection of field data. The authors thank Datong Coal Mine Group Co. Ltd, Tongxin coal mine, and Tashan coal mine for financial support.

REFERENCES

Gu Q.Z., Shi Y.W., Qi Q.X. Rules of roof movement in sub-level caving workings. *Journal of China Coal Society*. 1996, 21(1): 45–50.

Huang Z.Z. Study on laws of roof strata fracture of fully mechanized caving mining under goaf in ultra-thick seam. *Coal Science and Technology*. 2013, 41(7): 60–62, 66.

Jia X.R., Zhai Y.D., Yang S.S. Structure of roof rocks in sub-level caving face and calculation of rock weighting. *Journal of China Coal Society*. 1998, 23(4): 366–370.

Kang T.H., Chai Z.Y., Li Y.B., et al. Study on physical simulation of full-seam mining for a 20 m very thick and medium hard seam by sub-level caving mining with high bottom cutting height. *Chinese Journal of Rock Mechanics and Engineering*. 2007, 26(5): 1065–1072.

Li H.M., Jiang D.J., Syd S.P., et al. Analysis on loading features and suitability of hydraulic powered caving supports. *Coal Science and Technology*. 2015, 43(6): 22–28, 70.

Li H.M., Jiang D.J., Li D.Y. Analysis of ground pressure and roof movement in fully-mechanized top coal caving with large height in ultra-thick seam. *Journal of China Coal Society*, 2014, 39(10): 1956–1960.

Yan S.H. Theory study on the load on support of longwall with top coal caving with great mining height in extra thick coal seam. *Journal of China Coal Society*. 2009, 34(5): 590–693.

Yu B., Zhu W.B.,, Gao R., et al. Strata structure and its effect mechanism of large space stope for fully-mechanized sublevel caving mining of extremely thick coal seam. *Journal of China Coal Society*. 2016, 41(3): 571–580.

Yu B., Liu C.Y., Yang J.X., et al. Mechanism of strong pressure reveal under the influence of mining dual system of coal pillar in Datong mining area. *Journal of China Coal Society*. 2014, 39(1): 40–46.

Zhu S.S., Li H.C., Yang Z.F. The structure of overlying coal and rocks in sublevel caving faces. *Chinese Journal of Mechanics and Engineering*. 1996, 15(2): 150–154.

Innovation and Application of Engineering Technology – Oscar, Chen & Wang (Eds)
 ISBN 978-1-138-05057-0

Experimental study on the pressure compensation method for mine-used infrared methane sensor

J. Cao
State Key Laboratory of Gas Disaster Detection, Prevention and Emergency Control, Chongqing, China
Chongqing Research Institute Co. Ltd., China Coal Technology and Engineering Group, Chongqing, China

ABSTRACT: Accurate measurement of methane concentration is very important for the early warning and control of coal mine disasters. Aiming at the problem that the measurement precision and reliability of the existing infrared methane sensor are slightly affected by pressure, this paper analyzed the principle of the infrared methane sensor and the cause of the error. Furthermore, through the experiment of measuring methane concentration under different gas pressure conditions, the pressure compensation calculation model of the infrared methane sensor is obtained by using the least square method. Through the compensation calculation, the measurement error of the infrared methane sensor in the pressure range of 40–120 kPa can meet the requirement of various kinds of sensors in the coal industry, which provides an effective path for methane gas detection in coal mine.

1 INTRODUCTION

Methane (CH_4) is a flammable, explosive, colorless, and tasteless gas that is difficult to detect. It is also the main component of mine gas, accounting for 83–89%. Explosions occur when the methane concentration exceeds the standard permissible value, which have a huge impact on the safety of mine production (Liu & Qiao, 2006). Therefore, timely and accurate detection of methane concentration is of utmost importance for coal mine safety, personal safety, and the environment.

At present, there are several methane detection methods (Chai *et al.*, 2009, J & RP, 2012, Li & Zhou, 2009, Tong, 1997, Wu *et al.*, 2009, Yu, 2012, Zhang *et al.*, 2010) such as catalytic combustion, nondispersive infrared spectrometry, optical interferometry, and semiconductor method. The catalytic combustion method of methane measurement is the most widely used monitoring method in China, because of its simple device structure, nearly linear output signal, and insensitiveness to the impact of temperature and humidity. Sensors used in coal mines render them explosion-free; however, sulfide, phosphide, and other compounds are strongly adsorbed on the catalyst of the bridge and react with the lead elements in the catalyst to form new compounds, which will reduce the activity of the catalyst, resulting in sensor failure. Wang (2014) pointed out that sensors also have problems such as short life and short calibration cycle. Compared with the traditional catalytic combustion method, infrared sensors have a long lifespan, high reliability, long calibration period, and are sensitive to single detection gas, without being influenced by other interfering gases, which has developed rapidly in recent years. However, in practical application, subject to environmental temperature and pressure changes, the measurement accuracy and reliability of sensors will be reduced (Li, 2014, Wang *et al.*, 2013, Yu *et al.*, 2014, Zhang & Zhang, 2005). At present, a constant-temperature and constant-pressure compensation device and other measures are used to improve the measurement accuracy of sensors. This method increases the weight, size, and power consumption of sensors, and their structure is complex, not suitable for harsh underground coal mines (Liu *et al.*, 2012). However, the use of data processing to achieve the pressure compensation function is less expensive, efficient, and reliable.

Therefore, in this paper, different concentrations of standard methane gas were used as test objects, and experimental data of the same concentration of gas at different pressures were obtained. By analyzing the experimental data, the pressure compensation model was deduced and verified. Its principle and numerical analysis method can be applied to the existing infrared methane sensors.

2 FUNDAMENTALS OF INFRARED METHANE SENSORS

Infrared light is a kind of electromagnetic wave, with scattering, reflection, interference, absorption, and other properties. Passing through a gas medium, the methane gas molecules will absorb the specific wavelength of infrared light to make it attenuated, following the classic Lambert–Beer law (Paul & Austin, 2004):

$$I = I_0 \times \exp(-klc) \tag{1}$$

where I = transmitted light intensity; I_0 = incident light intensity; k = absorption coefficient; l = distance from the light source to the detector; and c = target gas concentration (g/L or mol/L).

When infrared light passes through gas, the intensity of the transmitted light becomes lower than that of the incident light, and the methane concentration can be detected by measuring the energy difference. However, when the gas pressure changes, as in the negative-pressure environment, the gas expansion per unit volume and the gas molecular spacing increase. That is, the total number of gas molecules per unit volume decreased, thereby reducing the infrared absorption of energy and the value measured by the sensor. However, the actual concentration of methane (in China, coal mine gas detection instruments often use volume concentration, VOL (%) or PPM ($\times 10^6$)) does not change. Therefore, when the gas pressure changes, there is a large deviation comparing the measured concentration with the true value.

3 EXPERIMENTAL STUDY OF THE PRESSURE COMPENSATION METHOD

In order to analyze the effect of gas pressure on the value of methane concentration measured by the sensor, a set of experiment platform for the pressure measurement of infrared methane sensor was set up, as shown in Figure 1. The platform is mainly composed of standard methane gas, flowmeter, absolute pressure gauge, concentration sensor, vacuum pump, and other parts, which can be used to carry out the methane concentration measurement under an absolute pressure of 20–130 kPa.

The experimental procedure is as follows. First, connect the standard methane cylinder and the other parts of the experimental platform. Then, control the gas pressure by adjusting

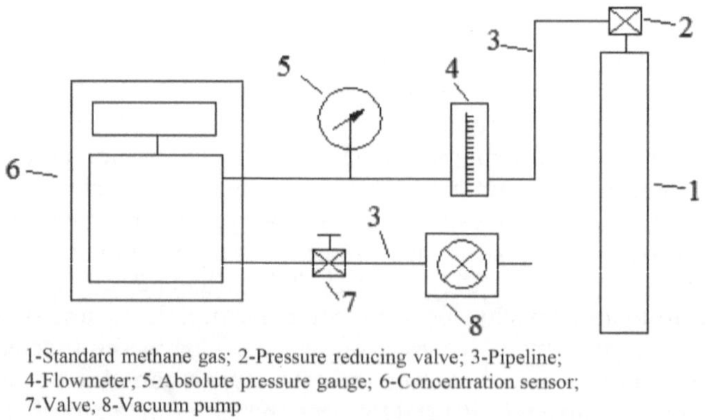

1-Standard methane gas; 2-Pressure reducing valve; 3-Pipeline; 4-Flowmeter; 5-Absolute pressure gauge; 6-Concentration sensor; 7-Valve; 8-Vacuum pump

Figure 1. Sketch map of infrared methane sensor pressure test platform.

the valve, starting from 40 kPa, record the gas pressure and its corresponding infrared methane sensor test results every 10 kPa. Repeat the above steps and test the pressure and its corresponding sensor concentration display value in 0.5%, 8.5%, 20%, 60%, and 85% CH_4 concentrations. To eliminate the effect of temperature on the sensor test data, all experiments are conducted at the same temperature. The test results are shown in Table 1 and Figure 2.

4 PRESSURE COMPENSATION MODEL AND ITS VERIFICATION

According to the experimental data, it can be seen that for the same concentration of methane gas, with the increase of environmental pressure, the sensor test concentration shows a linear increase. The greater the difference between ambient pressure and the pressure at which the instrument is calibrated, the greater the concentration error. On the basis of the experimental data, the pressure compensation model is constructed by least squares method. The least squares method is a mathematical optimization technique that finds the best function matching of data by minimizing the square of the error. By this method, a function relationship between the real concentration with the concentration display value and the gas pressure can be established.

Through the analysis, it can be concluded that the CH_4 concentration increases linearly with the increase of the sensor's measurement value under different environmental pressures. The model of the relationship is as follows:

$$C_0 = aC_h + b \tag{2}$$

where C_0 = the concentration of methane gas in the calibration state, %; C_h = the sensor concentration display value under different environmental pressure; and a, b = fitting parameters. The values of a and b can be obtained by fitting the linear slope of methane gas under

Table 1. Test data of infrared methane sensor before pressure compensation.

Concentration /%	Gas pressure/kPa 40	50	60	70	80	90	100	110	120
0.5	\	0.08	0.12	0.17	0.22	0.27	0.33	0.39	0.45
8.5	2.01	2.74	3.60	4.53	5.54	6.78	8.02	9.67	11.20
20.0	4.44	6.28	8.49	11.13	13.63	16.55	19.75	22.85	26.55
60.1	14.13	19.98	26.67	34.18	41.85	50.87	60.22	70.22	79.93
85.0	20.55	29.07	38.83	49.47	60.80	72.87	85.53	98.82	\

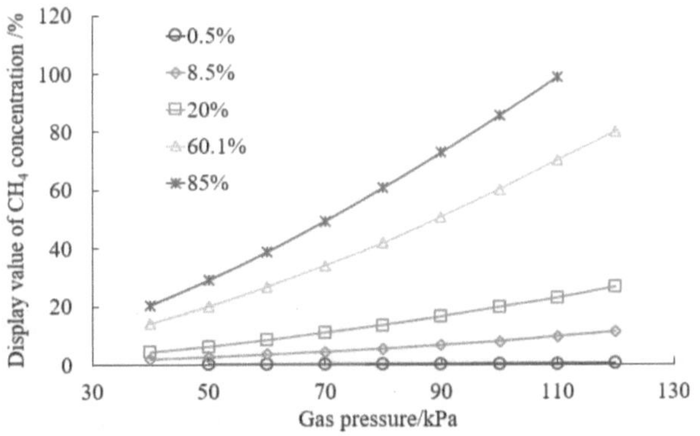

Figure 2. Test curve of infrared methane sensor before pressure compensation.

different environmental pressures through the least squares method. The results are shown in Table 2.

As shown in Table 2, fitting parameters a and b are different under different gas pressures; they showed a decreasing trend with increasing gas pressure. Similarly, a function of a, b, and gas pressure can be established:

$$\begin{cases} a = f(p_h) \\ b = g(p_h) \end{cases} \tag{3}$$

where p_h is the site environmental pressure, kPa. We can see from the analysis that the fitting parameter a and the gas pressure show a power function relationship: $f(p_h) = 1261.1 p_h^{-1.553}$. And because of different concentrations of methane, test error requirements are different, according to the China safety production industry standards (China State Administration of Work Safety, 2008). As shown in Table 3, the fitting parameter b needs to be approximated by piecewise fitting. Thus, by measuring the ambient gas pressure and concentration display value, the standard state of methane gas concentration can be calculated:

$$C_0 = f(p_h) \cdot C_h + g(p_h) \tag{4}$$

In order to verify the accuracy of the concentration model, using the same sensor, we adopted the pressure compensation calculation model and measured the methane concentration under the same temperature environment. The measured concentration values of five kinds of methane gas under different pressures are shown in Table 4. The pressure-corrected curve is shown in Figure 3.

According to the measured data after sensor pressure correction, for 0.5% methane gas, the error range is –0.03 to +0.06; for 8.5%, 20.0%, 60.1%, and 85% methane gas, the maximum

Table 2. Fitting parameters a and b under different environmental pressures.

Gas pressure/ kPa	40	50	60	70	80	90	100	110	120
a	4.115	2.921	2.187	1.716	1.398	1.164	0.990	0.856	0.750
b	1.087	0.855	0.825	0.670	0.709	0.534	0.391	0.232	0.119
R^2	0.9994	0.9995	0.9996	0.9998	0.9997	0.9999	1	1	1

Table 3. Basic error of measurement of nondispersive infrared methane sensor in Chinese coal mine.

Types	Measuring range/%CH_4	Basic error/%CH_4
A, B (for environmental monitoring)	0.00–1.00	±0.06
	1.00–100	±6% of true value
C (for gas drainage pipeline monitoring)	0.00–1.00	±0.07
	1.00–100	±7% of true value

Table 4. Test data of infrared methane sensor after pressure compensation.

Concentration /%	Gas pressure/kPa								
	40	50	60	70	80	90	100	110	120
0.5	0.51	0.47	0.51	0.54	0.56	0.54	0.52	0.50	0.49
8.5	8.01	8.89	8.85	8.68	8.55	8.62	8.11	8.17	8.25
20.0	19.54	19.37	20.02	19.49	19.04	19.19	19.40	19.31	19.57
60.1	58.83	58.52	58.59	58.91	58.43	58.99	59.15	60.35	58.91
85.0	85.53	85.12	85.34	85.27	84.89	84.61	84.03	83.55	83.35

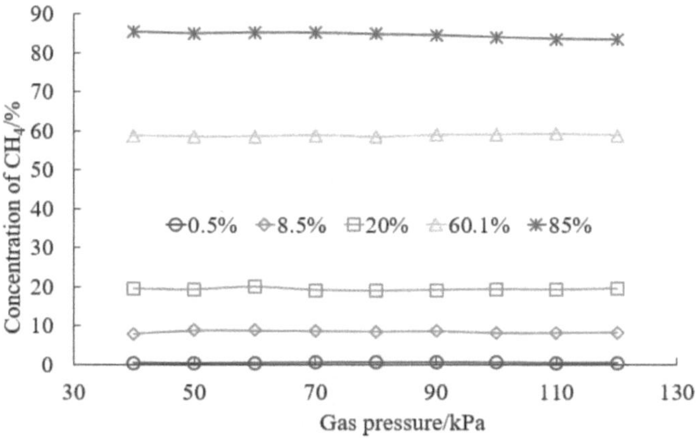

Figure 3. Test curve of infrared methane sensor after pressure compensation.

error range is −5.8% to + 4.6%. The greater the concentration, the smaller the measurement error. The range of experimental error can meet requirements of the various types of infrared methane sensor in Chinese coal industry. It is shown that the pressure correction method of the infrared methane concentration sensor is suitable for the measurement of methane gas in the coal industry.

5 CONCLUSIONS

In this paper, the principle of infrared methane sensor and its pressure compensation method have been described. On the basis of the experimental data, the pressure compensation calculation model of infrared methane sensor was obtained, which can reduce the influence of ambient gas pressure on the sensor. Through the compensation calculation, the measurement error of the infrared methane sensor in the pressure range of 40–120 kPa can meet the requirement of various kinds of sensors in the coal industry, which provides an effective path for methane gas detection in coal mines.

ACKNOWLEDGMENTS

This study was financially supported by Special Foundation of Par-Eu Scholars Program, National Natural Science Foundation of China (51374236, 51574280), Chongqing Research Program of Basic Research and Frontier Technology (cstc2015 jcyjBX0076), and Chongqing Technological Innovation Leading Talent Support Program (CSTCKJCXLJRC14).

REFERENCES

Chai Huapeng, Peng Feng, Bai Yunfeng, Tian Maozhong, Liang Wenjuan, Dong Chuan, Shuang Shaomin. 2009. Advances in methane sensor. *Journal of Shanxi Datong University (Natural Science)* 25(3):27–31.

China State Administration of Work Safety. 2008. AQ 6211–2008. *Coal mine non-dispersive infrared methane transducer.* Beijing.

Hodgkinson, J. R.P. Tatam. 2012. Optical gas sensing: a review. *Measurement Science & Technology* 24(1):111–123.

Liu Junjie, Qiao Deqing. 2006. Some aspects on the gas outburst/explosion hazard in coalmine of China. *Journal of China Coal Society* 31(1):58–62.

Liu Yan, Zhang Libin, Jiang Ze. 2012. Design of mine-used infrared methane sensor with compensation of temperature and pressure. *Industry and Mine Automation* (8):7–10.

Li Xiaogang. 2014. Pressure compensation method for mine-used infrared methane sensor. *Safety in Coal Mines* 45(7):7–14.

Li Zubin, Zhou Zongfu. 2009. Development and Application of Verification Instrument for Light Interference Methane Analyzer. *Acta Metrological Sinica* 30(4):385–388.

Paul C, Austin EAD. 2004. Theoretical analysis of a methane gas detection system, using the complementary source modulation method of correlation spectroscopy. *Measurement Science and Technology* (8):1629–1636.

Tong Minming. 1997. Study of a new kind of methane detector. *Chinese Journal of Scientific Instrument* 18(5):492–495.

Wang Hairong, Zhang Wei, You Liudong, Yuan Guoying, Zhao Yulong, Jiang Zhuangde. 2013. Back propagation neural network model for temperature and humidity compensation of a non-dispersive infrared methane sensor. *Instrumentation Science & Technology* 41(6):608–618.

Wang Liang. 2014. Principle and performance characteristics of methane detector using non-dispersive infrared method. *Gas & Heat* 34(11): B5–7.

Wu Suozhu, Shuang Shaomin, Zhang Yan, Li Zhongping, Martin M.F. Choi, Zhang Caihong, Dong Chuan. 2009. Study on mode-filtered light sensor for methane detection. *Chinese Chem Lett* 20: 210–212.

Yu Qing. 2012. Application of gas detection technology based on spectrum absorption in coal mine. *Mining Safety & Environmental Protection* 39(3):26–29.

Yu Xin, Lv RuiHong, Song Fang, Zheng ChuanTao, Wang YiDing. 2014. Pocket-Sized Nondispersive Infrared Methane Detection Device Using Two-Parameter Temperature Compensation. *Spectroscopy Letters* 47(1):30–37.

Zhang Jingchao, Liu Kui, Cao Yanpeng, Cao Suxia. 2010. Correlation based spectroscopy and differential absorption of methane detection method. *Chinese Journal of Sensors and Actuators* 23(2):168–170.

Zhang Liping, Zhang Fan. 2005. The mathematical modeling of infrared methane detector with real-time rectification of temperature influence. *Safety in Coal Mines* 36(7):1–3.

Innovation and Application of Engineering Technology – Oscar, Chen & Wang (Eds)
 ISBN 978-1-138-05057-0

Experimental study on the impact dynamic effect of coal-gas two-phase flow in the outburst process

Linchao Dai, Xusheng Zhao, Jie Cao & Dongling Sun
State Key Laboratory of Gas Disaster Detecting, Preventing and Emergency Controlling, Chongqing, China
Chongqing Research Institute Co. Ltd., China Coal Technology and Engineering Group, Chongqing, China

ABSTRACT: In order to explore the impact dynamic effect of coal-gas flow in underground equipment and facilities during the coal and gas outburst process, the coal-gas two-phase flow impact dynamic effect simulation experiment is carried out, and the impact of coal-gas two-phase flow on the impact damage of bracket, tramcar, air door, and other equipment and facilities is analyzed emphatically. The results show that under a pressure of 0.35 MPa, the gas pressure law of the bracket and air door is "rapid rise to peak, the pressure decreases and then increases, after slowing to reduce volatility lasts about five times, and ultimately tends to 0", and the tramcar, bracket, and other equipment and facilities caused serious damage under the joint action of the gas and coal flow. Meanwhile, the existence of shock wave and the impact dynamic effect of coal-gas two-phase flow have been verified.

1 INTRODUCTION

Coal and gas outburst is one of the most serious natural disasters that restrict the safe production of coal mine (Hu, 2012, and Zhang, 2014). In the event of an outburst accident, the gas and broken coal (rock) from the coal (rock) body are suddenly thrown into the tunnel space, accompanied by violent sound and strong dynamic effects (Cheng, 2004). The coal-gas two-phase flow with prominent shock wave property is the main cause of accidents. When the high-speed flow of coal gas encounters obstacles, the kinetic energy of the gas–solid two-phase fluid is converted into pressure by the potential energy acting on the blocking object. And if the yielding load of the barrier itself is less than the pressure on it, the barrier will be destroyed with the resulting dynamic instability, which is the main cause of damage for coal mine underground personnel, equipment, and facilities. Therefore, it is very important to study the migration law of coal-gas two-phase flow after outburst. Although several theoretical analyses and numerical simulations of coal-gas two-phase flow generation mechanism and its migration rule in the process of outburst have been made by scholars worldwide (Zhang, 2007, Cheng, 2008, Cheng, 2009, Zhang 2010, Wang, 2012, Sun, 2012 and Miao 2013), experimental studies are less (Wei, 2010, Wu, 2010, and Hu 2013). Therefore, this paper mainly uses the coal and gas outburst dynamic effect experimental device to carry out outburst coal-gas two-phase flow impact dynamic effect simulation experiment, focusing on the impact of coal-gas two phase-flow on bracket, tramcar, air door, and other equipment and facilities.

2 EXPERIMENTAL PROGRAM

2.1 *Experimental device*

Figure 1 shows the experimental device for the independent research and development of coal and gas outburst dynamic effect used in this study. The device is mainly composed of five modules (power system, air-charging system, pipeline system, data acquisition system, and dust removal system). It can be used to simulate the coal-gas two-phase flow movement

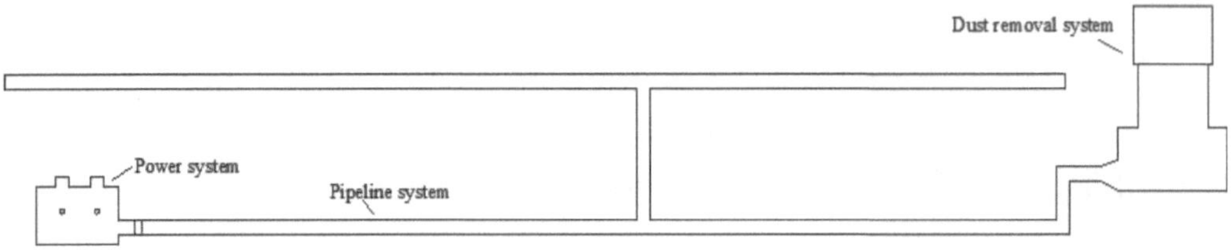

Figure 1. Experimental device for the simulation of coal and gas outburst dynamic effect.

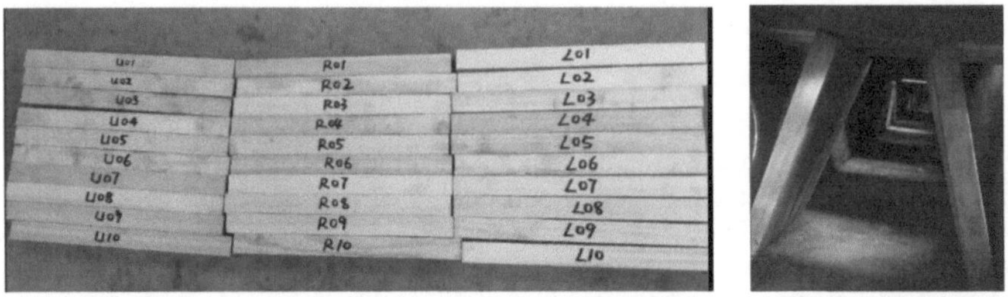

(a) bracket

(b) tramcar

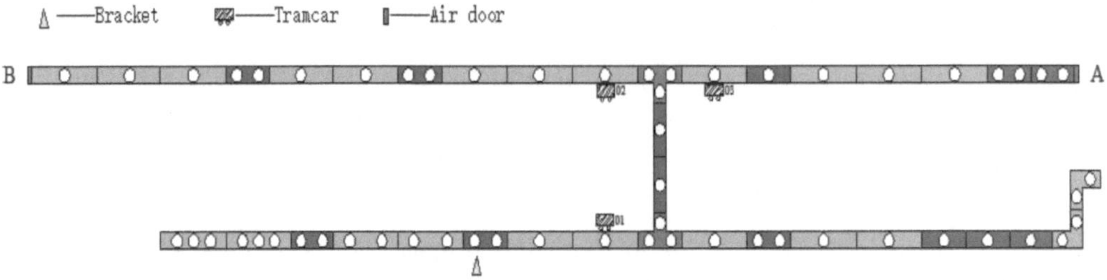

(c) arrangement position

Figure 2. Arrangement of equipment and facilities.

law, shock wave formation propagation law, and two-phase flow sorting accumulation characteristics under different experimental conditions, such as different outlet size, gas pressure, solid-to-gas ratio, coal particle size, and roadway network (different length and inclination).

1. The power system consists of a cylindrical high-pressure vessel tank (diameter, 1 m; length, 2 m; working pressure, 5 MPa; and total volume, 1.88 m^3), vessel installation platform, and a protruding mouth control device. The vessel tank is used to fill the coal particles and high-pressure gas, to simulate the outburst hole in the coal particles and determine the presence of gas and hole parameters.
2. The pipeline system consists of a straight tube, a right-angle elbow, a tee, a 10° inclined tube, and a 20° inclined tube, which can be used for different mine network simulation. It is square

shaped, with inner dimensions of 300 mm × 300 mm and design working pressure of 1.0 MPa. The observation window, clean window, and the sensor connection port are designed.

3. The data acquisition system is mainly used for collecting experimental data of gas pressure, temperature, gas concentration, wind speed, and so on, with a low-speed acquisition mode, which can automatically switch to a high-speed acquisition mode to ensure complete and reliable data.
4. The air-charging system is used for experimental gas fill and outlet opening control.
5. The dust removal system is mainly used to capture the escape of dust particles in the pipeline system, to keep the experimental site and the surrounding clean.

2.2 *Experimental arrangement and procedure*

In order to observe the migration and destruction of the simulation equipment and facilities in the pipeline, a wooden bracket and three simulation tramcars are arranged as shown in Figure 2.

The experimental steps are as follows:

1. Coal samples are filled into the power system of high-pressure tank. After sealing, use the connection adapter to connect the power system with the piping system and install a 0.3 MPa rupture disk in the middle.
2. Lay the pipeline system and install sensors, bracket, tramcar, and other equipment and facilities.
3. Use the air-charging system to gradually fill the high-pressure vessel tank up to 0.35 MPa.
4. When the gas pressure reaches the set value, the rupture disk is stimulated to make the protrusion suddenly open. At the same time, the high-speed acquisition mode of the data acquisition system is started to collect the experimental data until the experiment is finished.

3 EXPERIMENTAL RESULT AND ANALYSIS

3.1 *Bracket*

The position of the bracket in the pipeline is monitored by the gas pressure, and its impact is observed. The experimental results are shown in Figures 3–4.

As can be seen from the figures, the gas pressure in the bracket rapid rises to the peak value (0.057 MPa) and then the pressure decreases and then increases after slowing, to reduce volatility about five times and ultimately to 0. The change process verifies the existence of shock wave in outburst coal-gas flow. By the impact of shock waves in the outburst process, the simulation wood bracket in the main lane is greatly destroyed and migrated, which is moved about

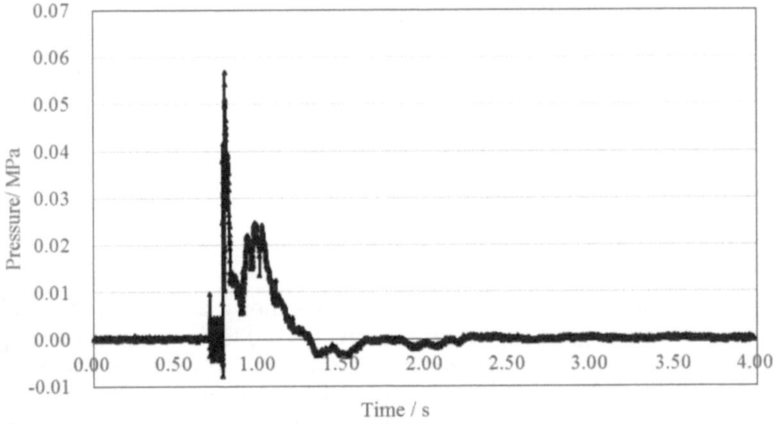

Figure 3. Gas pressure change.

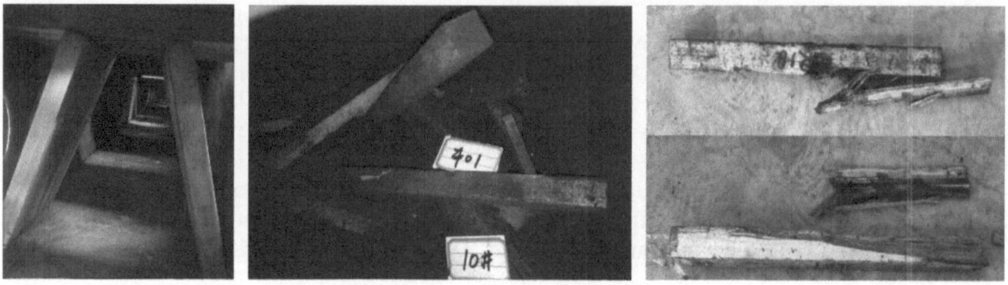

(a) (b) (c)

Figure 4. Damage of simulation bracket.

(a) No. 1 tramcar (b) No. 2 tramcar (c) No. 3 tramcar

Figure 5. Damage of simulation tramcar.

14–16 m in the right-angle turning assembly near the bottom dust removal device, and the surface of the wood bracket is filled with the pulverized coal particles. This shows that the wooden bracket is damaged by not only the role of gas but also the impact of pulverized coal flow.

3.2 *Tramcar*

The changes of three tramcars in the pipeline are shown in Figure 5. As can be seen from the figure, the No. 1 simulation tramcar in the main roadway is completely destroyed and the body is covered with pulverized coal. No. 2 simulation tramcar in the branch roadway experienced only rollover, and No. 3 simulation tramcar underwent only change of direction (deflection 45°). From the simulation tramcar change situation, the pulverized coal-gas flow plays the major role. When the kinetic energy of the two-phase fluid interacts with the simulation tramcar and produces a flow resistance, the tramcar rolls over or changes direction (No. 2 and No. 3 tramcars, respectively). However, when the Mach number of the two-phase fluid increases up to a certain extent (more than the long-time strength of the object), the tramcar will be permanently damaged (No. 1 tramcar). It shows that the effect of impact is obvious under the action of coal-gas flow.

3.3 *Air door*

The simulation of the pressure of air doors (A and B) is shown in Figure 6. It is evident from the figure that the pressure of the air door first increases suddenly and reaches the peak value around 0.85 s in the study area. Then, the pressure reduces and increases and continues to fluctuate about five times after slowing down. Meanwhile, the peak pressure of each fluctuation gradually decreases, indicating the gradual weakening of the impact of energy. At the same time, it can be seen that because the air door A is relatively far away from the contact

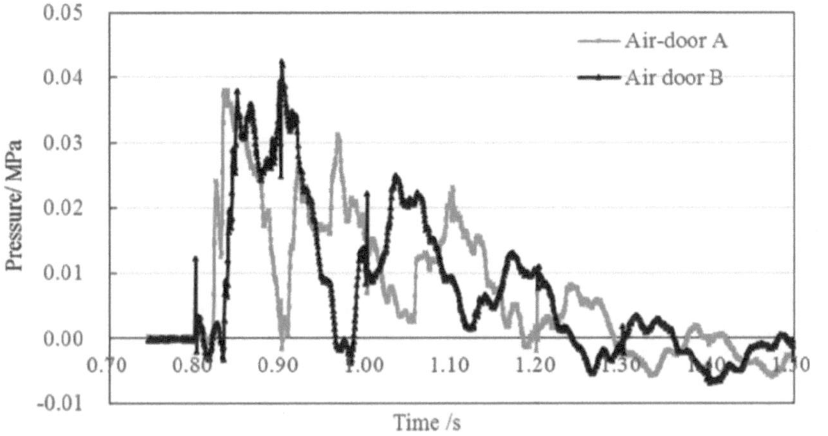

Figure 6. Comparison of air door pressure.

lane, the pressure increase of air door A is delayed compared with air door B; however, the peak pressure of air door A is slightly lower than that of air door B.

4 CONCLUSIONS

Under the pressure of 0.35 MPa, the gas pressure law of the bracket and air door is "rapid rise to peak, the pressure decreases and then increases, after slowing to reduce volatility lasts about five times, and ultimately tends to 0". The bracket and tramcars in the main roadway are completely destroyed, and the tramcars in the branch roadway experience only rollover and change of direction. The experimental study verifies the existence of shock wave and the impact dynamic effect of coal-gas flow. This study provides a reference for the coal and gas outburst impact dynamic effect research.

ACKNOWLEDGMENTS

This study was financially supported by National Key Research and Development Program (2016YFC0801402), National Natural Science Foundation of China (51374236, 51574280), Chongqing Frontiers and Application-based Research Program (cstc2015jcyjBX0076), Chongqing Technological Innovation Leading Talent Support Program (CSTCKJCXLJRC14), and the Science and Technology Innovation Fund of Coal Science Research Institute (2015ZYJ002).

REFERENCES

Cheng W.M., Wang G. & Zhang R. 2008. Formation of coal and gas outburst blast wave of anti-disaster damage. *Science Technology Review* 26(24): 61–65.

Cheng W.M., Wang G. & Zhou G. 2009. Numerical simulation of outburst prevention air door destruction by coal and gas outbursts. *Journal of Chongqing University* 32(3): 314–318.

Cheng W.Y., Liu X.Y & Wang K.J. 2004. Study on regulation about shock-wave-front propagating for coal and gas outburst. *Journal of China Coal Society* 29(1): 57–60.

Hu Q.T. & Zhao X.S. 2012. Present situation of coal and gas outburst accidents in China's coal mines and countermeasures and suggestions for their prevention. *Mining Safety and Environmental Protection* 39(5): 1–6.

Miao F.T., Sun D.L. & Hu Q.T. 2013. The formation mechanism of shock waves in the coal and gas outburst process. *Journal of China Coal Society* 38(3): 367–342.

Sun D.L, Hu Q.T. & Miao F.T. 2012. Motion state of coal-gas flow in the process of outburst. *Journal of China Coal Society* 37(3): 452–458.
Wang K., Zhou A.T. & Wei G.J. 2012. Study of the formation and propagation characteristics of shock wave and gas flow of outburst at straight roadway. *Journal of Mining & Safety Engineering* 29(4): 559–563.
Wei J.P., Zhu H.Q. & Wen Z.H. 2010. Coal and gas outburst shock wave propagation experiments. *Coal* 19(8): 11–13.
Wu A.J., Jiang C.L. & Tang J. 2010. Study on propagation laws of shock wave in coal-rock mass under the effect of gas outburst. *Journal of China Coal Society* 35(10): 1644–1648.
Zhang J.F., Wang K. & Wei C.P. 2010. Formation and propagation of shock waves during coal and gas outbursts. *Journal of Mining & Safety Engineering* 27(1): 67–71.
Zhang Q., Sun Y.R. & Wang X.Y. 2007. The shock wave propagating regulation of coal and gas outburst. *Mining Safety and Environmental Protection* 34(5): 21–23.
Zhang S.T. 2014. Review and prospect of coal and gas outburst simulation test. *Mining Safety and Environmental Protection* 41(1): 83–86.

Innovation and Application of Engineering Technology – Oscar, Chen & Wang (Eds)
© 2017 Taylor & Francis Group, London, ISBN 978-1-138-05057-0

Roof-caving regularity study of a fully mechanized large-cutting-height face

Shousong Guo
Chongqing Real Estate College, Chongqing, China
Chongqing Research Institute Co. Ltd., China Coal Technology and Engineering Group, Chongqing, China
State Key Laboratory of the Gas Disaster Detecting, Preventing and Emergency Controlling, Chongqing, China

Kun Zhong, Qingxiu Liu, Shuang Chen, Nana Liu & Guofu Wu
Chongqing Real Estate College, Chongqing, China

ABSTRACT: In this study, the roof-caving process of a 3207 fully mechanized large-cutting-height face is analyzed by calculating each bed load, the first caving step distance, the periodic caving step distance, and the other parameters of the 3207 fully mechanized large-cutting-height face roof and analyzing periodic weighting of the working face combining with the observed data. Test results show that: (1) the way of roof caving at mined-out areas of the 3207 working face is layer caving each time; (2) the pace of direct top collapsing initially is 14.1–16.0 m, the pace of basic top collapsing initially is 27.0–29.8 m, the cycle collapsing initially is about 11.0 m; and (3) the periodic roof weighting pace of the basic top decreases with increasing day advance progress, the difference of the periodic weighting distance among the upper, middle, and lower working faces is not big, but the middle dynamic load factor is 1.60, which shows a high weighting intensity. This study provides a basis for mining pressure control and safety mining of a 3207 fully mechanized large-cutting-height face.

1 INTRODUCTION

Shanxi Huoerxinhe coal mine is located in Zhangzi County, Changzhi city, China, with an area of about 71.3947 km^2. The No.3 coal seam, which is the main production unit of this coal mine, is located in the lower stratum of Shanxi formation, with a dip angle of about 5° and average thickness of 5.65 m. This coal seam is thick in the eastern part and thin in the western part. The 3207 fully mechanized large-cutting-height face is about 220 m in the inclined direction and 1389 m in the trending direction, which belongs to flat seam, with an average thickness of 5.3 m and coefficient of 0.5–0.6. The immediate roof of this working face is charcoal mudstone, with a complex geological condition. According to the roof classification standard of a gently inclined coal seam working face (MT554-1996), the roof of 3207 working face belongs to soft and broken roof thickness coal seam. Therefore, it is necessary to study the characteristics of roof caving of the 3207 fully mechanized large-cutting-height face in No.3 coal seam of Huoerxinhe coal mine in order to provide a basis for mining pressure control and safety mining of the face.

2 THEORETICAL ANALYSIS FOR ROOF ROCK LAYER CAVING LAW

Roof rock can be divided into immediate roof, basic roof, and a curve subsidence zone, according to the theory of roof structure. (1) Immediate roof is the rock formed in one or several layers above the coal seam and has no self-supporting capability after being self-broken. Immediate roof rock will fall along with the face advance, and it can sometimes form a hanging-arch structure. (2) Basic roof is the rock formed on top of the immediate roof (or on top of the coal seam sometimes). It is characterized by its thickness and hardness, and has an immediate impact on the working face after breakage to form a structure. The structure has the capability of balance rule, which has an effect on the movement rule of overlying strata, mine pressure, and control parameter of working face. (3) The curve subsidence zone moves as one structure and has no effect on the working face support generally.

Therefore, dropping characteristics of the roof rock layer adopted long-wall mining along strike and the mechanical effects induced are due to the surrounding rock control of the working face. Along with working face advanced from the open-off cuts, each roof rock layer will bend and sink under the effect of mine pressure, the roof-hanging distance of the gob roof increases continuously and collapses when the ultimate strength is reached. The way of large-area caving from the roof rock layer, whether synchronous or hierarchical, each time is determined by the horizon, thickness, strength, and size of load.

It is important to calculate the load of each rock stratum before calculating the breakage length. Because of the difference of thickness and lithology of each roof layer, the load on No.1 layer can be calculated according to the following formula:

$$(q_n)_i = \frac{E_i h_i^3(\gamma_i h_i + \gamma_{i+1} h_{i+1} + \cdots + \gamma_{i+n} h_{i+n})}{E_i h_i^3 + E_i h_i^3 + \cdots + E_{i+n} h_{i+n}{}^3} \tag{1}$$

where Ps: $(q_n)_i$ is the load from n layer upside to the i layer; $E_i \cdots E_{i+n}$ is the elasticity modulus of each layer; $h_i \cdots h_{i+n}$ is the thickness of each layer; and $\gamma_i \cdots \gamma_{i+n}$ is the unit weight of each layer; i=1,2,3….

When calculating $(q_{i+n-1})_i < (q_{i+n})_i, (q_{i+n})_i$ as the load applied to the i layer, the result of formula (1) can be used to calculate the breakage length as the load applied to the i layer at that moment.

According to the beam theory, the first and periodic collapsing initially of each roof layer can be calculated approximately as follows:

$$a_i = h_i \sqrt{\frac{2\sigma_{ti}}{(q_{i+n})_i}} \tag{2}$$

$$a_{ic} = h_i \sqrt{\frac{\sigma_{ti}}{3(q_{i+n})_i}} \tag{3}$$

where Ps: a_i, a_{ic} is the first and periodic collapsing initially of the i layer and σ_{ti} is the strength of extension of the i layer.

According to these formulas above, each rock formation caving synchronously or hierarchically can be confirmed by the following methods; the caving characteristics of the face roof are as follows: (1) when $a_i \geq a_{i+1}, a_i \geq a_{i+2}, \cdots, a_i \geq a_{i+j}$, caving occurs from the i+1 layer to the i+j layer, following the i layer (j=1,2,3…); (2) when $a_i < a_{i+j+1}$, the *i*+*j*+1 layer moves forward and collapses until it reaches its ultimate strength. By that analogy, the breakage length and the thickness of stratum caving can be confirmed simultaneously.

3 LOAD OF EACH ROCK FORMATION OF FACE ROOF

According to the distributing characteristics of the 3207 working face, the structure of the roof rock layer is complex, which is mainly made up of charcoal mudstone, sandy mudstone and sandstone, and coal streak partly. The roof is made up of charcoal mudstone, sandy mudstone, coal seam, charcoal mudstone, siltstone, and charcoal mudstone from bottom to up. Strata distribution and the physical and mechanical properties of No.3 coal seam roof are shown in Table 1.

Table 1. Strata distribution and physical and mechanical properties of No.3 coal seam roof.

No.	Roof rock layer	Thickness (m)	Poisson's ratio	Tensile strength (MPa)	Elasticity modulus (GPa)	Apparent density (kg/m^3)
1	Charcoal mudstone	3.8	0.34	2.97	8.67	2236
2	Sandy mudstone	3.6	0.31	2.88	13.8	2708
3	Coal seam	1.0	0.32	0.1	2.97	1390
4	Charcoal mudstone	1.0	0.34	2.97	8.67	2236
5	Siltstone	3.4	0.34	3.14	27.6	2656
6	Charcoal mudstone	0.5	0.34	2.97	8.67	2236
7	Sandy mudstone	3.7	0.31	2.88	13.8	2708
8	Coal streak	0.6	0.32	0.1	2.97	1390
9	Sandy mudstone	4.4	0.31	2.88	13.8	2708
10	Coal streak	0.4	0.32	0.1	2.97	1390
11	Sandy mudstone	1.9	0.31	2.88	13.8	2708
12	Siltstone	5.0	0.34	3.14	27.6	2656

Table 2. Load of each roof layer.

No.	1	2	3	4	5	6
Load (kN/m^2)	84.97	131.37	13.90	22.36	127.60	11.18
No.	7	8	9	10	11	12
Load (kN/m^2)	108.47	8.34	163.01	5.56	51.45	132.8

According to formula (1) and the physical and mechanical properties of each roof layer in Table 1, the load of each roof can be calculated as shown in Table 2.

4 CAVING CHARACTERISTICS OF EACH ROCK FORMATION OF FACE ROOF

The pace of collapsing initially is the main characteristic appraising roof stability, which can be calculated by "plate" and "beam" methods. The "plate" calculating method uses a "thin plate" located in the rectangular region above golf after mining of long-wall face. According to different boundary conditions, when calculating dimensions of the plate, the breakage length is the distance working face advancing when the plate breaks.

According to the assumption of "thin plate" theory, under the condition of $(\frac{1}{80}\sim\frac{1}{100})\leq\frac{h}{a(b)}\leq(\frac{1}{5}\sim\frac{1}{8})$, the maximum tensile stress can be confirmed by analyzing the stress states of different boundary conditions. When the maximum tensile stress exceeds the tensile strength of rock, the plate will be broken, and this is used to calculate the pace of collapsing initially. Generally, when the expected pace of collapsing initially a far less than the length of working face b, the error of calculation is less.

Using the recommended calculating method by "Strata control around long-wall face of China's coal mine", the pace of collapsing initially a1 is calculated by the following formula:

$$a_1=\begin{cases} b\cdot\sqrt[4]{\dfrac{l_m^2}{b^2-l_m^2}} & (l_m<b<\sqrt{2}l_m) \\ \dfrac{b}{\sqrt{2}l_m}\sqrt{b^2-\sqrt{b^4-4l_m^4}} & b\geq\sqrt{2}l_m \end{cases} \tag{4}$$

where Ps: $l_m=\frac{h}{1-\mu^2}\cdot\sqrt{\frac{2\sigma_t}{q}}$ and l_m is the limiting rupture length of slat when four edges rigid supports infinite.

Table 3. Calculation of the first and periodic collapsing initially of roof strata.

		First collapsing initially (m)		Periodic collapsing initially (m)
Horizon	Thickness (m)	by beam	by plate	by beam
12	5.0	34.4	38.9	14.0
11	1.9	20.1	22.2	8.2
10	0.4	2.4	2.7	1.0
9	4.4	26.2	28.9	10.7
8	0.6	2.9	3.3	1.2
7	3.7	27.0	29.8	11.0
6	0.5	11.5	13.0	4.7
5	3.4	23.9	27.0	9.7
4	1.0	16.3	18.4	6.7
3	1.0	3.8	4.2	1.5
2	3.6	10.6	11.7	4.3
1	3.8	14.1	16.0	5.8

According to the beam theory, the first and periodic collapsing initially of the each roof layer can be calculated using formulas (2) and (3), respectively.

The length of the 3207 fully mechanized large-cutting-height face in No.3 coal seam of the Huoerxinhe coal mine is 220 m. Its distribution characteristics and the physical and mechanical parameters of the roof rock are shown in Table 1. Because the structure of the roof rock layer is complex, and considering the influence of geological structure, joint fissure influence coefficient of the overlying strata is chosen 0.55, and stratified rock roof influence coefficient is chosen 0.38. The first and periodic collapsing initially of the 3207 fully mechanized large-cutting-height face in No.3 coal seam can be calculated by formulas (2) and (3), respectively, as shown in Table 3.

As shown in Table 3, (1) the pace of direct top collapsing initially is 14.1–16.0 m, the pace of basic top collapsing initially is 27.0–29.8 m, and the cycle collapsing initially is about 11.0 m; (2) the way of roof caving at mined-out areas of the 3207 working face is layer caving each time. The first layer, charcoal mudstone, caves first when working face advanced 14.1–16.0 m; the second layer, sandy mudstone, and the third layer, coal seam, caves synchronous. The caving height of roof is 8.4 m, with the working face advancing and the fourth layer, carbon mudstone, caving at the same time; the caving height of roof is 9.4 m, with the working face advancing to 23.9–27.0 m and the fifth layer, siltstone, and the sixth layer, carbon mudstone, caving at the same time; and the caving height of roof is 13.3 m, with the working face advancing to 27.0–29.8 m and the caving height of roof reaching 24.3 m.

5 ANALYSIS OF PERIODIC WEIGHTING OF WORKING FACE

The main roof of the 3207 working face occurred five times during the site observation. The periodic weighting changing curve of 100# hydraulic support is shown in Figure 1, and the periodic weighting of three parts of the working face are shown in Tables 4 and 5. From day 5.13 to 5.25, the average daily advance degree of the 3207 working face is 4.4 m, and from day 5.26 to 6.14, the average daily advance degree of the 3207 working face is 2.4 m.

As shown in Table 4, the periodic weighting of the main roof decreased with increasing daily advance degree. When the daily advance degree is 4.4 m, the periodic weighting of the main roof is 19.8 m, and when the daily advance degree is 2.4 m, the periodic weighting of the main roof is 14.7 m. As shown in Table 5, the difference of periodic roof weighting pace among bottom, middle part, and upside of working face is not large, but the dynamic load factor in the middle part is 1.60, which shows a high periodic pressure strength, followed by upside and bottom.

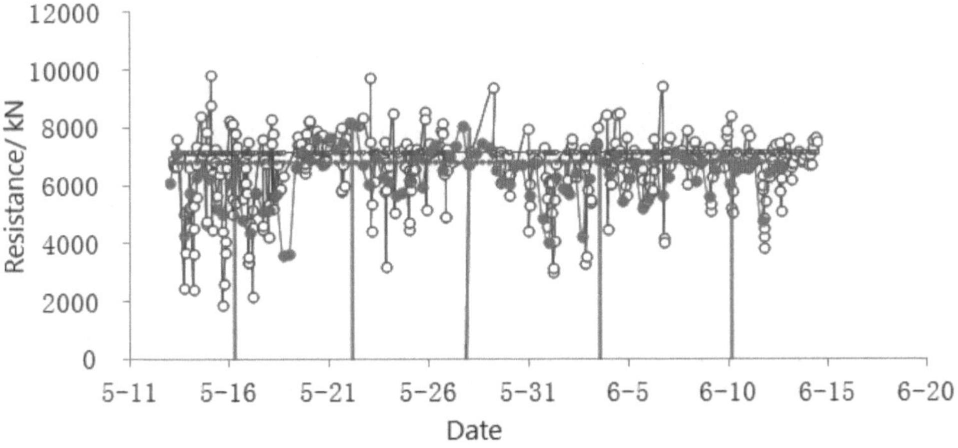

Figure 1. Periodic weighting changing curve of hydraulic support.

Table 4. Periodic roof weighting pace of working face.

Order of periodic roof weighting pace	Location of working face / serial number of hydraulic support	Bottom 8#	Bottom 30#	Middle part 53#	Middle part 77#	Upside 100#	Upside 122#	Average value of periodic weighting	Average value under different average daily advance degree
Average daily advance degree 4.4 m	1	17.6	17.2	20.2	21.6	19.2	20.2	19.3	19.8
	2	21	22.4	19.4	21	19.4	19	20.3	
Average daily advance degree 2.4 m	3	20	18.4	13.2	12	12	15.6	15.2	14.7
	4	16	20	14	11.6	13.6	13.2	14.7	
	5	/	/	14	12.4	16.4	/	14.3	

Table 5. Dynamic load factor during periodic weighting.

Order of periodic roof weighting pace	Location of working face / Serial number	Bottom 8#	Bottom 30#	Middle part 53#	Middle part 77#	Upside 100#	Upside 122#
Dynamic load factor	1	1.30	1.43	1.85	1.77	1.28	1.56
	2	1.47	1.44	1.45	1.51	1.37	1.38
	3	1.52	1.62	1.71	1.64	1.67	1.69
	4	1.32	1.38	1.69	1.3	1.31	1.26
	5	/	/	1.52	1.59	/	1.31
	Average value	1.43		1.60		1.36	
	General average	1.463					

6 CONCLUSION

1. The way of roof caving at mined-out areas of the 3207 working face is layer caving each time.
2. The pace of direct top collapsing initially is 14.1–16.0 m, the pace of basic top collapsing initially is 27.0–29.8 m, and the cycle collapsing initially is about 11.0 m.

3. The periodic weighting pace of the main roof increases with increasing daily advance degree. When the daily advance degree is 4.4 m, the periodic weighting pace ranges from 11.6 to 20 m, with an average value of 14.7 m; when the daily advance degree is 2.4 m, the periodic weighting pace ranged from 17.2 to 22.4 m, with an average value of 19.8 m. The difference of periodic roof weighting pace among bottom, middle part, and upside of the working face is not high, but the dynamic load factor in the middle part is 1.60, which shows a very high periodic pressure strength.

REFERENCES

Cao J.H. Liu H.J. (2015) Study on laws of stress and roof caving deformation in steeply inclined thin coal seam mining. *Zhongzhou Coal*, 2, 45–52.

Fu Y.P., Song X.M. and Xing P.W. (2012) Study on simulation of caving and evolution law of roof strata of large mining height workface in shallow thick coal seam. *Journal of China Coal Society*, 37(3), 366–371.

Tu M. Zhang D.S. (2014) Research on coal mining roof caving law under thick loose layer. *Mine Pressure and Roof Management*, 4, 8–11.

Wu Y.P., Yuan D.F. and Zhang M.F. (2000). Study on the elementary problems of full-mechanized coal mining in greater pitching seam. *Journal of China Coal Society*, 25(5), 465–468.

Yang Z.L., Yu X.Y. and Guo H.M. (2007) Study on catastrophe mechanism for roof strata in shallow seam longwall mining. *Chinese Journal of Geotechnical Engineering*, 29(12), 1763–1766.

Zhao Y.F., Zhang X.Y. and Tu M. (2007) Roof caving characteristic and strata behavior in exploiting steep coal seams. *Journal of Mining & Safety Engineering*, 24(2), 231–234.

Innovation and Application of Engineering Technology – Oscar, Chen & Wang (Eds)
© 2017 Taylor & Francis Group, London, ISBN 978-1-138-05057-0

Flotation mechanism of octylic hydroxamic acid on apatite

Anrong Liu
Guizhou Institute of Metallurgical and Chemical Engineering, Guiyang, China

Sanjun Liu & Wenming Yao
Guizhou Institute of Metallurgical and Chemical Engineering, Guiyang, China
School of Minerals Processing and Bioengineering, Central South University, Changsha, China

Dengpan Nie & Zhenjie Wang
Guizhou Institute of Metallurgical and Chemical Engineering, Guiyang, China

Qi Yue
School of Minerals Processing and Bioengineering, Central South University, Changsha, China

Liyuan Chai
School of Metallurgy and Environment, Central South University, Changsha, China

Dan Sun
Business School, Central South University, Changsha, China

ABSTRACT: Phosphate resources are widely spread in earth. However, effective utilization of this mineral is still lacking. The conventional collector for apatite mineral is oleic acid, which has three main drawbacks, insoluble at low temperatures, poor selectivity in apatite flotation, and low collecting ability, resulting in high dosage. In this study, octylic hydroxamic acid was synthesized, and its collecting ability toward two major minerals in phosphate ore apatite and dolomite was investigated. The results show that for the separation of dolomite and apatite, octylic hydroxamic acid had high selectivity and optimum pH of 6–8. Adding collectors can effectively decrease the surface tension of octylic hydroxamic acid and the zeta potential of these two minerals, and the decrease fordolomite was more apparent. By using Material Studio software and IR, it could be seen that the reaction between octylic hydroxamic acid and dolomite represents physical adsorption, while octylic hydroxamic acid cannot interact with apatite.

Keywords: octylic hydroxamic; apatite; dolomite; flotation

1 INTRODUCTION

China ranks second in phosphate resource reserves in the world. Most of them are present in the form of marine sedimentary phosphorite, which accounts for 85% of the total reserves. High magnesium content and low phosphorus content are the features of Chinese apatite (Wu Caibin et al. 2003, Li Wei et al. 2015 and Yan Zhenghua 2011). The average grade of phosphate of the phosphate ore is only 17%, lower than the world average of 30%, whereas the grade of magnesium of about 90% of the total phosphate ore is more than 1.5, which belong to low-grade phosphate ore with high magnesium content[1]. The key to deal with these resources is to improve the grade of phosphorus by flotation (Yu Yongfu et al 2008 and Young C. A. and Miller J.D. 2000). Diverse flotation, as one of two ways of flotation, is always applied to magnesium (Guo Fang et al 2010). Reagents play an extremely important role in the flotation and can greatly reduce the cost. Diverse flotation collectors were manufactured by many researchers (Houot Retal 1980 and Hsieh and Shuang Shi 1988).

However, reagents that belong to conventional collectors have many drawbacks such as high consumption, poor selectivity, and poor performance in low temperature. This problem has been focused by many researchers.

In this study, for the flotation separation of apatite and dolomite (Alfred W. Petersen 1972, Isgaque and Ahmed 1982), octylic hydroxamic acid (Wang 2013 and Che 2004) was synthesized. Meanwhile, the mechanism (Sreenivas and Padmanabhan 2005 and Guan et al 2009) was aided by the surface tension of collectors, zeta potential (Fuerstemau et al 2005), binding energies, and IR tests. It was found that octylic hydroxamic acid can be selective on the surface of dolomite, except on the surface of apatite. Thus, the separation of the dolomite and apatite was achieved.

2 EXPERIMENTAL

2.1 *Materials*

2.1.1 *Samples*

Minerals dolomite and apatite obtained from Wengfu, Guizhou Province, China, were ground by a porcelain ball mill. The 96 mm fractions were used in microflotation. The specific surface areas of dolomite and apatite are 8.21 and 8.73 $m^2\ g^{-1}$, respectively. The results of the chemical tests of the two samples are shown in Table 1.

2.1.2 *Reagents*

Hydrochloric acid (HCl) and sodium carbonate ($NaCO_3$) are of analytical purity. After being tested by IR, the purity of the synthesized hydroxamic acid was found to be more than 99%.

2.2 *Test methods*

Microflotation tests were performed in a 40 mL flotation cell. In each flotation test, 2 g of mineral powder was used in 40 mL of distilled water. In Figures 1 and 2, the concentrations can be regarded as different dosages in the flotation cell.

The pH of the solution was tested with pH meter pHS-3C.

The surface tension of collectors was tested with JK99D meter. Before measurement, the meter was preheated for 30 min. A platinum ring was burned red above the alcohol lamp after cleaning with flowing water and then hung on a tensiometer hook to be gradually cooled to room temperature. Octylic hydroxamic acid was configured to different concentrations, injected into a vessel, and then put into the sample set. The platinum ring was immersed into the liquid at a depth of 5 mm and then pushed up in order to form a liquid column. When the ring was disjointed from liquid surface, the liquid column disconnected instantly, and the surface tension is displayed on the screen.

Dolomite and apatite samples were ground to less than 2 mm in a corundum mortar. Each sample (20 mg) was added to distilled water until the final volume of the mixed medium became 50 mL. Then, the medium was dispersed by ultrasonic waves. Finally, the solution was transferred to test the zeta potential with a Delsa 440SX meter.

A three-dimensional molecular model was built using molecular simulation software Material Studio by the molecular simulation method, and a three-dimensional map of the

Table 1. Analysis results of apatite and dolomite.

	Ore grade/%		
Mineral names	P_2O_5	MgO	Purity/%
Apatite	36.71	1.22	91.75
Dolomite	0.03	19.29	91.64

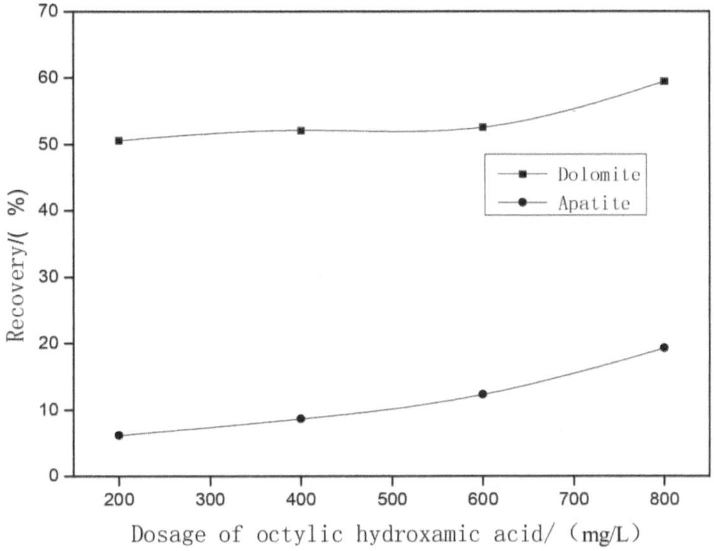

Figure 1. Recovery of apatite and dolomite versus dosage of octylic hydroxamic acid.

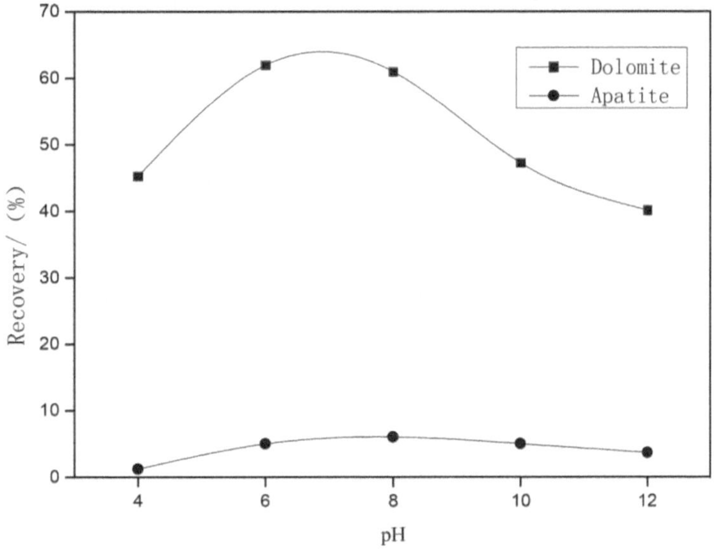

Figure 2. Recovery rate of apatite and dolomite as a function of pH.

interaction between flotation reagent and mineral crystal was constructed and analyzed. By comparing the energy value of the reagent and the mineral before and after the interaction, the action forms of the reagent and the mineral are studied.

Infrared absorption spectra analysis was performed using the KBr pellet method on the Fourier transform infrared spectrometer NEZUS 470-FTIR.

3 RESULTS AND DISCUSSION

3.1 *Flotation of the minerals*

Figures 1 and 2 show the flotation recoveries of the two minerals. It can be seen from the figures that in the same concentration of octylic hydroxamic acid and under the same pH conditions, the recovery of the dolomite is much higher than that of the apatite. When the octylic hydroxamic acid concentration was 200 mg/L and the pH value was 6, the recovery of dolomite will be 62%, higher than that of apatite, that is, 5%. The gap between the 57% is two. It is also indicated that the suitable pH values for the flotation of dolomite and apatite range from 6 to 8.

3.2 *Surface tension of octylic hydroxamic acid*

From Fig. 3, it can be seen that in the solution of a certain pH value, with increasing concentration of octylic hydroxamic acid, the surface tension of collectors decreased sharply at first, until it becomes constant. At pH = 6, the magnitude of surface tension reduction is maximum. It can be concluded that at pH = 6, the surface activity, solubility, and dispersion of octylic hydroxamic acid are high, and hence it is easily adsorbed on the surface of the mineral.

3.3 *Zeta potential*

The effect of pH on the zeta potential of the mineral surface was also tested (Fig. 4). It is evident from Fig. 4 that pH shows a strong influence on the zeta potential; at a concentration of 200 mg/L, the zeta potential of the minerals decreased with addition of octylic hydroxamic acid. Compared with the influence of the dolomite and apatite, it can be seen that adding octylic hydroxamic acid can effectively decease the zeta potential of these two minerals.

3.4 *Binding energies*

The three-dimensional map of the interaction of octylic hydroxamic acid and dolomite—and octylic hydroxamic acid and apatite—was drawn using Material Studio software. Furthermore, the energy value of the octylic hydroxamic acid and dolomite or apatite can also be calculated.

Table 2 shows that if octylic hydroxamic acid is adsorbed on the surface of dolomite, the total systematic energy will be reduced by 7.23 kcal·mol^{-1}. However, at an energy value of 761.42 kcal·mol^{-1}, octylic hydroxamic acid would adsorb on the surface of apatite. It could

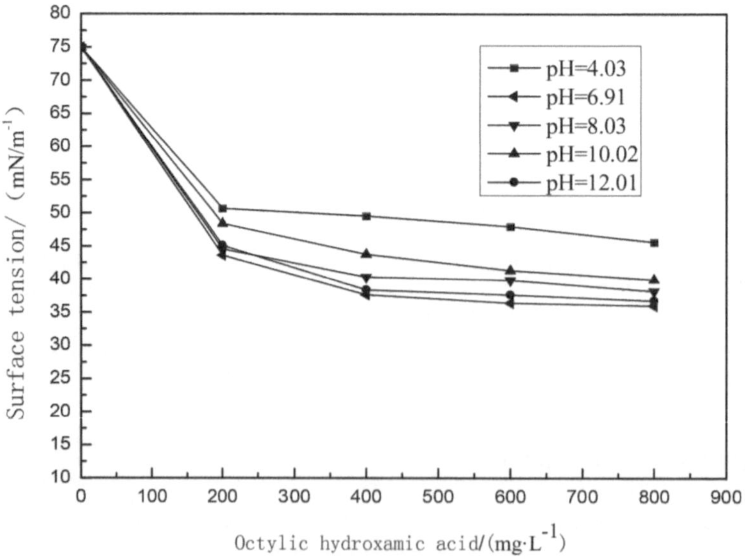

Figure 3. Surface tension of octylic hydroxamic acid.

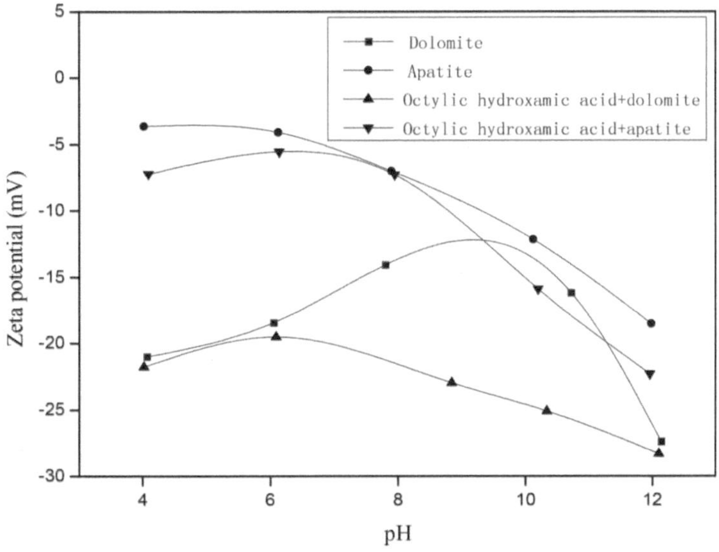

Figure 4. Effect of the pH of octylic hydroxamic acid on zeta potential.

Table 2. Effect of octylic hydroxamic acid on systematic energy/(kcal·mol^{-1}).

Reagent/ Octylic hydroxamic acid	Before the reaction		After the reaction		Binding energies	
	Apatite	Dolomite	Apatite +Reagent	Dolomite +Reagent	Apatite +Reagent	Dolomite +Reagent
12.400930	4177.522371	261.951792	4951.349980	267.101694	761.426679	–7.25103

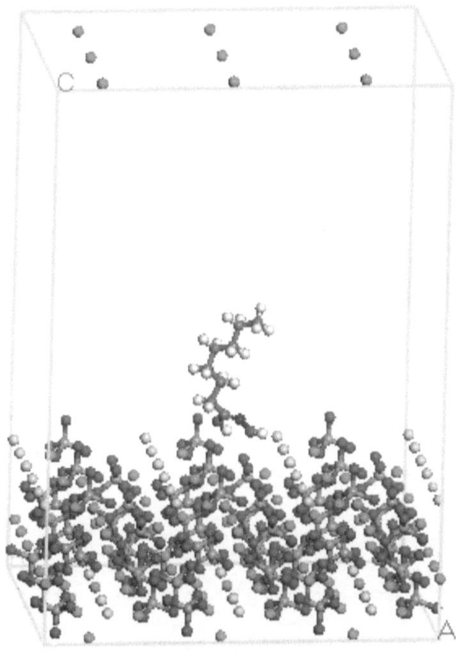

Figure 5. Interaction of octylic hydroxamic acid and apatite.

be concluded that octylic hydroxamic acid tends to be adsorbed on the surface of dolomite rather than apatite.

3.5 *IR test*

Figures 7 and 8 show the IR spectrum results for octylic hydroxamic acid. Figure 8 shows that in the IR spectrum after reaction with dolomite, the absorption peak becomes obvious in the

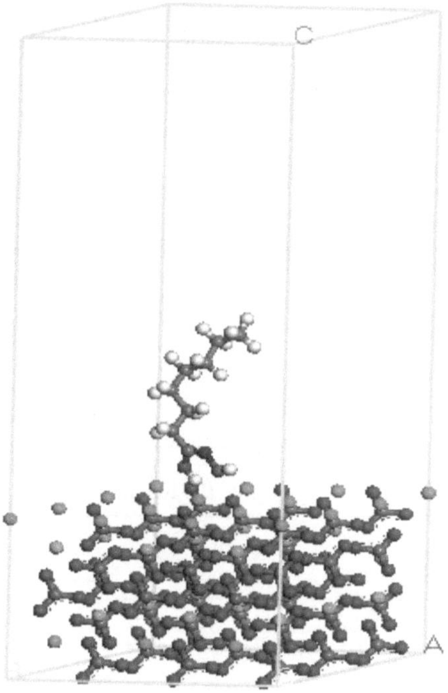

Figure 6. Interaction of octylic hydroxamic acid and dolomite.

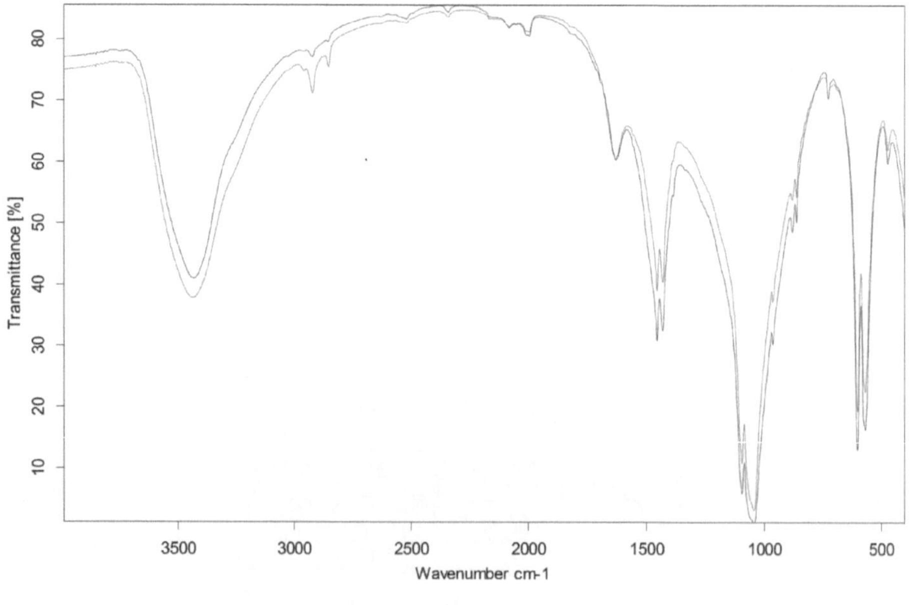

Figure 7. IR spectra of octylic hydroxamic acid and apatite.

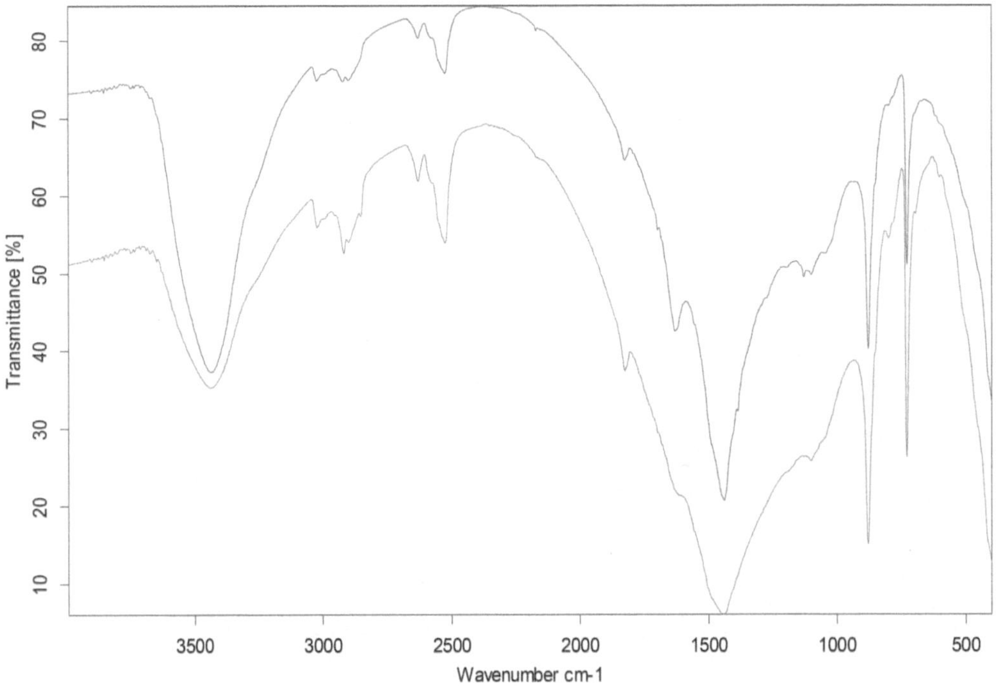

Figure 8. IR spectra of octylic hydroxamic acid and dolomite.

range of 800–3000 cm^{-1}, with characteristic sharp peaks around 2628 and 2918 cm^{-1}. It could be concluded that the reaction between octylic hydroxamic acid and dolomite represents the physical adsorption. However, from Fig. 7, no new characteristic peaks were found after reaction with dolomite, which shows that octylic hydroxamic acid cannot interact with apatite.

4 CONCLUSIONS

A series of flotation tests were performed to investigate the effectiveness of octylic hydroxamic acid as a collector. The results show that for separation between dolomite and apatite, octylic hydroxamic acid had a high selectivity, and the suitable pH values were within the range of 6–8. Adding collectors can effectively decrease the surface tension of octylic hydroxamic acid and the zeta potential of these two minerals; however, the decrease for dolomite was more apparent. By using Material Studio software and IR spectrum results, it could be seen that the reaction between octylic hydroxamic acid and dolomite represents physical adsorption, while octylic hydroxamic acid cannot interact with apatite.

ACKNOWLEDGMENTS

This study was supported by Guizhou Province Technology Hail and Guizhou academy of Science (LH[2015]); National Natural Science Foundation of China (51574282); Postdoctoral Science Foundation of Central South University (134707); and the Fundamental Research Funds for the Central Universities of Central South University (2015zzts269).

REFERENCES

Alfred W. Petersen. Process for Treating Phosphate Ore[J]. US. 1972(3):702–717.

Che Liping, Yu Yongfu, Pang Jinxing, et al. Synthesis, properties and role mechanism of hydroximic acid as collectors of rare mineral flotation[J]. Chinese Rare Earths, 2004(6):74–79.
Fuerstemau. D.W, Pradip. Zeta potentials in the flotation of oxide and silicate minerals[J]. Advanees in Colloid and Interface Science, 2005, 114–115:9–26.
Guan Feng, Zhong Hong, Liu Guangyi, et al. Flotation of aluminosilicate minerals using alkylguanidine collectors[J]. Transaction of Nonferrous Metals Society of China, 2009(1):228–234.
Guo Fang, Li Jun, Li Wei, et al. Improve the grade of a siliceous-calcareous phosphate ore with fine size by reverse flotation method[J]. Journal of University of Science and Technology Beijing, 2010(11):1388–1392.
Houot Retal. Inverseflotation Beneficiation of Phosphate Ore[J]. Proceeding Compounds, 1980, 4(21–25): 231–246.
Hsieh, Shuang shi. Beneficiation of a Dolomite Phosphate Pebble from Florida[J]. Ind. Eng. Chem. Res. 1988, 27(4):594.
Isgaque Muhammad, Ahmed Nasir. Removal of Magnesium by leaching from Lagarban Phosphate Rock Phosphoric Acid Manufacture[J]. Ferlitizer, 1982, 27(6):53–58.
Li Wei, Gao Hui, Luo Yingjie, et al. Status, trends and suggestions of phosphorus ore resources at home and abroad[J]. China Mining Magazine, 2015(6):6–10.
Sreenivas T, Padmanabhan N.P.H. Surface chemistry and flotation of cassiterite with alkyl hydroxamates [J]. Colloids and Surface A: Physicochemical and Engineering Aspects, 2005, 205(1/2):47–59.
Wang Peipei, Qin Wenqing, Ren Liuwei, et al. Solution chemistry and utilization of alkyl hydroxamic acid in flotation of fine cassiterite[J]. Transaction of Nonferrous Metals Society of China, 2013(6):1789–1796.
Wu Caibin, Duan Xixiang, Dai Huixin et al. New beneficiation method for mid-low-grade phosphate rock-dry electrostatic separation[J]. Industrial Minerals & Processing, 2003(9):7–9.
Yan Zhenghua. Review of development and ultilization of phosphate resources in China[J]. Mining & Metallurgy, 2011(3):21–25.
Young. C.A., Miller. J.D. Effect of temperature on oleate adsorption at a calcite Surface: an FT-NIR/IRS study and review[J]. International Journal of Mineral Processing, 2000, 58(2):331–350.
Yu Yongfu, Ge Yingyong, Pan Changlin. Progress and problems in beneficiation of phosphate ores[J]. Mining and Metallurgical Engineering, 2008(1):29–33.

Innovation and Application of Engineering Technology – Oscar, Chen & Wang (Eds)
 ISBN 978-1-138-05057-0

Evolutionary game and simulation analysis of rent-seeking behavior in real estate development

Kairong Hong & Dan Sun
Business School, Central South University, Changsha, China

Wenming Yao
School of Minerals Processing and Bioengineering, Central South University, Changsha, China
School of Metallurgy and Environment, Central South University, Changsha, China

Sanjun Liu
School of Minerals Processing and Bioengineering, Central South University, Changsha, China

Liyuan Chai
School of Metallurgy and Environment, Central South University, Changsha, China

ABSTRACT: In the real estate development process, rent-seeking behavior of the local government and developers is obvious; therefore, the regulatory authorities need to carry out its supervision. On the basis of the three-party rent-seeking game model, the dynamic evolution analysis is made on the choice of the behavior strategy of the local government and the regulatory authorities, and the evolutionary stable strategy is analyzed by using the dynamic equation and Jacobi matrix. At the same time, the evolutionary stable strategies of the three cases are simulated by using MATLAB, which simulates the ESS of the three cases and the specific impact of the endogenous variables on the final behavior strategy of the local government and the regulatory authorities. Finally, it not only analyzes the factors affecting the evolution of the game system, but also puts forward the management suggestions for the rent-seeking behavior of real estate development.

1 INTRODUCTION

The "rent-seeking" concept was first proposed by American economist Anne Kruger in 1974 in the discussion of policy formation of protectionism in international trade, in a paper called "Political economy of rent-seeking society" (Anne 1974). It extended the field of view of economic research from the productive activities to the nonproductive activities of rent seeking as well as distinguished people's behavior of the pursuit of the newly increased economic interests from the behavior of the pursuit of vested economic interests (Muzhe 2001). Later, Tullock (1967) and Buchanan (1980) conducted a systematic study on the theory of rent seeking through the book "The theory of rent-seeking society" (Yun 2007). Subsequently, the rent-seeking theory is used to describe and explain the trading behavior in the nonproductive field, especially the social phenomenon associated with privilege and corruption (Weizhou, 1999).

At present, the Chinese research literature has emerged largely, using the rent-seeking theory and the economic analysis tool for the interpretation and study of rent-seeking activities. These studies usually focus on bartering and corruption (Jianyun 2008, Yongping 2006 and Liping 2001). The research on the rent-seeking problem has gradually extended to government procurement rent seeking, commercial or enterprise rent seeking, land market rent seeking, construction project rent seeking, and rent seeking in the university science and technology achievement appraisal (Ping 2010, Xiaoxian 2007, Hong 2008 and Wenjuan 2009). Most of the research focused on the phenomena of rent seeking, cost of rent seeking, consequences of rent-seeking (Shigang 2014, Bin et al 2005 and Jianguo and Baojian 2012, Yu et al 2015 and Yu and Peng 2008), especially in the field of government procurement (Jianguo and Baojian 2012). In the process of rent-seeking research, the main research methods are empirical research and game theory. However, so far, there is only little research on rent-seeking activities in the real estate development process (Hong 2015). In the application of the game theory method, it is short of the verification of the nature and the practicality of the model. In other words, these issues need to be further supplemented and improved.

On the basis of the three-party rent-seeking models considering the opportunity cost, under the assumption of limited rationality, this paper will discuss on the evolutionary game equilibrium in view of the rent-seeking phenomenon in the field of real estate development. It also uses MATLAB simulation to carry out the depth test to the model, in order to find the factors affecting real estate development and seek the feasible method of controlling rent seeking.

2 ESTABLISHMENT OF MODEL

In the development of real estate in China, when the rent-seeking behavior is not easy to be found or is often difficult to pursue (rent-seeking low cost), the two key players—the local government and the real estate development enterprises—will make rent seeking for high returns (Gongyu et al 2009). The specific implementation of the real estate development process usually involves three behavior subjects: the central government supervision department, local officers, and property developers. Therefore, we put aside real estate developers to discuss the issue of rent seeking. This paper will further discuss the evolution process and the result of the game between the local government and the central government on the basis of the three-party rent-seeking game models.

2.1 *The tripartite game model of rent seeking*

For the three-party game of rent-seeking activities, the central government as the absolute owner of rights, it has a principal agent relationship with the local government (Xingyu and Laiyi 2001). The central government will entrust a piece of land variable X to the local government through contractual relations. It also has two strategic choices of audit and nonaudit, the results of which may or may not be positive. The local government officials and real estate developers also have two strategic options for rent seeking and nonseeking. When they rent, they may not succeed. Here, we describe the three-party's decision as well as the specific model, which is constructed as follows.

I. In the model, the market value of land variables X is V; the local government will transfer the land to the real estate enterprise with the price of P. Obviously, $V < P$ is clearly not possible. Of course, when $V = P$, local governments will fulfill their duties to the central government, without rent-seeking behavior. When $V > P$, local government officials and real estate enterprises will be in collusion with each other for rent seeking and the client suffered losses (V–P). The inspection probability of the central government, the probability of a successful audit, and the rent-seeking probability of the interest group are P_c, P_a, and P_b respectively.
II. When rent seeking occurs, the real estate business bribes agents through the interests of I. Set the direct economic income of local government officials to rent for I, W for the normal work of the wage function, C_z for audit costs of central government, which usually meets the condition $V–P–I > 0$. When the client finds the rent-seeking behavior of profit alliance, local government officials will suffer m ($0 < m < 1$) times of punishment and the real estate business will suffer n ($0 < n < 1$) times of punishment. The opportunity cost exists during the local government rent seeking; in other words, the punishment of the central government to local government officials often involves the official position and then it indirectly affects their wage income. In this model, the opportunity cost of rent seeking is a variable function of wages, set it to pay r ($0 < r < 1$) times of W, that is, $C_o = rW$. The size of the opportunity cost is largely determined by the size of the benchmark wage W, and the impact of r on C_o is extremely limited.

III. In this model, P, V, and I are exogenous variables and Cz, Pa, m, N, and W are endogenous variables to be explained by the model. When local governments and real estate developers found irregularities, the former will obtain revenue (1–m) I–rW, the latter will get income (1–n)(V–P)–I, and the central government will obtain mI+n(V–P)–C_z. When local governments and real estate developers did not find violation of regulations, local governments will obtain revenue I–rW, real estate developers will get income V–P–I; if the central government check is unsuccessful, it will get an income of –(V–P) –C_z; if the central government did not check, it will get an income of –(V–P). When the local government and real estate companies keep the rules and do not rent, both sides will not have income, and the central government pays the audit cost only when it checks. The game matrix is shown in Table 1.

2.2 *The general analysis of game equilibrium*

I. If the rent-seeking probability of interest alliance of P_b is given in advance, the expected returns of the client for checking and not checking are, respectively, given by the following expressions:

$$\pi_1 = P_b\{P_a[mI + n(V-P) - C_z]\} + (1-P_a)[-(V-P) - C_z] + (1-P_b)[P_a(-C_z) + (1-P_a)C_z] \quad (1)$$

$$\pi_2 = P_b[-(V-P)] + (1-P_b)\cdot 0 \quad (2)$$

When $\pi_1 = \pi_2$, the optimal solution for the rent-seeking probability of interest is:

$$P_b^* = \frac{C_z}{P_a[(1+n)(V-P)+mI]} \quad (3)$$

II. If the agents' interests are placed first, the probability of inspection of the clients of P_c is given in advance, the expected return of the agents (the local governments) of rent seeking and no rent seeking are, respectively, given by the following expressions.

$$\pi_3 = P_c\{P_a[(1-m)I - rW] + (1-P_a)(I - rW)\} + (1-P_c)(I - rW) \quad (4)$$

$$\pi_4 = 0 \quad (5)$$

When $\pi_3 = \pi_4$, the optimal solution for the probability of inspection of the clients is:

$$P_{c1}^* = \frac{I - rW}{mP_aI} = \frac{1-\frac{rW}{I}}{mP_a} \quad (6)$$

III. If the interests of the real estate business are in the first place, the probability of inspection of the clients of P_c is also given in advance, the expected return of rent seeking and no rent seeking of the real estate developers are, respectively, given by the following expressions:

$$\pi_5 = P_c\{P_a[(1-m)(V-P) - I] + (1-P_a)(V-P-I)\} + (1-P_c)(V-P-I) \quad (7)$$

$$\pi_6 = 0 \quad (8)$$

When $\pi_5 = \pi_6$, the optimal solution for the probability of inspection of the clients is:

$$P_{c2}^* = \frac{V-P-I}{nP_a(V-P)} \quad (9)$$

IV. Then, we get the mixed-strategy Nash equilibrium solution after comprehensive analysis of the above:

$$\left[P_b^* = \frac{C_z}{P_a[(1+n)(V-P)+mI]}, P_{c1}^* = \frac{I-rW}{mP_aI} = \frac{1-\frac{rW}{I}}{mP_a}\right]$$
$$\left[P_b^* = \frac{C_z}{P_a[(1+n)(V-P)+mI]}, P_{c2}^* = \frac{V-P-I}{nP_a(V-P)}\right]. \tag{10}$$

Formula (3) shows that the optimal solution for the rent-seeking probability of interest union P_b is proportional to the audit cost of central government C_z, and it is inversely proportional to P_a, n, and m. Therefore, to improve the quality and efficiency of the inspection of the central government (P_a), to reduce the cost of inspection (C_Z), and to improve the degree of punishment for offenders (n and m), we can effectively reduce the probability of occurrence of rent-seeking P_b.

Formulas (6) and (9) show that the optimal solution for the probability of inspection of the clients P_c* is inversely proportional to P_a, n, and m. In the actual decision, if $P_c > P_{c1}$* or $P_c > P_{c2}$, the central government will check, at this point, the optimal choice of real estate enterprises or the agent is not to rent. If $P_c < P_{c1}$* or $P_c < P_{c2}$*, the central government will not check. At the same time, the optimal choice of real estate enterprises or the agent is to rent. If $P_c = P_{c1}$* or $P_c = P_{c2}$*, the central government will not check, but at the same time, the real estate enterprises or the agent can choose to rent or not to rent randomly.

Through formula (6), we have another important discovery. The wage function W is inversely proportional to the central government's optimal inspection probability P_{c1}*. Therefore, we can reduce the optimal audit probability of the central government by appropriately increasing the value of W, which can help us to suppress the negative rent-seeking behavior to a certain extent. In other words, the appropriate increase in the wages of government officials (W) has a positive effect on reducing the corruption of government officials.

3 EVOLUTIONARY GAME ANALYSIS

The general game equilibrium analysis above shows that placing the interests of local governments in the first place or placing the interests of developers in the first place will lead to different mixed-strategy Nash equilibrium results. In this paper, we choose placing local government's interests first as the background and then explore the evolution process and the result of the game between local governments and the central government supervision department on the basis of the three-side game model. Besides, it is assumed that both sides of the game are bounded rational game groups. Then, according to Table 1, we can get the gain matrix of the two sides, which is shown in Table 2.

3.1 *Analysis of the local government*

Replication dynamics is one of the mechanisms of evolutionary game analysis, especially in the analysis of 2 * 2 symmetric games. Assuming that the dynamic equation of the replication

Table 1. Principal agent model of the three-party rent-seeking game.

Local government officials and real estate developers (Agents and real estate enterprises)	Central government (client)		
	Check (P_c)		
	Success (P_a)	Fail ($1-P_a$)	No check ($1-P_c$)
Rent seeking (P_b)	$(1-m)I-rW$, $(1-n)(V-P)-I$, $mI+n(V-P)-C_z$	$I-rW$, $V-P-I$, $-(V-P)-C_z$	$I-rW$, $V-P-I$, $-(V-P)$
No rent seeking ($1-P_b$)	0 0 $-C_z$	0 0 $-C_z$	0 0 0

Table 2. The game matrix of the central government and the local government.

	Central government	
Local government	Check (P_c)	Not check ($1-P_c$)
Rent seeking (P_b)	$I-rW-P_aP_cmI$, $P_amI+(P_an+P_a-1)(V-P)-C_z$	$I-rW$, $-(V-P)$
No rent seeking ($1-P_b$)	0, $-C_z$	$0, 0$

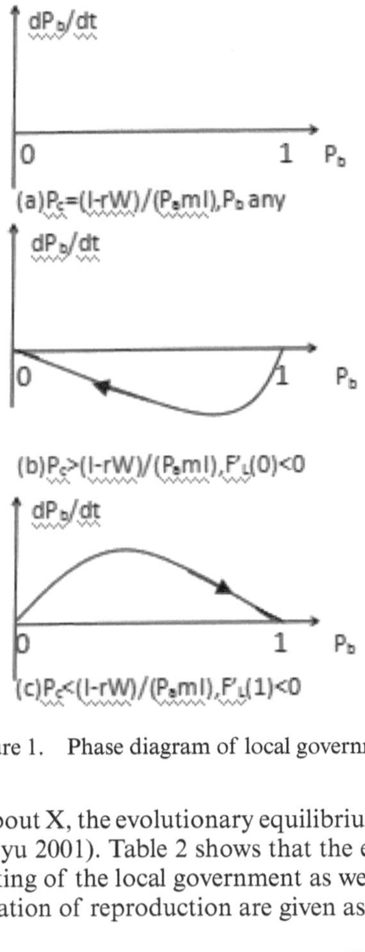

Figure 1. Phase diagram of local government.

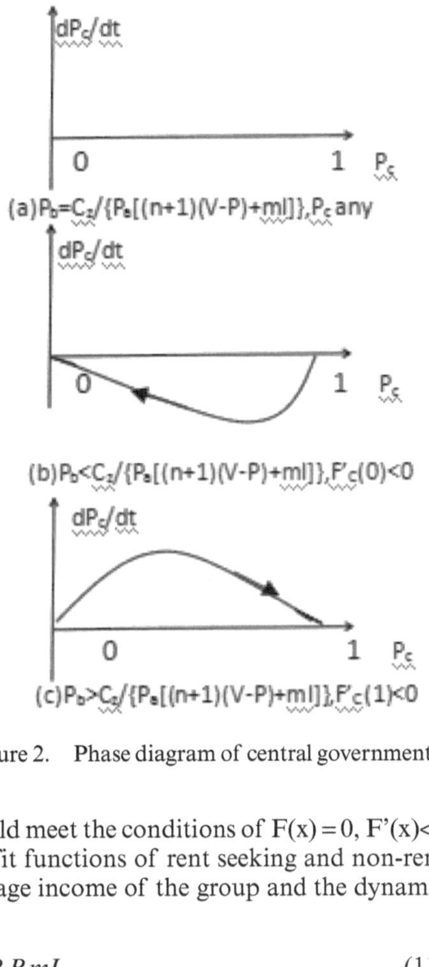

Figure 2. Phase diagram of central government.

is about X, the evolutionary equilibrium point should meet the conditions of $F(x)=0$, $F'(x)<0$ (Shiyu 2001). Table 2 shows that the expected profit functions of rent seeking and non-rent seeking of the local government as well as the average income of the group and the dynamic equation of reproduction are given as follows:

$$\pi_3 = I - rW - P_aP_cmI \tag{11}$$

$$\pi_4 = 0 \tag{12}$$

$$\bar{\pi}_L = P_b\pi_3 + (1-P_b)\pi_4 \tag{13}$$

$$F_L(P_b) = \frac{dP_b}{dt} = P_b(\pi_3 - \bar{\pi}_*) = P_b(1-P_b)(I - rW - P_aP_cmI) \tag{14}$$

Assuming $F_L(P_b)=0$, we can obtain $P_b=0$, $P_b=1$, and $P_c=(I-rW)/(P_amI)$. If the conditions meet $P_c=(I-rW)/(P_amI)$, all of the P_b will be in the stable state. If $P_c \neq (I-rW)/(P_amI)$, the evolutionary stable equilibrium point is required to satisfy this condition: $F(P_b)=0$, $F'_L(P_b)<0$. If $P_c>(I-rW)/(P_amI)$, then we will get $F'_L(0)<0$, so $P_b^*=0$ is the evolutionary stable strategy. If $P_c<(I-rW)/(P_amI)$, then we will get $F'_L(1)<0$, so $P_b^*=1$ is the evolutionary stable strategy. The phase diagrams of the above three cases are shown in Figure 1. It shows that when the inspection possibility of the central government is higher than a certain value, the rent-seeking strategy ratio of the local government and real estate developers group will gradually converge to 0. On the contrary, when the inspection proportion of the central government is lower than the reference value, the rent-seeking strategy proportion of the local government and real estate developers group will gradually converge to 1.

3.2 *Analysis of the central government*

Similarly, as is shown in Table 2, the expected return functions of the audit and nonaudit of the central government as well as the average income of the group and the dynamic equation of reproduction are also given as follows:

$$\pi_1 = P_b P_a mI + P_b(P_a n + P_a - 1)(V - P) - C_z \tag{15}$$

$$\pi_2 = -P_b(V - P) \tag{16}$$

$$\bar{\pi}_C = P_c \pi_1 + (1 - P_c)\pi_2 \tag{17}$$

$$F_C(P_c) = \frac{dP_c}{dt} = P_c(\pi_1 - \bar{\pi}_*) = P_c(1 - P_c)[P_b P_a mI + P_b P_a n(V - P) + P_b P_a(V - P) - C_z] \tag{18}$$

Assuming $F_C(P_b) = 0$, we can obtain $P_c = 0$, $P_c = 1$, and $P_b = C_z/\{P_a[(n+1)(V-P)+mI]\}$. If the conditions satisfy $P_b = C_z/\{P_a[(n+1)(V-P)+mI]\}$, all of the P_c will be in the stable state. If $P_b \neq C_z/\{P_a[(n+1)(V-P)+mI]\}$, the evolutionary stable equilibrium point is required to satisfy the condition: $F(P_c) = 0$, $F'_C(P_c) < 0$. If $P_b > C_z/\{P_a[(n+1)(V-P)+mI]\}$, then we will get $F'_C(1) < 0$, so $P_c^* = 0$ is the evolutionary stable strategy. If $P_b < C_z/\{P_a[(n+1)(V-P)+mI]\}$, then we will get $F'_c(0) < 0$, so $P_b^* = 1$ is the evolutionary stable strategy. The phase diagrams of the above three cases are shown in Figure 2. It shows that when the rent-seeking possibility of the local government and real estate developers is higher than a certain value, the inspection proportion of the central government will gradually converge to 1. On the contrary, when the rent-seeking possibility of the local government and real estate developers group is lower than the reference value, the inspection proportion of the central government will gradually converge to 0.

3.3 *Analysis of the overall system*

For the evolution of the local government and the central government in the field of real estate development, the system can be described by differential equations (14) and (18). Through the analysis of the replicated dynamic equation, we learn that there are five local equilibrium points in the entire game system in the plane of $\{(x,y);0 \le x,y \le 1\}$. They are $E_1(0,0)$, $E_2(0,1)$, $E_3(1,0)$, $E_4(1,1)$, and $E_5\left(\frac{C_z}{P_a[(n+1)(V-P)+mI]}, \frac{I - rW}{P_a mI}\right)$. However, not all of the local equilibrium point is the evolutionary stable point. According to the theory of Friedman, the local stability analysis of the Jacobi matrix of the system can be used to test the stability of the local equilibrium point of the system. If the determinant and trace of the Jacobi matrix corresponding to the dynamic system satisfy the $\det J > 0$ and $\mathrm{tr}J < 0$ synchronously, then the equilibrium point is the evolutionary stable equilibrium point (Paul 1991). Therefore, according to equations (11)-(18), we can calculate det. J and tr. J of equilibrium points E_1, E_2, E_3, E_4, and E_5, which are as follows:

$$E_1{:}\det.J = -C_z(I-rW) \quad \mathrm{tr}.J = I-rW-C_z$$

$$E_2{:}\det.J = C_z(I-rW-P_a mI) \quad \mathrm{tr}.J = I-rW-P_a mI+C_z$$

$$E_3{:}\det.J = (rW-I)\{P_a[(n+1)(V-P)+mI]-C_z\} \quad \mathrm{tr}.J = rW-I+P_a[(n+1)(V-P)+mI] - C_z$$

$$E_4{:}\det.J = (P_a mI-I+rW)\{C_z-P_a[(n+1)(V-P)+mI]\} \quad \mathrm{tr}.J = C_z+rW-I-P_a(n+1)(V-P)$$

$$E_5{:}\det.J = C_z(I-rW)(1-P_b)(1-P_c) \quad \mathrm{tr}.J = 0$$

Obviously, the stability of the five equilibria above is affected by $(I-rW-P_a mI)$ and $\{P_a[(n+1)(V-P)+mI] - C_z\}$ (Qingwen 2003). As $I > rW$ (if the opportunity cost is greater than the rent income, then the local government will not go to rent), for E_1, regardless of the relationship between I–rW and C_z, detj < 0 is determined, so E_1 is a saddle point. Because of this, it is not necessary to discuss the size of ($I-rW-C_z$). The process and results of the stability analysis of equilibrium points are shown in Table 3. The corresponding phase diagrams are given in Figures 3, 4, 5, and 6.

Table 3 shows that both E_3 and E_4 are likely to be the evolutionary stable strategy of the rent-seeking game.

I. If $P_a[(n+1)(V-P)+mI]-C_z > 0$ and $I-rW-P_a mI > 0$, $E_4(1,1)$ is an evolutionary stable strategy (ESS). The system evolution phase diagram is shown in Figure 3, where all arrows point to E_4 direction. This shows that regardless of the initial value of the game, the game equilibrium will eventually converge to (1,1), that is, the local government will take rent-seeking strategy and the central regulatory authorities will check.

II. If $P_a[(n+1)(V-P)+mI]-C_z < 0$ and $I-rW-P_a mI > 0$, $E_3(1,0)$ is an evolutionary stable strategy (ESS). The system evolution phase diagram is shown in Figure 5, where all arrows point to E_3 direction. This shows that regardless of the initial value of the game, the game equilibrium will eventually converge to (1,0), that is, the local government will take rent-seeking strategy and the central regulatory authorities will not check.

III. If $P_a[(n+1)(V-P)+mI]-C_z < 0$ and $I-rW-P_a mI < 0$, $E_3(1,0)$ is an evolutionary stable strategy (ESS). The system evolution phase diagram is shown in Figure 6, all arrows point to

Table 3. Judgment process and results of the stability of evolutionary equilibrium points.

Categories	Condition	Number	Equilibrium point	Determinant symbol	Trace symbol	Conclusion	Phase diagram
Case 1	$P_a[(n+1)(V-P)$ $+mI]-C_z>0$, $I-rW-P_a mI>0$	4	E_1	–	Uncertain	Saddle point	Fig. 3
			E_2	+	+	Unstable	
			E_3	–	Uncertain	Saddle point	
			E_4	+	–	ESS	
Case 2	$P_a[(n+1)(V-P)$ $+mI]-C_z>0$, $I-rW-P_a mI<0$	5	E_1	–	Uncertain	Saddle point	Fig. 4
			E_2	–	Uncertain	Saddle point	
			E_3	–	Uncertain	Saddle point	
			E_4	–	Uncertain	Saddle point	
			E_5	+	0	Core	
Case 3	$P_a[(n+1)(V-P)$ $+mI]-C_z<0$, $I-rW-P_a mI>0$	4	E_1	–	Uncertain	Saddle point	Fig. 5
			E_2	+	+	Unstable	
			E_3	+	–	ESS	
			E_4	–	Uncertain	Saddle point	
Case 4	$P_a[(n+1)(V-P)$ $+mI]-C_z<0$, $I-rW-P_a mI<0$	4	E_1	–	Uncertain	Saddle point	Fig. 6
			E_2	–	Uncertain	Saddle point	
			E_3	+	–	ESS	
			E_4	+	+	Unstable	

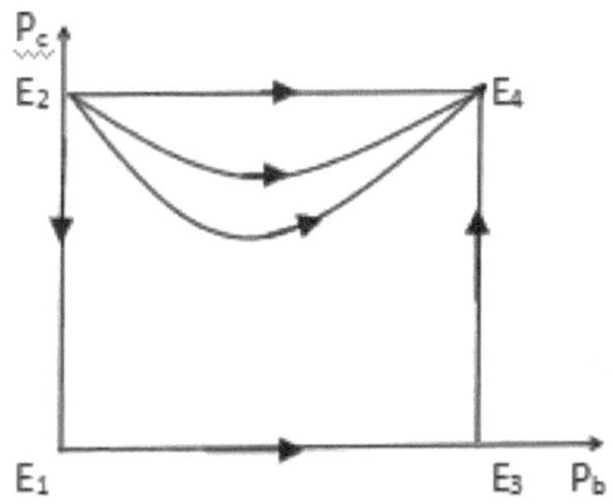

Figure 3. System evolution phase diagram of case 1.

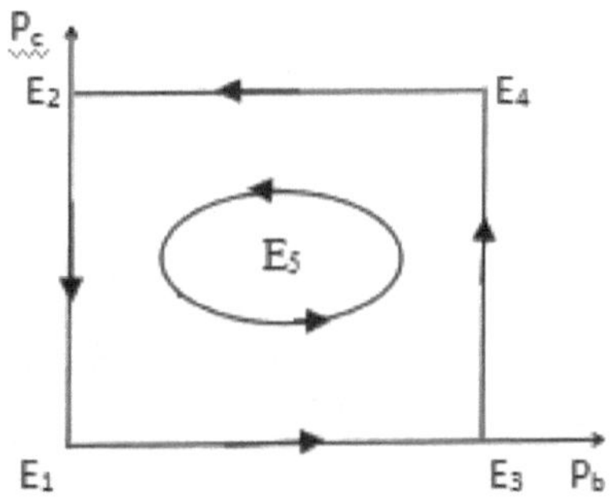

Figure 4. System evolution phase diagram of case 2.

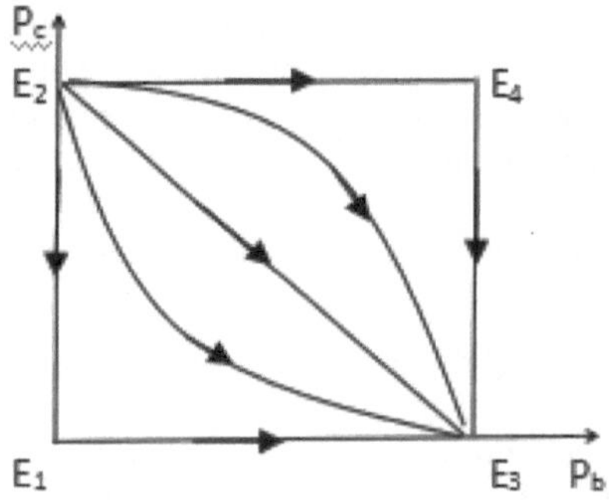

Figure 5. System evolution phase diagram of case 3.

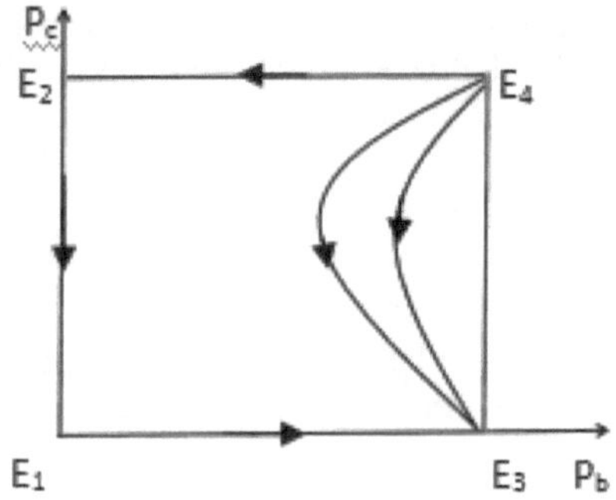

Figure 6. System evolution phase diagram of case 4.

Table 4. Different assignments of numerical simulation.

V	P	I	r	W	m	n	P_a	C_z	Number	Description
10	6	3	0.8	2	0.5	0.5	0.6	2	①	Fig. 7→ESS of case 1
10	6	3	0.8	2.5	0.5	0.5	0.6	2	②	①②③→Fig. 10
10	6	3	0.8	3	0.5	0.5	0.6	2	③	①②③→Fig. 10
10	6	3	0.8	2	0.8	0.8	0.6	2	④	①④→Fig. 12
10	6	3	0.8	2	0.5	0.5	0.8	2	⑤	①⑤→Fig.11
10	6	3	0.8	2	0.5	0.5	0.6	5	⑥	Fig. 8→ESS of case 3
10	6	3	0.8	3	0.5	0.5	0.6	5	⑦	Fig. 9→ESS of case 4

E_3 direction. This shows that regardless of the initial value of the game, the game equilibrium will eventually converge to (1,0), that is, the local government will take rent-seeking strategy and the central regulatory authorities will not check.

IV. If $P_a[(n+1)(V-P)+mI]-C_z > 0$ and $I-rW-P_a mI < 0$, then there is no evolutionary stable strategy in the whole evolution system. The phase diagram of the system evolution is shown in Figure 4, where the arrow is centered around the center point (E_5). That is, the local government may or may not choose the rent-seeking strategy and the central government audit departments may or may not select audit. If the time is infinite, the two sides of the game process will be permanently carried on.

V. In a word, the evolution trajectory and equilibrium results of the rent-seeking game between local governments and the central government ultimately depend on the size of $(I-rW-P_a mI)$ and $\{P_a[(n+1)(V-P)+mI]Cz\}$. In the four cases shown in Table 3, as long as the initial value is not (0,0), the stable equilibrium result will not possibly be (0,0). In other words, as long as the proportion of local government groups choosing rent seeking is not zero, the Pareto optimal equilibrium (no rent seeking, no audit) cannot be achieved.

4 SIMULATION ANALYSIS

In order to investigate the nature and practicality of the model, the variables in the rent-seeking game model are specified as real values. Then, we use MATLAB to simulate the ESS of the three cases and those changes of endogenous variables (W, P_a, m, n, C_z) in the model influencing the behavior strategy choice of the two parties. Numerical simulation of the whole system evolution is assumed under the condition of initial value (0.2,0.2)[1], and throughout the whole simulation process, we maintain the exogenous variables $V = 10$, $P = 6$,

[1]The simulation results of other initial values are exactly the same as the initial values (0,0). This paper takes the initial value (0.2,0.2) as an example to simulate the effects of ESS and simulation-related variables on the equilibrium.

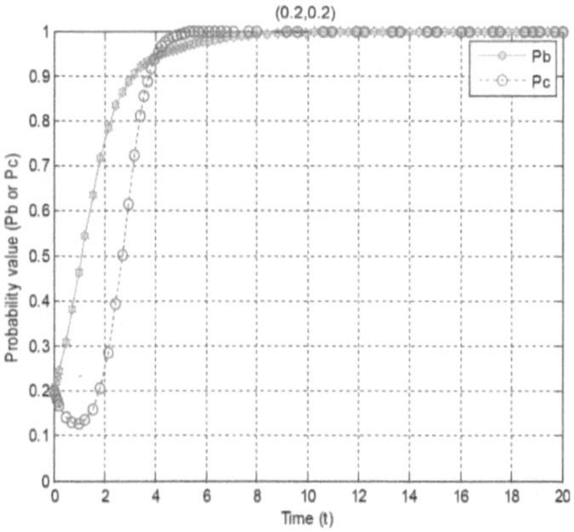

Figure 7. ESS of case 1.

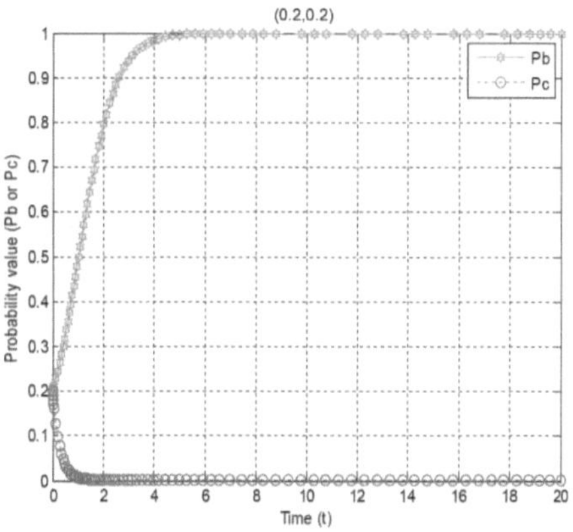

Figure 8. ESS of case 3.

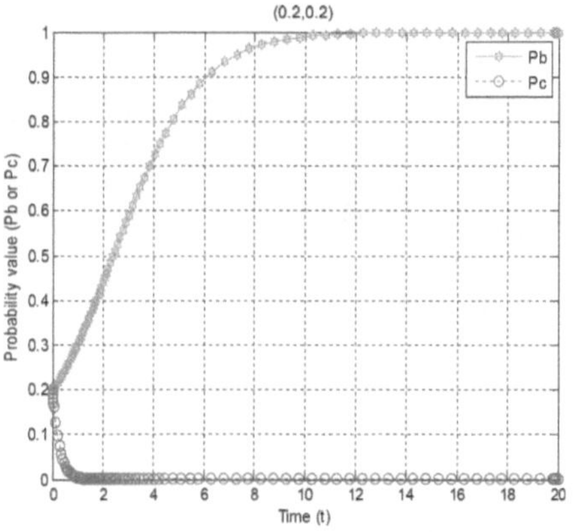

Figure 9. ESS of case 4.

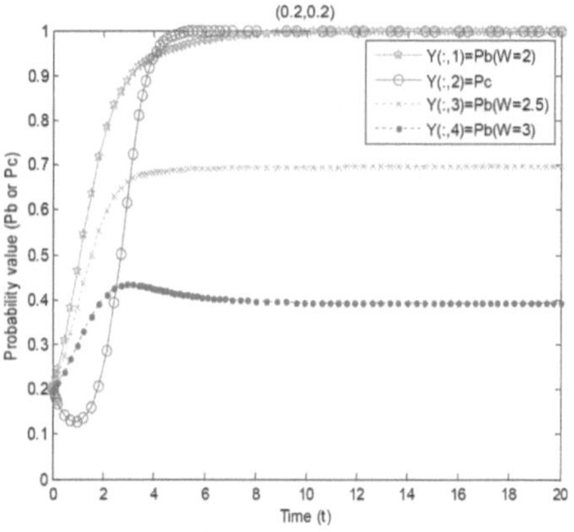

Figure 10. Effect of W on P_b.

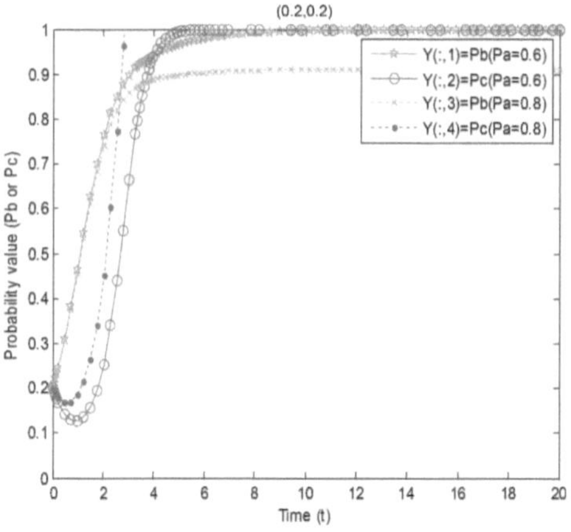

Figure 11. Effect of P_a on P_b,P_c.

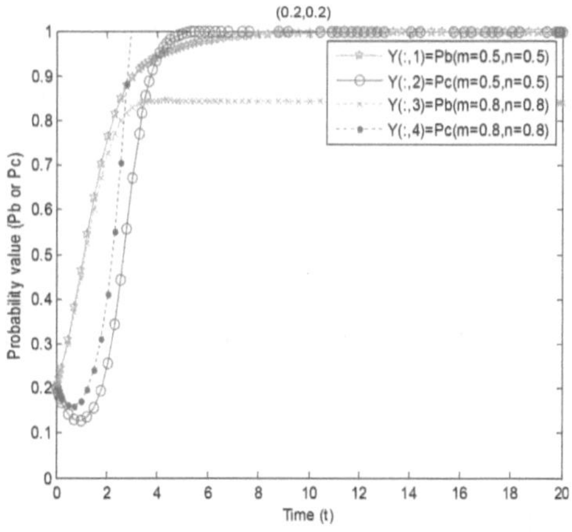

Figure 12. Effect of m and n on P_b,P_c.

I = 3 invariably, and the different values of the endogenous variables are given in Table 4. In addition, in the simulation diagram, the horizontal coordinates represent the simulation time and the longitudinal coordinates represent the proportion of different strategies.

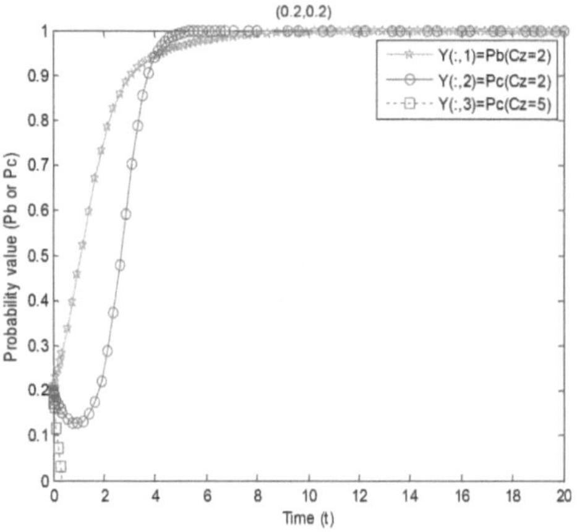

Figure 13. Effect of C_z on P_c.

Situation 1: $P_a[(n+1)(V-P)+mI]-C_z > 0$ and $I-rW-P_amI > 0$. Enter the data of $r = 0.8$, $W = 2$, $n = 0.5$, $m = 0.5$, $P_a = 0.6$, and $C_z = 2$ into the MATLAB system[2] to get the output shown in Figure 7. The green line represents P_b, and the red line represents P_c. From 0.2 to 1, P_b tends to be stable at about $t = 10$ and P_c tends to be stable at about $t = 6$. The final simulation result is (1,1), which is in complete agreement with the results of the dynamic analysis above.

Situation 2: $P_a[(n+1)(V-P)+mI]-C_z < 0$ and $I-rW-P_amI > 0$. Enter the data of $r = 0.8$, $W = 2$, $n = 0.5$, $m = 0.5$, $P_a = 0.6$, $C_z = 5$ into the MATLAB system to get the output shown in Figure 8. From 0.2 to 1, P_b tends to be stable at about $t = 6$. From 0.2 to 0, P_c tends to be stable at about $t = 2$. The final simulation result is (1,0), which is in perfect agreement with the results of the dynamic analysis above.

Situation 3: $P_a[(n+1)(V-P)+mI]-C_z < 0$ and $I-rW-P_a mI < 0$. Enter the data of $r = 0.8$, $W = 3$, $n = 0.5$, $m = 0.5$, $P_a = 0.6$, $C_z = 5$ into the MATLAB system to get the output shown in Figure 9. From 0.2 to 1, P_b tends to be stable at about $t = 14$. From 0.2 to 0, P_c tends to be stable at about $t = 2$. The final simulation result is (1,0), which is in perfect agreement with the results of the dynamic analysis above.

Situation 4: On the basis of situation 1, the value of W is changed to 2.5 and 3. Enter the three sets of values ①②③ into the MATLAB system and the output is shown in Figure 10. The evolution path of P_c is not changed, and the evolution of P_b is changed with the change of W. From the evolution route of P_b, we can obtain that P_b will reduce gradually, with the increase in the wage level W of civil servants. That is, there is a negative correlation between W and P_b, but there is no relationship between W and P_c.

Situation 5: On the basis of situation 1, the value of P_a is changed to 0.8. Enter the two sets of values ①⑤ into the MATLAB system, and the output is shown in Figure 11. After P_a increased, the evolution result of P_b changes from 1 to about 0.9 and at the same time the evolution result of P_c is still 1 with the only change of increasing speed, it reaches the steady state at about $t = 3$. The results show a negative correlation between P_a and P_b and a positive correlation between P_a and P_c.

Situation 6: On the basis of situation 1, the values of m and n are changed from 0.5 to 0.8. Enter the two sets of values ①④ into the MATLAB system, and the output is shown in Figure 12. With increasing n and m, the evolution result of P_b decreases to about 0.85, and synchronously, the evolution result of P_c is still 1; with the only change of increasing speed, it reaches the steady state at about $t = 3$. This implies a negative correlation between n, m and P_b and a positive correlation between n, m and P_c.

[2] The corresponding programming statement of the function is saved as a functional M file, then it can be run to output the corresponding graphics in the MATLAB software system.

Situation 7: On the basis of situation 1, the value of C_z is changed from 2 to 5. Enter the two sets of values ①⑥ into the MATLAB system, and the output is shown in Figure 13. With increasing C_z, no changes occurred in P_b; on the contrary, there is a serious change in P_c, as shown by the red line. With increasing C_z, P_c changes from 1 to 0 rapidly. This result shows a negative correlation between C_z and P_c and no relationship between C_z and P_b.

After the analysis above, it is concluded that the results of MATLAB simulation and the conclusions of dynamic equation analysis are identical. The equilibrium results of (1,1) and (1,0) are two evolutionary stable strategies (ESS) under certain conditions. In general, the study shows that W, P_a, n, and m all have a negative relationship with P_b, but C_z has no relationship with P_b. And P_a, n, and m all have a positive relationship with P_c, and C_z has a negative relationship with P_c, but W has no relationship with P_c. The results indicate that adjusting the value of the variables in the model appropriately can resolve the prisoner's dilemma.

5 CONCLUSIONS

On the basis of the three-party rent-seeking model, this paper studies the evolutionary game analysis between local governments and the central government and discusses the factors influencing the behavior of both sides by using the dynamic equation and MATLAB, thus drawing the following conclusions.

I. The behavior strategy choice of the local government and the central government supervision department of the real estate market is a dynamic game. The proportion of local government's choice of rent seeking is affected by local government officials' wage level, the probability of successful audit of the central government supervision departments, and the degree of punishment on the local government and real estate developers for their rent-seeking behavior. Moreover, this study also finds that the proportion of local government's choice of rent seeking having a negative relationship with local government officials' wage and the probability of successful check of the central government departments as well as with the degree of punishment on the local government and real estate developers. The proportion of the central government's choice of inspection is also affected by audit costs of the central government, the probability of successful audit of the central government supervision departments, and the degree of punishment on the local government and real estate developers for their rent-seeking behavior. Moreover, the inspection possibility of the central government has a negative relationship with audit costs of the central government and a positive relationship with the probability of successful audit of the central government supervision departments and the degree of punishment on the local government and real estate developers. Therefore, in order to reduce the occurrence of rent-seeking activities, we can think about changing these factors.

II. To reduce the rent-seeking proportion of local government, increasing the wages of local government officials appropriately, improving the inspection ability of the central government, and increasing the punishment of rent-seeking behavior may play a positive role. Opportunistic behavior is easy to breed by the low salaries of civil servants, and the low salaries of civil servants can also increase the difficulty of restraining rent seeking, so the central government may properly adjust the level of civil servants wages. Local governments and real estate developers need to establish a sense of social responsibility and a concept of social honor, that is, "to rent for the shame, not to rent for the honor", and they also need to be strictly bound by his/her actions. The supervision department of the central government should establish the internal self-discipline mechanism and the external supervision mechanism at the same time and then further improve their audit ability through their own improvement and social public supervision. Besides, the central government should also severely punish local government officials and real estate developers for rent seeking, as serious violations of local government officials can cancel their duties and serious violations of the real estate developers can cancel the qualification of the market bidding. Necessarily, the central government can formulate relevant laws and policies that specifically manage rent-seeking behavior in the real estate market.

REFERENCES

Anne Krueger, The Political Economy of the Rent-seeking Society, American Economic Review, 1974(64), 291–303.

Bin Wang, Yanfeng Xu, Zhimin Li. Dynamic game analysis of incomplete information on the supervision and control of rent seeking phenomenon[J]. Systems engineering, 2005(10):81–85.

Gongyu Hou, Yaqi Wei, Fei Zhao. Analysis on the three party game model of the theory of rent seeking in engineering project[J]. Journal of Beijing radio and TV University, 2009(02):62–64.

Hong Jia, Game analysis of rent seeking behavior in construction industry[J]. Journal of Beijing Jiaotong University, 2008(03):80–84.

Hong Jiang, Real estate development and local government rent seeking: an extended game model of strategy choice[J]. Journal of Zhejiang Business Technology Institute, 2015(02):20–24.

Jianguo Li, Baojian Yang. Rent seeking games and welfare analysis in government procurement[J]. Shandong Social Science, 2012(11):178–181.

Jianyun Bao, An economic analysis on the rights of the lessor and the lessee in the commercial rent seeking[J]. Journal of central university of finance and economics, 2008(01):59–64.

Liping Zheng. The analysis on the economic rationality and expected utility of corruption[J]. Chinese Social Sciences, 2001(01):91–99.

Muzhe Yang, The three party decision-making model of corruption: the analysis of the principal, the agent and the rent-seeking behavior and the suggestion of the anti corruption policy[J]. Economic Science, 2001(05):35–43.

Paul Burrows. Compensation for Compulsory Acquisition[J], Land Economics, 1991(01):49–63.

Ping Cang, Junmin Wu, Hong Wang. Evolutionary game analysis of the government procurement rent seeking regulation under the three party model[J]. Journal of Guizhou College of Finance and Economics, 2010(06):64–68.

Ping Sun, Hua Ding. Analysis of rent seeking game in the evaluation of scientific and technological achievements in Colleges and Universities[J]. Journal of Social Science of Hunan Normal University, 2016(04):106–109.

Qingwen Sun, Liu Lu, Guangle Yan, Hongan Che. Asymptotic Stability of Evolutionary Equilibrium under Imperfect Knowledge[J]: 2003(07):11–16.

Shigang Li, Heng Yin. How Much is the social cost of the allocation of talent which is caused by Rent-seeking behavior?[J]. Economic Research, 2014(07):56–66.

Shihai Li, Xiaolong Li, Xianzhong Song. Property right, market competition and corporate social responsibility behavior—an analysis based on the perspective of political rent-seeking [J]: 2015(01):162–169.

Shiyu Xie. Evolutionary Game Theory Under Bounded Rationality[J]. Journal of Shanghai University of Finance and Economics, 2001(05):3–9.

Weizhou Zhong, The analysis and policy implications of the rent seeking behavior of public power[J]. Journal of shanxi university of finance and economics, 1999(03):5–9.

Wenjuan Song, Jianxin Hao. Game analysis and prevention of the phenomenon of rent seeking in the land market[J]. Commercial economy, 2009(09):26–27.

Xiaoxian Hong, The construction of government procurement anti rent seeking target system based on Game Theory[J]. Economic Theory and Business Management, 2007(06):30–34.

Xingyu Wang, Laiyi Xue. Analysis on the three party game model of rent seeking theory[J]. Research on Financial and Economic Issues, 2001(11):14–17.

Yongping Ruan, Chen Liu. Rent-Seeking, Rent-Protection and Optimal Financial Contracts[J]. Journal of university of electronic science and technology (Social Science Edition), 2006(01):10–13.

Yu Chen, Peng Xu. The incomplete information static game model of the rent seeking problem in the state owned enterprises with foreign capital merger and acquisition[J]. Industrial technology economy, 2008(04):99–101.

Yun Wang, The three party game analysis of the government investment project under the agent construction system based on rent seeking behavior[J]. Science Technology and Engineering, 2007(07):1419–1421.

Yu Shen, Lili Fu, Jingmei Zhao. An empirical study on the influence of the replacement of Party committee secretary on the rent seeking of enterprises[J]. China Industrial Economy, 2015(09):37–52.

Innovation and Application of Engineering Technology – Oscar, Chen & Wang (Eds)
© 2017 Taylor & Francis Group, London, ISBN 978-1-138-05057-0

Preparation of activating MnO_2 from light manganese

Sanjun Liu
School of Minerals Processing and Bioengineering, Central South University, Changsha, China

Wenming Yao
School of Minerals Processing and Bioengineering, Central South University, Changsha, China
School of Metallurgy and Environment, Central South University, Changsha, China

Qi Yue
School of Minerals Processing and Bioengineering, Central South University, Changsha, China

Liyuan Chai
School of Metallurgy and Environment, Central South University, Changsha, China

Dan Sun
Business School, Central South University, Changsha, China

Anrong Liu
Guizhou Institute of Metallurgical and Chemical Engineering, Guiyang, China

ABSTRACT: The architecture and mechanism of modification of light MnO_2 are studied in the process of activating manganese dioxide by roasting, lixiviating, and oxidizing. It is concluded that different structures of Mn_2O_3 produce different discharge capacities of activating MnO_2 and that the sample is compact and the tap density of the sample is improved by controlling the concentration of MnO_2 in the solutions.

1 INTRODUCTION

At the end of the 20th century, about 25 billion batteries have been consumed worldwide, 70% among which are Zn-Mn cells. For this, over 400,000 tons of MnO_2 were used (Yang 1997). China is the biggest producer and market of batteries in the world (Chen et al 2003 and Fu et al 2014); Chinese battery manufacturers comprise one-third of the world market (Yang et al 2009). Over 1.6×10^8 tons of manganese minerals were imported by China every year, which occupy 60% of the Chinese market (Lei et al 2013). This shows that activating MnO_2 using Zn-Mn cells has a bright future.

On the one hand, by considering the reserves and discharge capacity of manganese dioxide, it is unpractical to use manganese dioxide for depolarizing the cell material. Although EMD and CMD are highly activating, the process is complicated and pollutes the environment. The tap densities of the AMD and CMD are low. These factors restrict the applications of EMD, CMD, and AMD to industrial production.

On the other hand, high Mn content, high tap density of MnO_2, and low content of useless ingredients are demanded in manganese dioxide used for depolarizing the cell material. There are abundant manganese minerals in China, but 90% is the unavailable lean ore, the content of which is <30%. Manganese dioxide with MnO_2 content of >90% is lacking. Natural manganese minerals that are directly used for depolarizing cell material are rare (1%) (Li 1994). More than 70% of manganese minerals have high tap density and high contents of useless ingredient (e.g., Daxin manganese minerals of Guangxi), whereas the remaining have a light tap density and low contents of useless ingredient (e.g., Mugui manganese minerals of Guangxi). The process of the former is mainly to get rid of useless ingredients. The process of the latter is mainly to improve the tap density. On the condition that China is one of the leading producers and exporters of batteries in the world (Wen et al 1988), it is wasteful that the latter is used for steel making or manganese production. Therefore, improving the tap density by a new technique becomes the key of availably utilizing this kind of manganese minerals.

This paper discusses the mechanism of improving tap density and discharge capacity of activating MnO_2. And it is concluded that the different structure of Mn_2O_3 results in a different discharge capacity of activating MnO_2 on the basis of thermodynamics, kinetics, and crystallography.

2 PROPERTIES OF SAMPLE

Natural high-content and light manganese minerals are used as the sample. By crushing and grinding, the granularity of 90% particle is less than 0.15 μm. The result of the analysis of the content of the components in the sample is shown in Table 1. The tap density of the sample is 2.02 g/cm³.

Fig. 1 shows the XRD patterns of the sample. The XRD pattern of the sample shows that the sample mainly contains MnO_2 with little amount of SiO_2, Fe_2O_3, and Al_2O_3. Although the content of MnO_2 is high, MnO_2 is mostly in the form of α-MnO_2 without discharge capability. If this kind of manganese mineral is used for depolarizing the cell material of cell, it must be activated. The tap density is so low that it cannot be used for depolarizing the cell material by the general technique of activating. Considering the special features of sample, the experiment focuses on the factors of roasting affecting the tap density and discharge capacity of activating MnO_2.

Table 1. Content of the component of sample (%).

Item	MnO_2	Fe	SiO_2	Al_2O_3	CaO	MgO
Content (%)	73.05	3.84	10.13	1.39	1.50	0.072

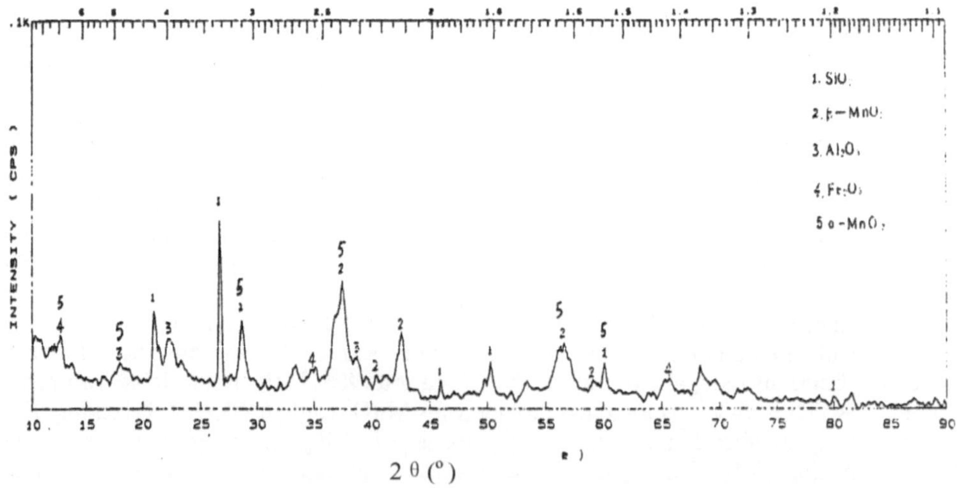

Figure 1. X-ray diffraction patterns of the sample.

3 THEORIES AND EXPERIMENT

3.1 *Theories of roasting*

MnO_2 in natural manganese minerals mostly is α-MnO_2 without discharge capability. The first stage of activating is changing MnO_2 into Mn_2O_3 by roasting. It is a process of deoxidization. The reaction is as follows:

$$MnO_2 \xrightarrow{\Delta} Mn_2O_3 + O_2$$

The reaction shows that the temperature and the partial pressure of O_2 are the main factors. In real production, the roasting reaction is conducted in the open environment, so the partial pressure of O_2 is invariable. Temperature is a decisive factor.

On the basis of the theories of thermodynamics, equation (1) about the temperature and the partial pressure of O_2 is as follows:

$$\lg p_{o_2} = -\frac{8277.7}{T} + 11.2045 \quad (1)$$

There is a reaction between MnO_2 and Mn_3O_4. Equation (2) about the temperature and the partial pressure of O_2 is as follows:

$$\lg P_{O_2} = -\frac{11380.8}{T} - 1.339\lg T - 0.65\times10^{-3}T + 1.113\times10^{5}T^{-5} + 13.9 \quad (2)$$

In the open environment, the partial pressure of O_2 is 21.278 kPa. The temperature of equilibrium of equation (1) is 424°C and that of equation (2) is 889°C. Therefore, the existing temperature of Mn_2O_3 ranges from 424 to 889°C.

3.2 *Theories of crystallography*

In essence, crystal growth is the process of regulating the out-of-order particle array, namely it is the process of forming a lattice. Crystal growth involves two stages: (1) development of crystal seeds and (2) growth of crystal seeds (Kaldis and Emanuel 1985).

Over saturation concentration, temperature, and surface tension are the factors affecting the rate of crystal formation. The most important factor is the oversaturation concentration of the crystal component. Crystal seeds will develop when the concentration of the crystal component exceeds the critical concentration (Brice and Laudise 1973).

The particles of component in solution are adsorbed on the surface of the crystal seeds in the second stage. The oversaturation concentration of the crystal component in solution motivates the growth of crystal seeds. The shape of the crystal is determined by the growth speed ratio of different crystal surfaces (Zehe et al., 2006).

The oversaturation concentration of the crystal component in solution motivates the process of crystal formation, which greatly affects the growth rate and the character and shape of the crystal. Therefore, controlling the oversaturation concentration of the crystal component in solution is the key technique of controlling the character and growth rate of the crystal. On the condition of low oversaturation concentration of the crystal component, the low growth rate helps to develop crystal surface and compact crystal. The high oversaturation concentration of the crystal component facilitates self-purification and crystal formation (Wilcox 1983).

3.3 *Experiment*

In the experiment, natural manganese minerals are spread upon the roasting apparatus. The thickness of materials is 2 cm. After the sample is roasted in the roasting apparatus at the given temperature, the contents of all Mn and MnO_2 are assayed. The transform ratio of MnO_2 is calculated by the contents.

The different roasted sample is processed under the same activating conditions. The tap density and discharge capability of MnO_2 processed with different roasted samples are tested in order to show the influence of roasting factors on the discharge capacity and tap density.

Tap density of MnO_2 is the major parameter that demonstrates the content of MnO_2 in the given volume. The method of measuring tap density is shown in as follows.

MnO_2 (40 cm^3) is put in the measuring cylinder, which is slightly tapped until the volume of MnO_2 in the measuring cylinder becomes constant. The weight-to-volume ratio of MnO_2 in the measuring cylinder is called tap density of MnO_2, and the equation (Yuan et al 2000) is as follows:

$$\text{Tap density}\left(\frac{g}{cm^3}\right)=\frac{\text{the weight of } MnO_2}{\text{the volume of } MnO_2}\left(\frac{g}{cm^3}\right)$$

4 RESULTS AND DISCUSSIONS

4.1 *Roasting factors*

In order to get the optimal roasting temperature, the roasting time is set to 3 h. The percent of conversion, tap density, and discharge capability of MnO_2 are shown in Figs. 2 and 3 at four different roasting temperatures.

Figures 2 and 3 show that the percent of conversion of MnO_2 increases when the roasting temperature increases. The percent of conversion of MnO_2 is 93.56% at 780°C. The percent of conversion of MnO_2 increases slowly when the roasting temperature is over 780°C.

Figures 2 and 3 also show that the tap density decreases with increasing roasting temperature. The reason is that when the temperature increases, more MnO_2 in the natural manganese minerals will change into Mn_2O_3 and thus more MnO_2 will be dissolved and go into the solution. It makes the structure of the grains porous and destroys the compactness of the structure. Although the MnO_2 coming into being by oxidation will fill into the grains, the grains lose compactness. Therefore, the tap density decreases with increasing roasting temperature.

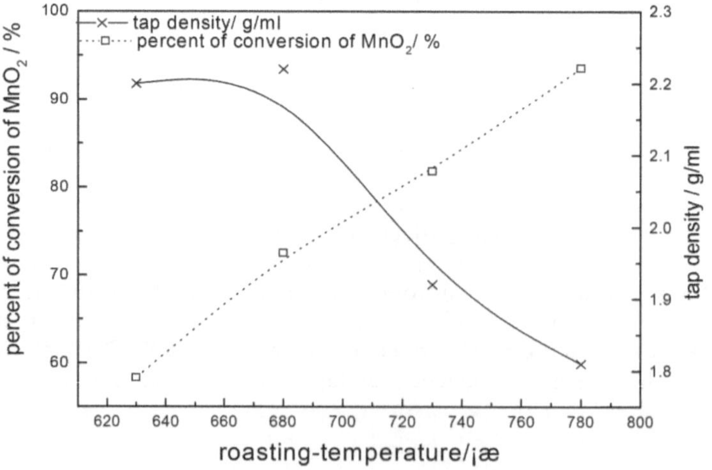

Figure 2. Roasting temperature versus percent of conversion of MnO_2 versus tap density. Condition: Roasting time is 3 h.

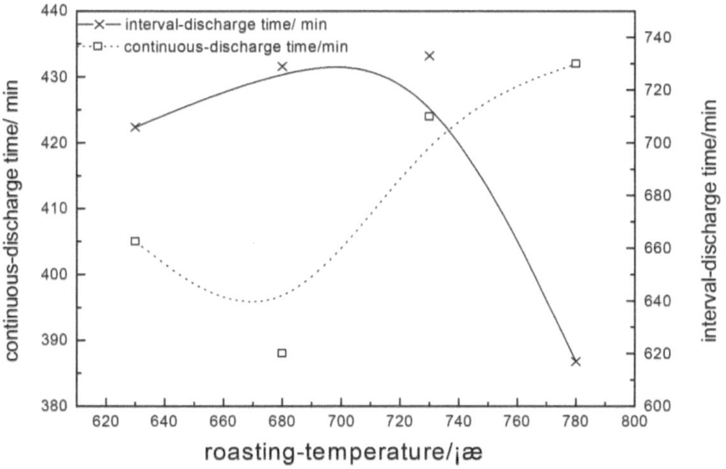

Figure 3. Temperature versus discharge of MnO_2.
Condition: Roasting time is 3 h.

As the discharge capability of MnO_2, the continuous-discharge capability of MnO_2 reverses interval discharge capability of MnO_2. When the temperature is less than 680°C, interval discharge capability of MnO_2 increases, but the continuous discharge capability of MnO_2 decreases with increasing temperature. When the temperature is over 680°C, the interval discharge capability of MnO_2 decreases, but the continuous discharge capability of MnO_2 increases with increasing temperature. The reason is that the crystal structure of Mn_2O_3 will be different at different roasting temperatures. Different crystal structures of Mn_2O_3 produces different crystal structures of MnO_2. Different crystal structures of MnO_2 show different continuous discharge capabilities and interval discharge capabilities. γ-MnO_2 shows better continuous discharge capability and interval discharge capability; however, β-MnO_2 shows better continuous discharge capability and worse interval discharge capability. MnO_2 is composed of different contents of γ-MnO_2 and β-MnO_2 at different roasting temperatures. Therefore, the continuous discharge capability of MnO_2 reverses the interval discharge capability of MnO_2 when the roasting temperature is changed. This result is proved by the X-ray diffraction patterns of active MnO_2.

At the optimal temperature, the effect of different roasting times on the percent of conversion of MnO_2 is shown in Fig. 4.

Fig. 4 shows that the percent of conversion of MnO_2 increases with increasing roasting time at the optimal temperature. The increasing of roasting time greatly affects the percent of conversion of MnO_2 when the roasting time is less than 2 h. The effect of the increase of roasting time on the percent conversion of MnO_2 is not evident when the roasting time is more than 2 h. The increasing of the percent of conversion of MnO_2 is low with increasing roasting time. It is believed that most MnO_2 in the natural manganese minerals changed into Mn_2O_3 in 2 h.

On the basis of the above experiments, the increase of roasting temperature and roasting time can help increase the percent of conversion of MnO_2 between 424°C and 889°C, if only the high percent of conversion of MnO_2 is considered. However, overall, contrasting the tap density with discharge capability, the appropriate roasting temperature and roasting time are selected.

4.2 *Concentration of the oxidant*

Under conditions that the time of oxidizing reaction is 12 h and the volume of H_2SO_4 (98%) is 30 cm^3, the volume of water added is changed in order to change the concentration of oxidant. The data are presented in Table 2.

The result is that the tap density of MnO_2 increased when the concentration of oxidant increased. It is believed that the high concentration of oxidant leads to a large number of

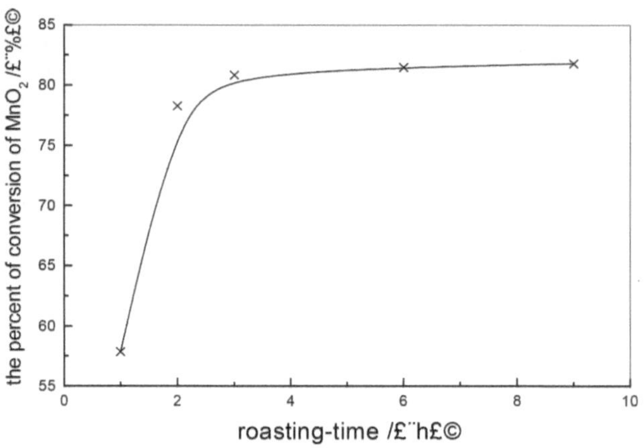

Figure 4. Time versus percent of conversion of MnO_2.

Table 2. Tap density as a function of the concentration of oxidant.

Weight of $NaClO_3$ (g)	Volume of water added (cm^3)	Tap density (g/cm^3)
30	300	1.628
30	500	1.412

Conditions: time of oxidizing reaction is 12 h, and the volume of H_2SO_4 (98%) is 30 cm^3.

MnO_2 being developed in a short time, and a high concentration of MnO_2 helps to shorten the time of crystal seeds being developed. By the way, the formation and growth of crystal seeds is separated as much as possible, and the compacting of MnO_2 is enhanced. As a result, the tap density of MnO_2 increases.

4.3 *The way of adding oxidant*

According to the theory of crystallography, it is known that the formation and growth of crystal seeds are the two main factors affecting the crystallite. If the formation and growth of crystal seeds can be separated, compacting of MnO_2 will be enhanced. The data presented in Table 3 are obtained by changing the way of adding oxidant in this experiment.

It is displayed that the way of adding oxidant affects the tap density of MnO_2 greatly.

First, the tap density of MnO_2 will be different when the oxidant is added by different times. Contrasting twice with once, it increases when the oxidant is added twice. Contrasting twice with more times, the tap density of MnO_2 decreases when the oxidant is added for more times, but contrasting whole with more times, the tap density of MnO_2 increases when the oxidant is added for more times.

The reason is that the way of addition of oxidant affects the formation and growth of crystal seeds.

When the oxidant is added twice, the first addition can make the concentration of produced MnO_2 reach the critical concentration of crystal seeds formed. The second addition of oxidant provides enough MnO_2 for the growth of crystal seeds, thereby increasing the tap density of MnO_2. When the oxidant is added once, the way is helpless to growth of crystal seeds. When the oxidant is added more times, the concentration of produced MnO_2 is less than the critical concentration of crystal seeds formed, so that the tap density of MnO_2 decreased.

Moreover, it is also displayed that the interval time of oxidization affects the tap density of MnO_2. When the oxidant is added three times, the tap density increased with increasing interval time of oxidizations.

Table 3. Tap density as a function of the number of oxidants added.

Weight of $NaClO_3$ (g)	Count	Oxidation times (h)	Tap density (g/cm^3)
30	1	10	1.20
30	2	4, 6	1.69
30	2	10, 10	1.408
30	3	2, 3, 5	1.24
30	3	4, 4, 6	1.28
30	3	6, 6, 6	1.413
30	6	1, 1, 1,1,1,5	1.22

Conditions: time of oxidizing reaction is 12 h; volume of H_2SO_4 (98%) is 30 cm^3; weight ratio of liquid to MnO_2 is 5.

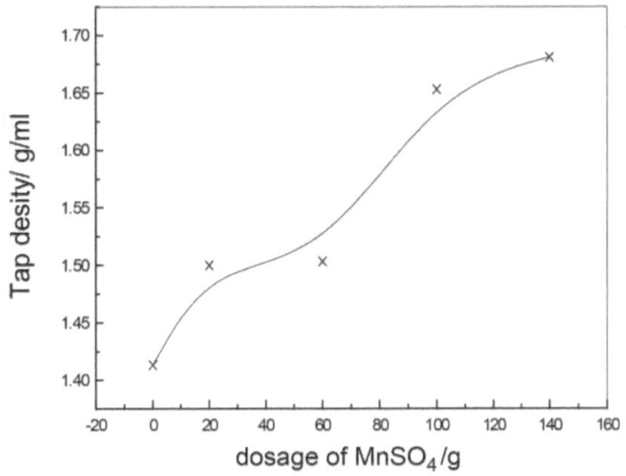

Figure 5. Tap density versus dosage of $MnSO_4$.
Conditions: Time of oxidizing reaction is 12 h; volume of H_2SO_4 (98%) is 30 cm^3; weight ratio of liquid to MnO_2 is 5.

4.4 *Mix $MnSO_4$*

Another method of increasing the concentration of MnO_2 in the solution is adding $MnSO_4$ into the solution when the oxidant is added. Adding $MnSO_4$ results in the increase of the concentration of Mn^{2+} in the solution. Therefore, the concentration of MnO_2 produced is high. On the one hand, it results in the formation and growth of crystal seeds be separated as soon as possible; one the other hand, it provides enough MnO_2 for the growth of crystal seeds. The addition of $MnSO_4$ increases the tap density of MnO_2 and decreases the effect of acid on oxidation. The corresponding results are drawn from the data of experiment.

Fig. 5 shows the relationship between the tap density and the dosage of $MnSO_4$.

5 CHARACTER OF ACTIVATING MNO_2

Figure 6 shows the XRD patterns of roasted manganese minerals at the optimal temperature (710°C) and time (2 h). Figure 5 shows that the major component is Mn_2O_3.

Figure 7 shows the XRD patterns of active MnO_2. The sample contains MnO_2 with little amount of SiO_2 and Fe_2O_3. MnO_2 contains γ-MnO_2 and β-MnO_2. The discharge capabilities of γ-MnO_2 and β-MnO_2 are better than that of α-MnO_2.

The result of the analysis of the content of the components in the sample is presented in Table 4. The discharge capability of the sample is shown in Table 5.

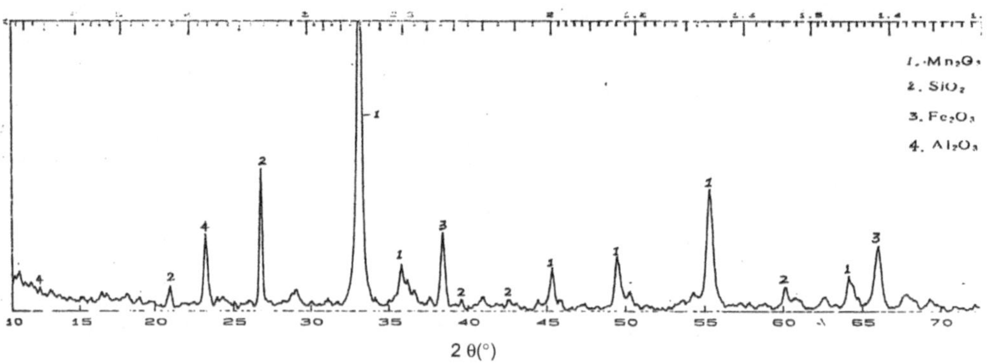

Figure 6. X-ray diffraction patterns of roasted-natural manganese minerals.
Conditions: Optimum temperature (710°C) and time (2 h).

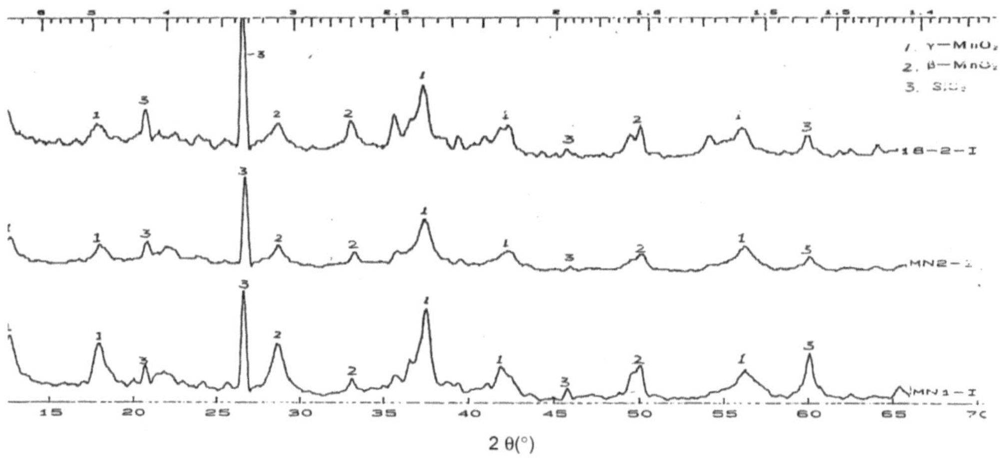

Figure 7. X-ray diffraction patterns of Active-MnO_2.

Table 4. Content of the component of sample (%).

Component	MnO_2	Fe	Al_2O_3	SiO_2	MgO	CaO	Pb	As
Content (%)	74.19	3.74	0.08	7.63	0.06	0.08	无	无

Table 5. Discharge time of the sample (min).

Condition	Continuous discharge (3.9 Ω)	Interval discharge (3.9 Ω)	Continuous discharge (2 Ω)
Time (min)	479	622	185

6 CONCLUSIONS

1. The roasting temperature and roasting time affect the percent of conversion of MnO_2 as well as the tap density and discharge capability.
2. The crystal structure of Mn_2O_3 is different when the roasting—temperature is different, which causes the MnO_2 to be composed of different ratios of γ-MnO_2 to β-MnO_2.

3. The concentration of MnO_2 in solution is the key factor affecting the tap density of MnO_2. Controlling the concentration of MnO_2 in solution helps to control the formation and growth of crystal seeds and increase the tap density of MnO_2.
4. Active MnO_2 shows the character of discharge because α-MnO_2 changed into γ-MnO_2 and β-MnO_2.
5. The optimal roasting factors are 710°C roasting temperature and 2 h roasting time.
6. Optimal parameters activating light MnO_2 (50 g) are the following:

Light MnO_2	50%
Acid	2 5 ml H_2SO_4 (98%) + 5 ml HNO_3 (96%)
Water	500 ml
$NaClO_3$	35%
$MnSO_4$	50%
Temperature	90–95°C
Reaction time	24 h
Times of adding oxidant	3 times

ACKNOWLEDGMENTS

This work was supported by the National Natural Science Foundation of China (No. 51574282).

REFERENCES

Brice J C, Laudise R A. The Growth of Crystals from Liquids [M]. North-Holland Pub. Co., 1973: 50.

Chen Wei-liang, Chai Li-yuan, Min Xiao-bo, Peng Bing, Zhuang Chuan-fu, Dai Yong-nian. Recycling of valuable metals from spent zinc-manganese batteries by vacuum metallurgy [J]. Transactions of Nonferrous Metals Society of China, 2003, 13(5): 1213–1216.

Fu Yong, Xu Zhi-gang, Fei Hao-xiang, et al, Study on metallogenic regularity of manganese ore deposits in China [J]. Acta Geologica Sinica, 2014, 88(12): 2192–2193. (In Chinese)

Kaldis, Emanuel. Crystal growth of electronic materials [M]. Nederland, North-Holland, 1985.

Lei Xiao-li, Hu Yong-da, Du Yi-lun, Zhang Fu-liang, Wang Dan. Thinking of status and development of manganese ore resource utilization [J]. China Mining Magazine, 2013, 37(3): 382–383. (In Chinese)

Li Xu-fu, New methods of exploitation and utilization to the lean, inferior and scrap manganese resources [J], China's Manganese Industry, 1994, 12(1): 8–12. (In Chinese)

Wen Li, Xu Bao, Femg Yan-fu, Production status and development trend of world battery [J]. Battery Bimonthly, 1988(1): 63. (In Chinese)

Wilcox W R. Influence of convection on the growth of crystals from solution [J]. Journal of Crystal Growth, 1983, 65(1–3): 133–142.

Yang Hong. Application of Manganese Compound in Battery Idustry [J]. China Manganese Industry, 1997, 3: 49. (In Chinese)

Yang Lin, Lai Yong, Zheng Wen-xiang, Fan Yin-liang. The present conditions and prospects for primary Zn-MnO_2 battery industries [J]. Chinese Battery Industry, 2009, 14(6): 413–415. (In Chinese)

Yuan Ming-liang, Chen Ke, et al, the research of preparation of ANMD by activating NMD [J], Utilization of Mineral Resources, 2000(1): 19.

Zehe A, Crystal Growth and Characterization of Advanced Materials [J]. Acta Botanica Brasilica, 2006, 20(1): 649–655.

Innovation and Application of Engineering Technology – Oscar, Chen & Wang (Eds)
 ISBN 978-1-138-05057-0

Steklov eigenvalue problem and finite-element approximation

LingLing Sun
School of Biology and Engineering, Guizhou Medical University, Guiyang, China

ABSTRACT: In this paper, we study a reconstructional posteriori error estimate of a special Steklov eigenvalue problem on the basis of the interpolation postprocessing. First, on the basis of the works of LingLing Sun (Adv. Material Research (2012) pp 2081–2086) and Mingxia Li (Adv. Com put. Math., 2010, pp 25–44), we conducted a postprocessing research. Second, given a posteriori error indication, its validity and reliability are proved. Finally, numerical experiments are carried out using MATLAB to confirm the theoretical analysis results.

Keywords: Steklov eigenvalue problem; Reconstructional posteriori error estimates; Finite element; Interpolation postprocessing

INTRODUCTION

In recent years, the Steklov eigenvalue problem has occurred in quite a few physical and electromagnetic systems and received increasing attention.

It is well known that the interpolation postprocessing, which was established by Lin and Yang (Chen and Huang 2000, Brandts 1995, Li and Lin 2006 and Zhu and Lin 1989), is an efficient procedure to increase the accuracy of finite elements. In this paper, we studied a special Steklov eigenvalue problem:

$$-\Delta u = 0 \; in\,\Omega, \quad \partial u/\partial n = \lambda u \; on\,\partial\Omega$$

Bermudez and Rodriguez (Bermudez et al 1995) have probed its prior error estimation, and Houde Han (Han 1994) have studied its boundary element approximation. LingLing Sun (2012) has studied its residual posteriori error estimate.

On the basis of the above-mentioned works, in Section 2, we provide a finite-element approximation (Sun 2012). Graded meshes are introduced in Section 3 by means of the method in Li et al. (2010), and some related results are presented. Moreover, in Section 4, by virtue of Bramble-Hilbert Lemma in Li et al. (2008), the technique of interpolation postprocessing in Li et al. (2010) and the Rayleigh quotient acceleration, we obtain a higher-accuracy posteriori error indicator. Finally, in Section 5, we present a numerical result to validate the theoretical result computationally.

2 FINITE-ELEMENT APPROXIMATION

Consider the following Steklov eigenvalue problem:

$$\begin{cases} -\Delta u = 0 & in\,\Omega \\ \dfrac{\partial u}{\partial n} = \lambda u & on\,\partial\Omega \end{cases}. \tag{1}$$

where $\Omega \subset R^2$ is a polygonal domain with the largest inner angle ω and $\partial u/\partial n$ is the unit normal derivative of u. The variation problem associated with (1.1) is given as follows: Find $\lambda \in R$ and $u \in V = H^1(\Omega)$, $b(u,u) = 1$, $u \neq 0$ satisfying

$$a(u,v) = \lambda b(u,v) \tag{2}$$

where $a(u,v) = \int_\Omega \nabla u \cdot \nabla v + \int_{\partial\Omega} uv,\ \ b(u,v) = \int_{\partial\Omega} uv$. Obviously, if (λ,u) is an eigenpair of (2), then $(\lambda - 1,u)$ is an eigenpair of (1.1); if (λ,u) is an eigenpair of (1), then $(\lambda + 1,u)$ is an eigenpair of (2).

From the differential equation theory, we know that $u \in H^{1+r}(\Omega)$, where $r = 1$ if $\omega < \pi$ and $r < \pi/\omega$ but close to the π/ω sufficiently if $\omega > \pi$.

Let π_h be a family of regular triangulations of Ω, h be the largest diameter of the triangle, $h = \max_{T\in\pi_h} h_T$, and h_T be the diameter of the triangle T. Assuming $h_T/\rho_T \le \sigma, \rho_T$ is the diameter of inscribed circle of the triangle unit.

Let $V_h = \{v \in V : v|_T \in P_1, \forall T \in \pi_h\}$ be a finite-element space and P_1 denote the space of linear polynomials. Then, the finite-element approximation of problem (2) is as follows: find $\lambda_h \in R, u_h \in V_h, b(u_h, v_h) = 1$, such that:

$$a(u_h,v) = \lambda_h b(u_h,v),\ \forall v \in V_h \tag{3}$$

Obviously, $a(u,v)$ and $\|\cdot\|_a = \sqrt{a(\cdot,\cdot)}$ are the inner product and norm of V, respectively. According to Friedrich's inequality, $\forall v \in H^1(\Omega)$, we obtain:

$$\|v\|_{1,\Omega} \le C_\Omega \left\{ |v|_{1,\Omega} + \left| \int_{\partial\Omega} v \right| \right\} \le C\|v\|_a.$$

It is obvious that $\|v\|_a \le \|v\|_{1,\Omega}$, so the two norms $\|\cdot\|_a$ and $\|\cdot\|_1$ are equivalent.

In problem (1), when $\Omega \in R^2$ is rectangular, the length and width correspond to the x and y axes of the coordinate system. At the moment, we consider the problem in the standard Sobolev space.

Define the norm and seminorm in standard space, $W^{m,p,\Omega}(\Omega)(m \in N)$, respectively, as follows:

$$\|\psi\|_{m,p,\Omega} = \left(\sum_{0\le|\alpha|\le m} \|D^\alpha \psi\|^p_{L^p(\Omega)} \right)^{\frac{1}{p}},\quad |\psi|_{m,p,\Omega} = \left(\sum_{|\alpha| = m} \|D^\alpha \psi\|^p_{L^p(\Omega)} \right)^{\frac{1}{p}}.$$

In fractional Sobolev space, $H^\nu(\Omega)$, $\nu \in R^+ \setminus N$, defining the norm as follows:

$$\|\psi\|_{\nu,\Omega} = \left(\|\psi\|^2_{m,\Omega} + \sum_{|\alpha|=m} \int_\Omega \int_\Omega \frac{\left(\partial^\alpha \psi(x) - \partial^\alpha \psi(y)\right)^2}{|x-y|^{2+2\theta}} dxdy \right)^{\frac{1}{2}}$$

At this point, there is still

$$a(u,v) = \int_\Omega \nabla u \cdot \nabla v + \int_{\partial\Omega} uv,\ \ b(u,v) = \int_{\partial\Omega} uv.$$

Obviously, problem (2) has a sequence of pairs $(\lambda_j, u_j) \in R \times V$, where $\lambda_j > 0, j = 1,2,3,\cdots$, and $\lambda_j \to +\infty$, we assume the eigenvalues to be increasingly ordered: $0 < \lambda_1 \le \lambda_2 \le \cdots \le \lambda_j \le \cdots$. Let $\{T_h\}$ be a family of finite-element partition of Ω into rectangles and V_h denote the bilinear finite-element space. Then, the corresponding finite-element approximation associated with problem (2) is given by:

Find $(\lambda_h, u_h) \in R \times V_h$ and $\lambda_{h,j} > 0 (1 \le j \le N_h)$, where N_h denotes the number of vertices on $\delta\Omega$. Without loss of generality, we assumed that the eigenvalues are increasingly ordered: $0 < \lambda_{h,1} \le \lambda_{h,2} \le \lambda_{h,3} \le \cdots \le \lambda_{h,N_h}$.

3 THE GRADED MESH

Let ω be the largest inner angle of domain Ω. For the eigenvalue problem (1), when singular points appear to the solution domain of Ω from the differential equation theory, we know

$u \in H^{1+r}$ but close to the r_0 sufficiently, where $0 < r < r_0 (r_0 = \pi/\omega)$. Therefore, when we use the method of finite element, we need to introduce locally refined mesh, singular odd function, and other skills to reach the ideal accuracy. For the eigenvalue problem (2.1), assuming that the domain Ω is a square domain or an L-shaped domain. For the square domain, we divide Ω into fives parts (see Figure 1): the four corner subdomains $\Omega^i (i = 1,2,3,4)$ containing singular points $O_i (i = 1,2,3,4)$ and the remainder $\Omega \setminus \bigcup_{i=1}^{4} \Omega^i$. Similarly, adopting the same methods to refine the domain $\Omega^i (i = 2,3,4)$ and the partition satisfies the following grading conditions (4):

The partition of L-shaped domain is similar to the squared domain. First, we still find the corner areas containing singular points, where there are two types of domains, with one of them having the largest inner angle ($\pi/2$) and the other having the largest inner angle ($3\pi/2$). For the first-type corner, the partition is the same to the squared area, and for the second corner domain with inner angle $3\pi/2$, the partition is as follows (see Figure 2). By virtue of $u \in H^{1+r}(\Omega)$, the eigenvalue function of problem (2) is analytical on $\Omega \setminus \bigcup_{i=1}^{4} \Omega^i (i = 1,2,3,4)$ and the function has singularity on the points $O_i (i = 1,2,3,4)$. The following is selecting a weight space.

If (ρ, θ) is a polar coordinate, where $\rho = \rho(x_1, x_2) = \sqrt{x_1^2 + x_2^2}$, then the solution is $u \approx \rho^{r_0}$. For the L-shaped domain with the largest inner angle ($3\pi/2$), we know that $r_0 = 2/3$, therefore $|\partial^j u| \le \rho^{r_0 - |j|}$. From integration theory, it is obvious that the integration of $(\rho^{r_0-1})^2$ exists when $j = 1$, but the integration value of $(\rho^{r_0-1})^2$ does not exist when $j = 2$, that is to say $u \notin H^2(\Omega)$. Therefore, it is necessary to pursue a weight coefficient $\rho^{\alpha} (\alpha \in R)$ on the basis of ρ^{r_0-2}. Then, $\rho^{\alpha} \rho^{r_0-2} = \rho^{r_0-2+\alpha}$. Only at $\alpha > 1 - r_0 = 1/3$ did the integration value $(\rho^{r_0-2+\alpha})^2$ exist.

In the same way, it must pursue a weight coefficient $\rho^{\beta} (\beta \in R)$ when we solve u'''. That is to say, if $\rho^{\beta} \rho^{r_0-3} = \rho^{\beta + r_0 - 3}$, then we obtain $\beta > 4/3$. Only in this way did the integration exist.

From what has been discussed above, we select a weight space $H_\tau^3(\Omega)$ for the L-shaped domain, where $\tau = 2\rho > 4/3$. For the squared domain, we select a weight space $H_\tau^3(\Omega)$, $\tau = 2\rho > 0$.

Define the norm and seminorm in weight space $H_\tau^3(\Omega)$, respectively, as follows:

$$\|u\|_{3,\tau,\Omega} = \left(\sum_{i=1}^{n} \sum_{|j| \le 3} \int_{\Omega^i} \left(|r_i|^{\tau - 3 + |j|} |\partial^j u|^2 \right) dxdy + \sum_{|j| \le 3} \int_{\Omega \setminus \bigcup_{i=1}^{n}} |\partial^j u|^2 dxdy \right)^{\frac{1}{2}}$$

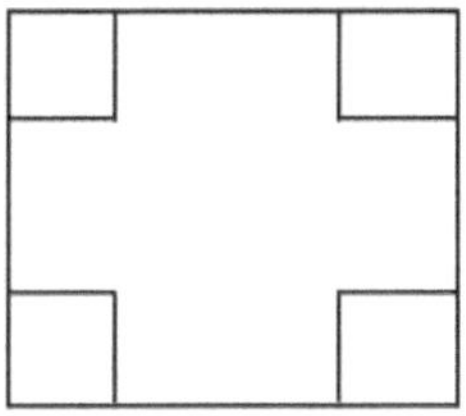

Figure 1. Five parts.

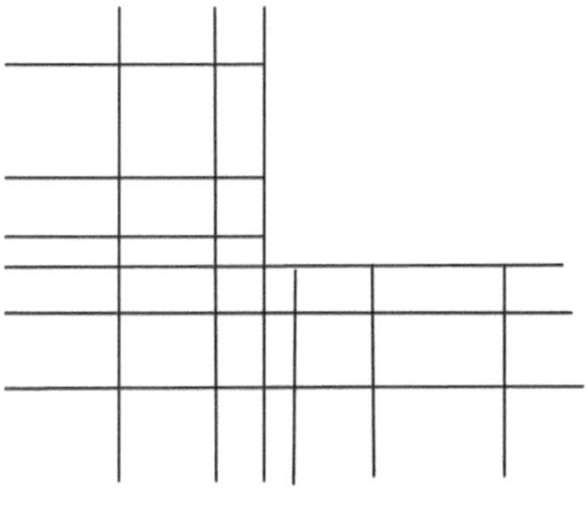

Figure 2. Partition satisfying construction.

$$|u|_{3,\tau,\Omega} = \left(\sum_{i=1}^{n} \sum_{|j|=3} \int_{\Omega^i} \left(|r_i|^{\tau-3+|j|} |\partial^j u|^2 \right) dxdy + \sum_{|j|=3} \int_{\Omega \setminus \bigcup_{i=1}^{n}} |\partial^j u|^2 \, dxdy \right)^{\frac{1}{2}},$$

Here, $n = 4$ for the squared domain and $n = 6$ for the L-shaped domain, respectively, and r_i denotes the distance between the point (x, y) and the singular points $O_i (i = 1,2,3,4)$. For the solution on $\Omega \setminus \bigcup_{i=1}^{4} \Omega^i$, which have enough regularity, asymptotic error expansions for bilinear finite elements have been obtained in [11]. Therefore, we focus our attention on one of the four corner subdomains, say Ω^1. Without loss of generality, we set $\Omega^1 = [0,1]^2$ with the singular point $(0,0)$. Let T_h^1 be the partition T_h over Ω^1. Let e denote an arbitrary element in T_h^1 and $2h_e, 2k_e$ denote its widths in x– and y– directions.

Set $d_e = \max\{h_e, k_e\}, h = \max\{d_e, e \in T_h^1\}, d_0 = \max\{d_e, (0,0) \in e\}$.

Let $r_{1,e} = \inf_{(x,y)\in e}\{|r_1|\}$ denote the distance between e and the corner point $(0,0)$, where $r_1 = \sqrt{x^2 + y^2}$. Then, Ω^1 can be divided into two parts as follows:

$$\Omega_s^1 = \{e \in T_h^1, r_{1,e} < d_0\}, \Omega_r^1 = \{e \in T_h^1, r_{1,e} \geq d_0\}.$$

Let $q = \dfrac{1}{1-\rho}$ be the grading parameter. For the future need, we assume that the mesh satisfies the following grading conditions:

$$d_0 \leq Ch^{\frac{1}{1-\sigma}}, de \leq \left(\frac{1+i}{n}\right)^q - \left(\frac{i}{n}\right)^q = \frac{q}{n}\left(\frac{i+\theta}{n}\right)^{q-1} \leq Chr_{1,e}^{\sigma} \quad i = 0,1,2,\cdots,n-1.\Omega^1 \tag{4}$$

The partition satisfying (4) is shown in Figure 2.

The projection operator $R_h : H^1(\Omega) \to V_h$ is defined by:

$$a(R_h u, v_h) = \lambda b(u, v_h) \ \forall v_h \in V_h.$$

Then, taking $u_h = u_h/(u, u_h)$, we obtain:

$$b(u, u_h) = b\left(u, \frac{u_h}{b(u, u_h)}\right) = \frac{b(u, u_h)}{b(u, u_h)} = 1.$$

Moreover, we have for the exact eigenvalue λ that $\lambda = \lambda b(u, u_h) = \lambda_h b(R_h u, u_h)$. Thus, the approximate eigenvalue λ_h can be expressed as:

$$\lambda_h = \lambda_h b(u, u_h) = \lambda_h b(u - (R_h u, u_h)) + \lambda_h b(R_h u, u_h) = \lambda_h b(u - (R_h u, u_h)) + \lambda.$$

4 POSTPROCESSING AND RECONSTRUCTIONAL POSTERIORI ERROR ESTIMATES

4.1 *Weak estimate of interpolation*

In this part, we focus our attention on the weak estimate of interpolation. Thus, we introduce *Bramble– Hilbert* Lemma first.

Lemma 4.1 denoted by $B(u)$ a bounded linear function from $K^{k+1}(s)$ to R. If for a polynomial P_k of k, $B(P_k) = 0$, there exists a constant C independent of u such that

$$|B(u)| \leq C|u|_{k+1,s}.$$

For arbitrary element $e \in \Omega_r^1$, let $\hat{e} = [-1,1] \times [-1,1]$ be the reference element of e. Define an affine transformation

$$F: e \rightarrow \hat{e}, \text{ with } \qquad F:(x,y) \rightarrow (\hat{x},\hat{y}), \hat{x} = \frac{x-x_e}{h_e}, \hat{y} = \frac{y-y_e}{k_e}.$$

Consider the following function for $\forall \hat{u} \in H^3_{2\sigma}(\hat{e}), \hat{v}_h \in Q_1(\hat{e})$,

$$B_1(\hat{u},\hat{v}_h) = \int_{\hat{e}} (\hat{u}_I - \hat{u})_{\hat{x}} (v_h)_{\hat{x}} \, d\hat{x}d\hat{y} + \int_{\hat{e}} (\hat{u}_I - \hat{u})_{\hat{y}} (v_h)_{\hat{y}} \, d\hat{x}d\hat{y}.$$

It is obvious that we can verify $\forall \hat{u} \in P_2(\hat{e}), \hat{v}_h \in Q_1(\hat{e}), B_1(\hat{u},\hat{v}_h) = 0$. As $\hat{u} \in P_2(\hat{e}) = span\{\hat{x},\hat{y},\hat{x}^2,\hat{y}^2\}, v_h \in Q_1(\hat{e}) = span\{1,\hat{x},\hat{y},\hat{x}\hat{y}\}$, we obtaining the results presented in Tables 1 and 2.

From Tables 1–2, we know $B_1(\hat{u},\hat{v}_h) = \int_{\hat{e}} (\hat{u}_I - \hat{u})_{\hat{x}} (v_h)_{\hat{x}} \, d\hat{x}d\hat{y} + \int_{\hat{e}} (\hat{u}_I - \hat{u})_{\hat{y}} (v_h)_{\hat{y}} \, d\hat{x}d\hat{y} = 0$

$$\forall \hat{u} \in P_2(\hat{e}), \hat{v}_h \in Q_1(\hat{e}).$$

Now, we need to verify the boundary character of $B_1(\hat{u},\hat{v}_h)$. From *Bramble– Hilbert* Lemma, we know that:

$$|B_1(\hat{u},\hat{v}_h)| \le C|\hat{u}|_{3,2\sigma,\hat{e}} |\hat{v}_h|_{1,\hat{e}} \ \forall \hat{v}_h \in Q_1(\hat{e}).$$

Adapting coordinate transfer, we can obtain $B_1(\hat{u},\hat{v}_h)$ translation from $\hat{e}$ to e.

From the definition of F, the following equality has been used:

$$\hat{u}(\hat{x},\hat{y}) = u(x,y), \hat{u}_I(\hat{x},\hat{y}) = u_I(x,y), \hat{u}_{\hat{x}} = heu_x, \hat{u}_{\hat{y}} = keu_y, d\hat{x} = dx/he, d\hat{y} = dy/ke$$

$$\int_e (u_I - u)_x (v_h)_x \, dxdy + \int_e (u_I - u)_y (v_h)_y \, dxdy = O(h^2)|u|_{3,2\sigma,e} (v_h)_{1,e} \quad \forall v_h \in V_h$$

That is to say

$$\int_e \nabla (u_I - u)(\nabla v_h) dxdy = \int_e (u_I - u)_x (v_h)_x \, dxdy + \int_e (u_I - u)_y (v_h)_y \, dxdy$$
$$= O(h^2)|u|_{3,2\sigma,e} (v_h)_{1,e} \le Ch^2 |u|_{3,2\sigma-1,e} \|v_h\|_{1,e}.$$

Let $a_{\Omega_s^1}(u,v_h) = \int_{\Omega_s^1} \nabla u \cdot \nabla v_h + \int_{\partial\Omega_s^1} uv_h$, $\quad a_{\Omega_r^1}(u,v_h) = \int_{\Omega_r^1} \nabla u \cdot \nabla v_h + \int_{\partial\Omega_r^1} uv_h$.

Table 1. Results.

$\hat{v}_h$	1	$\hat{x}$	$\hat{y}$	$\hat{x}\hat{y}$
	0	1	0	$\hat{y}$
	0	0	1	$\hat{x}$

Table 2. Results.

$\hat{u}$	1	$\hat{x}$	$\hat{y}$	$\hat{x}^2$	$\hat{y}^2$
$\hat{u}_I$	1	$\hat{x}$	$\hat{y}$	1	1
$\hat{u}_I - \hat{u}$	0	0	0	$1-\hat{x}^2$	$1-\hat{y}^2$
$(\hat{u}_I - \hat{u})_{\hat{x}}$	0	0	0	$-2\hat{x}$	0
$(\hat{u}_I - \hat{u})_{\hat{y}}$	0	0	0	0	$-2\hat{y}$
$\int_{\hat{e}} (\hat{u}_I - \hat{u})_{\hat{x}} (\hat{v}_h)_{\hat{x}} d\hat{x}d\hat{y}$	0	0	0	0	0
$\int_{\hat{e}} (\hat{u}_I - \hat{u})_{\hat{y}} (\hat{v}_h)_{\hat{y}} d\hat{x}d\hat{y}$	0	0	0	0	0

Theory 4.1 $\forall v_h \in V_h$, the inequality $a(u_I - u, v_h) \le Ch^{\frac{3}{2}} \|u\|_{3,2\sigma,\Omega} \|v_h\|_{1,\Omega}$ has been obtained.
Proof: under the conditions of (4),
We have $\int_{\Omega_r^1} \nabla(u_I - u)(\nabla v_h) dxdy \le Ch^2 |u|_{3,2\sigma,\Omega_r^1} \|v_h\|_{1,\Omega_r^1}$.
As $b_{\Omega_r^1}(u_i - u, v_h) = \int_{\partial\Omega_r^1} (u_I - u) v_h \le C\|u_I - u\|_{0,\partial\Omega_r^1} \|v_h\|_{0,\partial\Omega_r^1} \le Ch^{\frac{3}{2}} \|u\|_{3,2\sigma,\Omega} \|v_h\|_{1,\Omega}$.

$$a_{\Omega_r^1}(u_I - u, v_h) = \int_{\Omega_r^1} \nabla(u_I - u) \cdot \nabla v_h + \int_{\partial\Omega_r^1} (u_I - u) v_h \le Ch^{\frac{3}{2}} \|u\|_{3,2\sigma,\Omega} \|v_h\|_{1,\Omega}.$$

Which, together with Lemma 4.1, the equivalence of norm and condition (4.1), leads to $a_{\Omega_s^1}(u_I - u, v_h) \le C\|u_I - u\|_{a,\Omega_s^1} \|v_h\|_{a,\Omega_s^1} \le Ch^2 \|u\|_{2,2\sigma-1,\Omega} \|v_h\|_{1,\Omega}$.
Thus,

$$a(u_I - u, v_h) \le Ch^{\frac{3}{2}} \|u\|_{3,2\sigma,\Omega} \|v_h\|_{1,\Omega} \qquad \forall v_h \in V_h, \tag{5}$$

Therefore, we complete the proof of this theory.
Theory 4.2 Assume that $u \in H^3_{2\sigma}(\Omega)$ and on each corner subdomain Ω^i $(i = 1,2,3,4)$ the mesh satisfies (4). Then, we have:

$$\|u - R_h u\|_{0,\Omega} + \|u - u_h\|_{0,\partial\Omega} + |\lambda_h - \lambda| \le Ch^{\frac{3}{2}} \|u\|_{3,2\sigma,\Omega}.$$

Proof: As $a(u - R_h u, v_h) = 0 \quad \forall v_h \in V_h$,
We have $a(u - R_h u, v_h) = a(u - u, v_h) \quad \forall v_h \in V_h$.
From (5), we can derive that $a(u - R_h u, v_h) = a(u - u, v_h) \le Ch^{\frac{3}{2}} \|u\|_{3,2\sigma,\Omega} \|v_h\|_{1,\Omega} \quad \forall v_h \in V_h$.
Taking $v_h = u - R_h u$ in the above inequality, we obtain $\|u_I - R_h u\|_{a,\Omega} \le Ch^{\frac{3}{2}} \|u\|_{3,2\sigma,\Omega}$.
That is to say, $\|u_I - u\|_{a,\Omega} \le Ch^{\frac{3}{2}} \|u\|_{3,2\sigma,\Omega}$.
Therefore, $\|u_I - R_h u\|_{0,\Omega} \le \|u_I - R_h u\|_{a,\Omega} = \|u - u_I + u_I - R_h u\|_{a,\Omega} \le Ch^{\frac{3}{2}} \|u\|_{3,2\sigma,\Omega}$.

$$\|u - u_h\|_{0,\partial\Omega} \le C\|u - u_h\|_{1,\Omega} \le \|u - u_h\|_{a,\Omega} \le Ch^{\frac{3}{2}} \|u\|_{3,2\sigma,\Omega},$$

$$|\lambda_h - \lambda| = |a(u_h, u_h) - a(u,u)| = \left|(\|u_h\|_{a,\Omega} + \|u\|_{a,\Omega})(\|u_h\|_{a,\Omega} - \|u\|_{a,\Omega})\right| \le C\|u_h - u\|_{a,\Omega} \le Ch^{\frac{3}{2}} \|u\|_{3,2\sigma,\Omega}$$

Together with above inequalities, we have $\|u - R_h u\|_{0,\Omega} + \|u - u_h\|_{0,\partial\Omega} + |\lambda_h - \lambda| \le Ch^{\frac{3}{2}} \|u\|_{3,2\sigma,\Omega}$.

4.2 *Postprocessing and reconstructional interpolation*

In this section, we will employ *Rayleigh* quotient acceleration technique and the interpolation postprocessing method to get the superconvergence for the eigenvalue finite-element solution of the Steklov eigenvalue problem.

We assume that the rectangular partition T_h has been obtained from T_{2h} with mesh size $2h$, by subdividing each element of T_{2h} into four small congruent rectangles. Let $\tau = \bigcup_{i=1}^4 e^i$ with $e^i \in T_h$. We define the biquadratic interpolation operator I^2_{2h} associated with T_{2h} according to the following conditions:

$$I^2_{2h}\omega|_\tau \in Q_2(\tau), I^2_{2h}\omega(z_i) = \omega(z_i) \quad i = 1,2,\cdots,9.$$

where z_i $(i = 1,2,\cdots,9)$ are the nine vertices of the four small element $e^i (i = 1,2,3,4)$. In addition, it is easy to check that

$$I^2_{2h}\omega_I = I^2_{2h}\omega, \|I^2_{2h}v_h\|_{a,\Omega} \le C\|v_h\|_{a,\Omega} \quad \forall v_h \in V_h$$

And $\|I^2_{2h}\omega - \omega\|_{a,\Omega} \le Ch^{\frac{3}{2}} \|\omega\|_{3,2\sigma,\Omega} \quad \forall \omega \in H^3_{2\sigma}(\Omega)$

Theory 4.3 assumes that $u \in H^3_{2\sigma}(\Omega)$ and $(\lambda,u),(\lambda_h,u_h)$ are the solutions of (1) and (2), respectively. On each corner subdomain, the mesh satisfies (4). Then, we have the following super convergence results:

$$\|I^2_{2h}u_h - u\|_{a,\Omega} \leq Ch^{\frac{3}{2}}\|u\|_{3,2\sigma,\Omega}, \lambda^r - \lambda \leq Ch^3,$$

where I^2_{2h} is the biquadratic interpolation operator defined above and λ^r is the *Rayleigh* quotient:

$$\lambda^r = \frac{a(I^2_{2h}u_h, I^2_{2h}u_h)}{b(I^2_{2h}u_h, I^2_{2h}u_h)}.$$

Remark 4.1 Form Theory 4.3, we can obtain the following a posteriori error estimates:

$$\|u - u_h\|_{a,\Omega} = \|I^2_{2h}u_h - u\|_{a,\Omega} + O\left(h^{\frac{3}{2}}\right), \lambda_h - \lambda = \lambda_h - \lambda^r + O(h^3)$$

This remark demonstrates that $\|I^2_{2h}u_h - u\|_{a,\Omega}$ and $\lambda_h - \lambda^r$ are the asymptotic and accurate reconstructional posteriori error indications of $\|u - u_h\|_{a,\Omega}$ and $\lambda_h - \lambda$, respectively.

5 NUMERICAL RESULTS

In this section, we consider problem (1), where $\Omega = [0,1]^2$ or $\Omega = [0,1] \times [0,1] - [1/2,1] \times [1/2,1]$. The exact eigenvalues of Ω have been obtained by Sun L.L (2012). By programming with MATLAB, those results are presented in Tables 3–6.

Comparing Tables 3–6 with Tables 1–2 in Sun L.L (2012), we have that the reconstructional posteriori error indications are better than the residue posteriori error indications.

Table 3. Numerical result of the second eigenvalue on square domain.

h	λ^r_2	$\lvert\lambda_{2,h} - \lambda^r_2\rvert$	$(\lambda_{2,h} - \lambda^r_2)/(\lambda_{2,h} - \lambda_2)$
$\sqrt{2}/8$	1.37651136079875	$6.645975219234947e-005$	2.37151581281720
$\sqrt{2}/16$	1.37650585239339	$5.653840315369073e-005$	2.07393197740623
$\sqrt{2}/32$	1.37650550766252	$4.843724702224447e-005$	1.90529266145341
$\sqrt{2}/64$	1.37650548610988	$3.786958614718925e-005$	1.85600558457778
$\sqrt{2}/128$	1.37650548476372	$3.692392333012389e-005$	1.83654923539452

Table 4. Numerical result of the fourth eigenvalue on square domain.

h	λ^r_2	$\lvert\lambda_{2,h} - \lambda^r_2\rvert$	$(\lambda_{4,h} - \lambda^r_4)/(\lambda_{4,h} - \lambda_4)$
$\sqrt{2}/8$	2.00000000000000	$1.042658601888893e-004$	2.78898765505838
$\sqrt{2}/16$	2.00000000000001	$8.269403510947444e-006$	2.40401638029279
$\sqrt{2}/32$	2.00000000000006	$6.342403306270674e-006$	2.16041665089905
$\sqrt{2}/64$	2.00000000000025	$3.641905102513831e-006$	1.92109097522537
$\sqrt{2}/128$	2.00000000000099	$3.349050637699805e-006$	1.76298112600987

Table 5. Numerical result of the second eigenvalue on L-shaped domain.

H	λ_2^r	$\lvert\lambda_{2,h} - \lambda_2^r\rvert$	$(\lambda_{2,h} - \lambda_2^r)/(\lambda_{2,h} - \lambda_2)$
$\sqrt{2}/8$	0.77682507300506	$4.202095483944036e-004$	1.92495908242756
$\sqrt{2}/16$	0.77537848934079	$8.772070676429814e-004$	1.65305982111731
$\sqrt{2}/32$	0.77480665361259	$3.283591674687836e-004$	1.80025243590090
$\sqrt{2}/64$	0.77458001496668	$1.243298746974864e-004$	1.47240255762176
$\sqrt{2}/128$	0.77449011966902	$1.146122908801886e-004$	1.35732003782266

Table 6. Numerical result of the third eigenvalue on L-shaped domain.

h	λ_3^r	$\lvert\lambda_{2,h} - \lambda_2^r\rvert$	$(\lambda_{3,h} - \lambda_3^r)/(\lambda_{3,h} - \lambda_3)$
$\sqrt{2}/8$	1.59197111519033	$4.036538448339841e-005$	2.00244362105619
$\sqrt{2}/16$	1.59190634714951	$7.454589336042257e-006$	2.16752175192487
$\sqrt{2}/32$	1.59189769317003	$3.353652358484286e-006$	1.81723569066303
$\sqrt{2}/64$	1.59189643633955	$2.479708585934048e-006$	1.54732003782266
$\sqrt{2}/128$	1.59189624568945	$2.893728126749005e-006$	1.56389426786050

REFERENCES

Armentano, M.G. (2004): The effect of reduced integration in the Steklov eigenvalue problem. Math. Model. Numer. Anal.(MAN): 38,27–36.

Armentano, M.G., C, Padra. (2008): A posteriori error estimates for the Steklov eigenvalue problem, Applied Numerical Mathematics.Vol. 58, p. 593–601.

Babuska, I., J. Osborn (1991): Handbook of Numerical Analysis, vol. 2, pp. 640–787 (El-sevier Science Publishers B.V, North-Holland.

Bermudez, A., R. Rodriguez, D. Santamarina. (1995): A finite element solution of an added mass formulation for coupled fluid-solid vibrations, Numer. Math Vol. 87(2000), p. 201–227.

Brandts, J. (1995): Superconvergence phenomena in finite element methods. phD. Thesis, Utrecht Univ.

Chen, C., Y. Huang: High Accuracy Theory for Finite Element Methods. Hunan Scientific and technology, Hunan.

Han. H.D. (1994): The boundary element approximation of Steklov eigenvalue problems. App Math Ser A-A Journal of Chinese Universities Vol. 9(2), p. 128–135.

Li, M.X., Q. Lin, S. Zhang (2010): Extrapolation and superconvergence of the Steklov eigenvalue problem. Adv. Comput. Math. (2010) pp. 25–44.

Li, Q., H.T. Huang, Z.C. Li, (2008): New expansions of Numerical eigenvalues for $-\Delta u = \lambda \rho u$ by nonconforming elements. Mathematics of computation, Volume 77, Number 264, pp. 2061–2084.

Li, Q., J. Lin. (2006): Finite Element Methods: Accuracy and Improvement. Science, Beijing.

Sun, L.L., Y.D. Yang, (2012): A poteriori error estimates for a Steklov eigenvalue problem. Advanced Material Research Vols. 557–559, pp. 2081–2086.

Zhu, Q., Q. Lin. (1989): Superconvergence Theory of Finite element methods. Hunan Science, Hunan.

Innovation and Application of Engineering Technology – Oscar, Chen & Wang (Eds)
© 2017 Taylor & Francis Group, London, ISBN 978-1-138-05057-0

Growth and nutrient degradation characteristics of an aerobic denitrifying bacterium

Ling Sun
School of Environment Engineering, Xuzhou Institute of Technology, Xuzhou, China
School of Environment Science and Spatial Informatics, China University of Mining and Technology, Xuzhou, China

Wen-ping Cao & Shao-hui Huang
School of Environment Engineering, Xuzhou Institute of Technology, Xuzhou, China

Hui-fang Zhang
School of Environment Engineering, Xuzhou Institute of Technology, Xuzhou, China
School of Environment Science and Spatial Informatics, China University of Mining and Technology, Xuzhou, China

Yan Li
School of Environment Science and Spatial Informatics, China University of Mining and Technology, Xuzhou, China

ABSTRACT: An efficient aerobic denitrifying bacterium, QX1, identified as Arthrobacter sp. by morphological observations, physiological-biochemical tests, and 16S rDNA sequence analysis, was isolated from burdock fields. A phylogenetic tree for this strain was built based on the 16S rDNA sequence, and the effects of nitrogen source, carbon source, pH value, and temperature on the growth of the strain, in addition to its denitrification characteristics, were evaluated. The results showed that under conditions of a gas bath with shaking at 160 rpm, suitable growth conditions for this strain were LB medium plus 1% acetic acid sodium, 30°C, and pH 8. With these conditions, the strain grew to the logarithmic phase after about 4 h, the logarithmic phase was maintained for 16 h, steady growth was observed for 24 h, and growth began to decline thereafter. QX1 exhibited alkali resistance and could tolerate low (15°C) and high temperatures (40°C). At 30°C, pH 8, and shaking at 160 rpm, strains could degrade nitrate, with a removal rate of 89.92% within 48 h in synthetic sewage. Thus, strain QX1 could have applications as an alkaline-tolerant, broad-spectrum temperature-tolerant, efficient denitrifying bacterium in the field of denitrification repair.

Keywords: aerobic denitrification; Arthrobacterium sp.; isolate; bacterial identification; bacterial characteristics

1 INTRODUCTION

Traditional biological denitrification processes include autotrophic nitrification and anaerobic denitrification (Li et al 2014, Shi et al 2013 and Ren et al 2014). In recent years, heterotrophic nitrification and aerobic denitrification have become hot research topics in the field of biological denitrification. Several types of bacteria have been shown to function as aerobic denitrifying bacteria, including Pseudomonas sp. (Li 2014, Miyahara et al 2012 and Yang et al 2011), Acinetobacter sp. (Ren et al 2014, Wang et al 2011 and Ren et al 2014), Paracoccus sp. (Shi et al 2013 and Liu et al 2010), Rhizobium (Xiao et al 2012), Alcaligenes sp. (Wang et al 2011 and Makoto and Yoichi 2014), Bacillus sp. (Li et al 2011), and Vibrio sp. (Duan et al 2015). A strain of aerobic denitrifying bacteria, Pseudomonas GL19, selected by Wang et al. (Wang et al 2014) showed resistance to low temperatures and could use a different nitrogen source for fast and efficient denitrification; this strain could also be applied for denitrification of wastewater and may have applications in wastewater denitrification treatment during winter. Diaphorobacter sp. PDB3 (Wang et al 2015) has also been shown to remove ammonia nitrogen through cell assimilation and heterotrophic nitrification/aerobic denitrification, with a total organic carbon removal rate of up to 90.8%. The NO3--N removal rate by Pitt Acinetobacter selected by Huang et al. (Huang et al 2015) reached 78.89% at 36 h under aerobic conditions with NO3--N as the sole nitrogen source. Additionally, the NH4+, total nitrogen, and total organic carbon removal rates of Pitt Acinetobacter at 48 h were 95.25%, 80.42%, and 98.30%, respectively, when NH4+ was the only nitrogen source. Thus, biological denitrification technologies could be developed for use in practical applications.

In this study, we obtained a strain, designated QX1, from burdock planting field. We then aimed to analyze the morphological, physiological, and biochemical characteristics of this strain and to assess 16S rDNA sequences and denitrification characteristics in water to provide a theoretical basis for future applications of this strain in biological denitrification.

2 MATERIALS AND METHODS

2.1 *Bacterium source*

The bacterium in this study was isolated from burdock field of Feng County, Xuzhou City, China, and saved at 4°C in environmental engineering laboratory of Xuzhou Institute of Technology.

2.2 *Culture medium*

The following media were prepared: 1) Enrichment medium, containing 2.0 g/L KNO_3, 5.0 g/L $C_6H_{12}O_6$, 1.0 g/L K_2HPO_4, 0.5 g/L KH_2PO_4, 0.2 g/L $MgSO_4{\cdot}7H_2O$, 2 mL/L trace elements (pH 7.0–7.2). 2) LB medium, containing 5.0 g/L yeast extract, 10 g/L peptone, 5.0 g/L NaCl (pH 7.0–7.2; optionally, 25 g gelatin was added to make the medium solid). 3) BTB medium, containing 3.0 g/L beef extract, 10.0 g/L peptone, 15.0 g/L sucrose, 20.0 g/L NaCl, 0.01 g/L bromothymol blue (pH 7.0–7.2). 4) Denitrification medium, containing 5 g/L sodium citrate, 1 g/L KNO_3, 1 g/L KH_2PO_4, 0.5 g/L K_2HPO_4, 0.2 g/L $Mg_2SO_4{\cdot}7H_2O$, 2 mL/L trace elements (pH 7.2). 5) BTB-denitrification medium, containing BTB medium plus denitrification medium at 4 mL/L (pH 7.0–7.2). 6) Selective medium for liquor, containing 5 g/L $CH_3COONa{\cdot}3H_2O$, 0.05 g/L K_2HPO_4, 0.2 g/L KH_2PO_4, 0.2 g/L $MgSO_4{\cdot}7H_2O$, 0.5 g/L $CaCl_2$, 2 mL/L trace elements (pH 7.0–7.2). The trace element solution contained 1.5 g/L $FeCl_3{\cdot}6H_2O$, 0.15 g/L H_3BO_3, 0.03 g/L $CuSO_4{\cdot}5H_2O$, 0.03 g/L KI, 0.06 g/L $Na_2MoO_4{\cdot}2H_2O$, 0.12 g/L $MnCl_2{\cdot}4H_2O$, 0.12 g/L $ZnSO_4{\cdot}7H_2O$, and 0.12 g/L $CoCl_2{\cdot}2H_2O$ (pH 7.0).

2.3 *Enrichment and selection of bacterial species*

One gram of soil sample was collected in a 100 mL conical flask containing enrichment broth and shaken at 120 rpm at 26°C. The enrichment cycle was as follows: every 3 days, the bottle was stood for a moment for the precipitate to settle, after which 20 mL of the liquid was removed and fresh enrichment broth was added to a total volume of 100 mL. This method ensured that the denitrifying bacteria could continue to grow. Griess reagent was used to detect NO_2^--N, and diphenylamine-sulfuric acid reagent was used to detect the removal of NO_3^--N. When the solution turned colorless, the enrichment domestication process was considered complete.

Next, 0.5 mL bacteria after enrichment and domestication was removed and subjected to a gradient dilution (10^{-1}–10^{-9}). Twenty microliters of bacteria from the diluted samples was used to coat LB tablets, with three replicates for each dilution. Samples were maintained for

2–3 days in a biochemical incubator at 26°C. Mature colonies were transferred and streaked on plates; pure colonies were obtained after repeating this step 6–8 times. Pure colonies were inoculated in BTB-denitrification medium and developed for 2–3 days at 26°C; strains showing blue color were considered denitrifying bacteria. Selected denitrifying bacteria were cultured in 100 mL denitrifying medium at 26°C for 24 h with shaking at 160 rpm. The absorbance of NO_3^--N and NO_2^--N was determined from the liquid after centrifugation. The denitrification rate was calculated according to the formula: $\eta = (A - A_t - B_t)/A \times 100\%$, where η is the denitrification rate (%), A is the initial absorbance of NO_3^--N, A_t is the absorbance of NO_3^--N in the solution after a certain time, B_t is the absorbance of NO_2^--N in the solution after a certain time. The strain with the highest denitrification rate was chosen for follow-up experiments.

2.4 *Analysis of the morphological, physiological, and biochemical characteristics of the strain*

Analysis of the morphological, physiological, and biochemical characteristics of the strain was carried out according to previously described methods (Dong and Cai 2001 and CASIM-BCG 1978).

2.5 *Amplification of the 16s rDNA gene by polymerase chain reaction (PCR) [19] and construction of the phylogenetic tree*

The 16s rDNA sequence was amplified from bacterial DNA using universal primers as follows: upstream (27FM), 5′-AGTTTGATCMTGGCTCAG-3′; downstream (1492), 5′-GGTTACCTTGTTACGACTT-3′. PCR was carried out using 50 ng/μL template (0.5 μL), 5 × buffer (with Mg^{2+}) 2.5 μL, 2.5 mM dNTPs (1 μL), 10 μM each primer (0.5 μL), 1 μM Taq-TM DNA polymerase (1 μL), with double distilled water to 25 μL. The PCR protocol included denaturation at 98°C for 3 min; 30 cycles of 98°C for 25 s, 55°C for 25 s, and 72°C for 1 min; and a final extension at 72°C for 10 min. Agarose gel electrophoresis was used to analyze PCR products, and sequencing was performed by Sangon Biotechnology Co., Ltd. (Shanghai, China).

The 16S rDNA gene sequences were submitted to NCBI BLAST to retrieve similar sequences, based on the principles of maximum homology sequence alignment, with Mega5.0 software and Clustal X1.83 software were used to implement the neighbor-joining method for construction of the phylogenetic tree. The stability of the branches was analyzed by the bootstrap method with 1000 repeats.

2.6 *Effects of different conditions on growth*

Fresh bacterial suspensions were inoculated into 250 mL conical flasks (150 mL) at an OD_{600} of 0.3. The cultures were then shaken at 160 rpm for 24 h, the OD_{600} was measured, and cultures were used for experiments below.

2.6.1 *Different nitrogen sources*

Selective medium was added to mother liquor peptone (1%) and yeast extract (1%), with urea, $NaNO_2$, KNO_3, $NaNO_2$, or KNO_3 as the nitrogen source (molar ratio of C/N = 10). In addition, when the composite nitrogen source (LB) medium was prepared, the above six types of medium were used separately. Cultures were incubated with shaking at 26°C.

2.6.2 *Different carbon sources*

The suitable nitrogen source was selected as described above. Then, sodium citrate $CH_3COONa{\cdot}3H_2O$ as the sole carbon source was replaced with 1% (v/v) sodium acetate, glucose, fructose, ethanol, potassium or sodium tartrate. This medium was then added to the selective medium in the mother liquor, and cultures were incubated at 26°C with shaking.

2.6.3 *Different temperatures*

The optimal temperature was determined using the optimized sources of nitrogen and carbon. Temperatures of 15, 20, 25, 30, and 35°C were examined after incubation at 40°C.

2.6.4 Different pH values

The optimal pH was determined by testing growth using the above optimized conditions at pH values of 5, 6, 7, 8, 9, and 10.

2.6.5 *Growth curves*

A suitable amount of the bacterial suspension was selected to inoculate into 200 mL optimized culture medium. The culture was adjusted to an OD_{600} of 0.1 and incubated at the optimal temperature and pH. Samples were collected at certain times, and the absorbance at OD_{600} was determined to create bacterial growth curves.

2.7 *Analysis of aerobic denitrification*

A suitable amount of bacterial suspension was inoculated into 200 mL synthetic sewage (0.6 g/L glucose, 0.1 g/L peptone, 0.01 g/L yeast powder, 0.5 g/L sodium acetate, 0.05 g/L NaCl, 0.09 g/L $KH_2PO_4 \cdot 3H_2O$, 0.4 g/L $MgSO_4 \cdot 7H_2O$, 0.2 g/L KNO_3). The OD_{600} was adjusted to 0.3, and bacteria were incubated with shaking under suitable conditions, with sampling every 4 h.

2.8 *Analytical methods*

Analyses were performed using the NO_3^--N: phenol disulfonic acid method and NO_2^--N: alpha-naphthylamine spectrophotometry. Each experiment was repeated three times, and the results were averaged.

2.9 *Data processing and statistics*

Data were analyzed using Microsoft Excel software with single-factor analysis of variance and multiple comparisons. Differences with P values of less than 0.05 were considered statistically significant.

3 RESULTS

3.1 *Strain screening*

From the selection procedures, 18 strains were initially picked and replated in LB culture medium. From these colonies, a total of seven single colonies were finally selected, named QX1–8. The screened strains were used for further analysis of denitrification rates, as shown in Table 1, and the strain QX1, which exhibited the highest denitrification rate, was used for further analyses.

3.2 *Morphological, physiological, and biochemical characteristics of QX1*

QX1 colonies were pale yellow and neat, with a smooth, moist surface. For QX1, H_2S analysis, transparent circle experiments, catalase assays, V-P experiments, nitrate reduction assays,

Table 1. Strain denitrification rates under aerobic culture conditions after 24 h.

	Strain							
Project	QX1	QX2	QX3	QX4	QX5	QX6	QX7	QX8
Denitrification rate (%)	72.3	65.3	45.6	60.1	59.2	68.4	56.4	62.3

and nitrate and oxidation origin gas assays were positive, whereas gram staining, semisolid AGAR piercing experiments, methyl red assays, and ammonia production were negative (Table 2).

3.3 *16S rDNA gene PCR amplification and phylogenetic tree construction*

The 16S rDNA sequence of strain QX1 was 1464 bp. An NCBI BLAST search showed that similarity was more than 99% compared with the sequences of *Arthrobacter nicotianae* SB42, *Arthrobacter arilaiti*, and *Arthrobacter nicotianae* and more than 98% compared with the sequences of *Arthrobacter arilaitensis*, *Arthrobacter* DSM12798, *Arthrobacter protophormiae* SSCT48, *Arthrobacter* sp. TVH. 003, and *Arthrobacter* sp. ZJUTW. The constructed phylogenetic tree is shown in Figure 1. Taken together, our data showed that this bacterium was within the genus *Arthrobacter*.

3.4 *Effects of different conditions on the growth of QX1*

3.4.1 *Effects of nitrogen source on growth*

The optimum culture medium for strain QX1 was LB culture medium (Fig. 2a), with the following order of growth based on the nitrogen source: LB medium > peptone > yeast extract > urea > KNO_3 > $NaNO_2$ (OD_{600} values of 1.30, 1.24, 1.21, 1.13, 1.06, and 0.57, respectively). Strains grew better in the presence of organic nitrogen than in the presence of inorganic nitrogen, with highest growth seen for the composite organic nitrogen. Moreover, the differences in growth with different types of nitrogen sources were significant, with

Table 2. Physiological and biochemical characteristics of QX1.

Experiments	Phenomenon
Gram staining	–
Transparent circle experiment	+
Catalase experiment	+
Semi-solid AGAR puncture	–
V-P experiment	+
Methyl red experiment	–
Nitrate reduction experiments	+
Nitrate as a native gas	+
Ammonia production experiment	–
Oxidation of H_2S	+

+: positive, –: negative

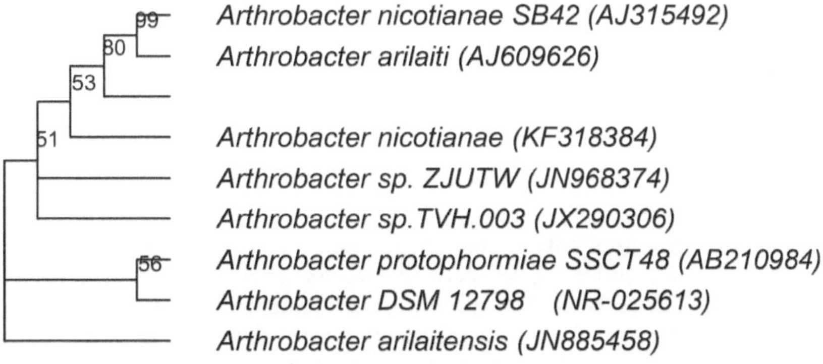

Figure 1. Phylogenetic tree of QX1 based on analysis of the 16S rDNA gene.

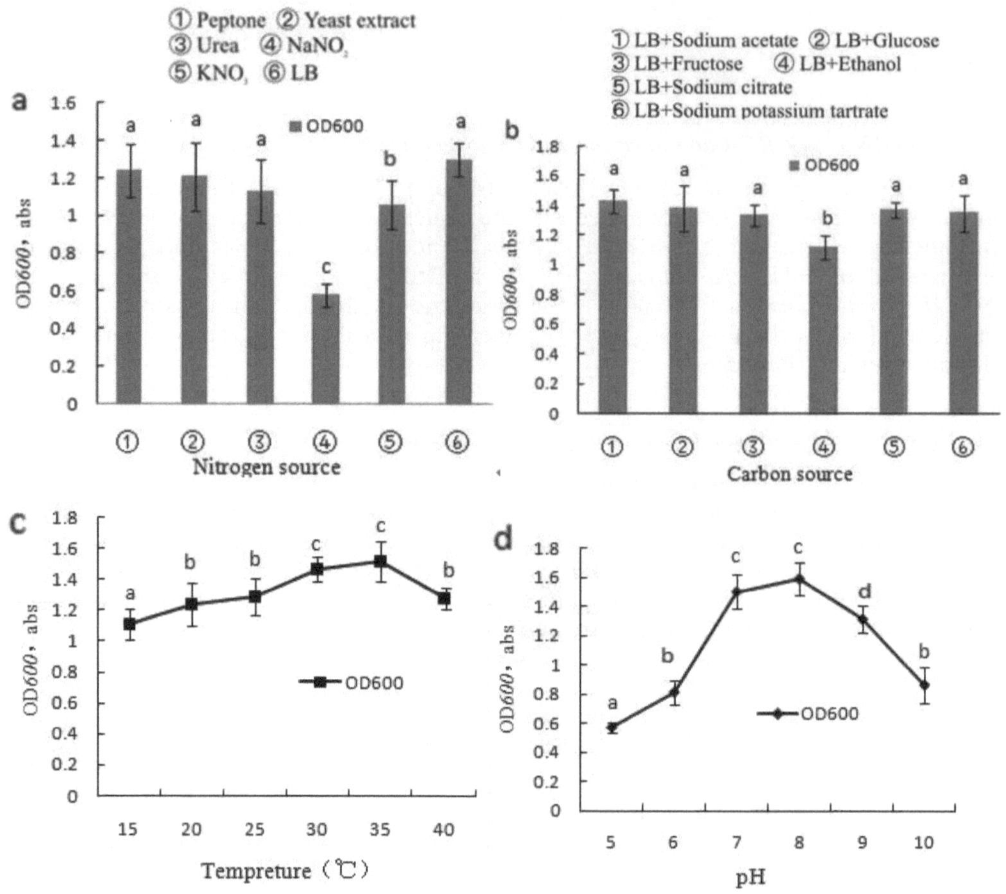

Figure 2. Effects of external conditions on the growth of QX1. (Different letters represent significant differences, $p < 0.05$ or $p < 0.01$.)

growth in LB medium significantly better than that in yeast extract medium ($P < 0.05$) or the three other types of culture medium ($P < 0.01$). Therefore, LB medium was used for subsequent analyses.

3.4.2 *Effects of carbon source on growth*

The optimum carbon source was sodium acetate, with growth under conditions of different the carbon sources following the order of sodium acetate > glucose > sodium citrate > sodium potassium tartrate > fructose > ethanol (OD_{600} values of 1.43, 1.38, 1.37, 1.35, 1.33, and 1.12) after 24 h (Fig. 2b). Growth was significantly lower when ethanol was the carbon source compared with that of the other carbon sources ($P < 0.05$). Therefore, sodium acetate was used as the carbon source for subsequent analyses.

3.4.3 *Effects of temperature on carbon source*

The growth of QX1 increased with increasing temperature from 15 to 35°C (OD_{600} values of 1.11, 1.24, 1.29, 1.47, and 1.52, respectively). However, at a temperature of 40°C, the OD_{600} decreased to 1.28 after 24 h (Fig. 2c). Thus, temperature affected the growth of QX1, and growth at 30 and 35°C was significantly better than that at other temperatures ($P < 0.05$ or $P < 0.01$, respectively). Therefore, we chose a temperature of 30°C for further experiments.

3.4.4 *Effects of pH value*

As shown in Fig. 2d, pH affected the growth of QX1 as well. Optimal growth was observed at a pH of 8. Reduced growth was observed at lower pH, whereas strong growth was observed when the pH was in the range of 6–9. Statistical analysis showed that growth was significantly better at pH 7 and 8 than that at other pH values ($P < 0.05$). Moreover, growth was significantly reduced when the pH was 5 ($P < 0.01$) or when the pH was 6 or 10 ($P < 0.05$), indicating that strains could grow in a weakly acidic, neutral, or alkaline environment.

3.4.5 *Growth curve*

Growth curves were created for QX1 based on the optimal culture conditions determined in the previous experiments (Fig. 3). Cells grew slowly during the adaptation phase from 0 to 4 h. However, after 4 h, QX1 cells grew quickly and entered the logarithmic phase, which lasted for approximately 16 h. After 20 h, the growth rate plateaued, exhibiting stable growth at 24 h. During this period of stable growth, the OD_{600} value was approximately 1.5–1.8, with bacterial counts of 1.756×10^{13}–3.018×10^{13} cfu/mL. After 44 h, growth began to decrease. These differences in growth over time were significant.

3.5 *Denitrification performance of strain QX1*

QX1 was inoculated into 200 mL synthetic sewage at pH 8, with shaking at 160 rpm at 30°C. Rocking bed shaking culture was performed for 48 h, with sampling every 4 h. Nitrate nitrogen, changes in nitrate nitrogen concentrations, and nitrate nitrogen removal rates are shown in Fig. 4. The content of nitrate nitrogen was reduced gradually, from 200 to 20.36 mg/L, with a maximum removal of 9.18 mg/L (20–24 h) and a maximum removal rate of 89.92%. The content of nitrite nitrogen did not accumulate and was maintained at low levels (0–0.06 mg/L). No ammonia nitrogen was detected during the entire process, similar to the results of physiological and biochemical identification. These data further verified that nitrate nitrogen gas was also present in air products. As shown in Fig. 4, QX1 played a role in denitrification mainly during the logarithmic bacteria growth period; in this period, bacteria are the most active in terms of growth and reproduction, and nutrition, energy, and power of cell synthesis are critical during this period. Therefore, denitrification was mainly observed during this period.

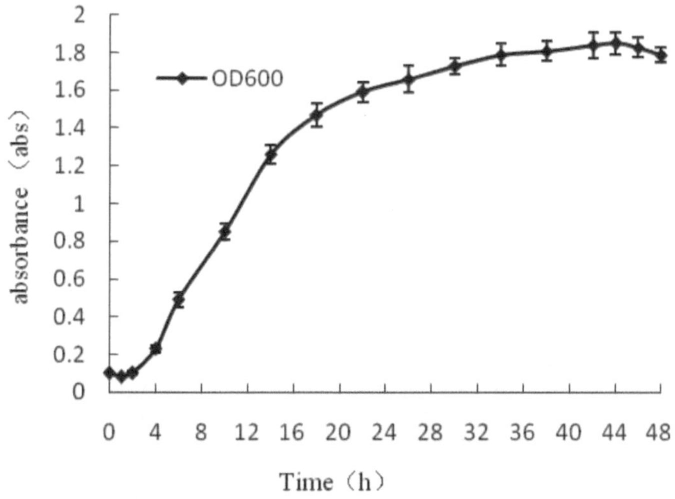

Figure 3. Growth curve for strain QX1.

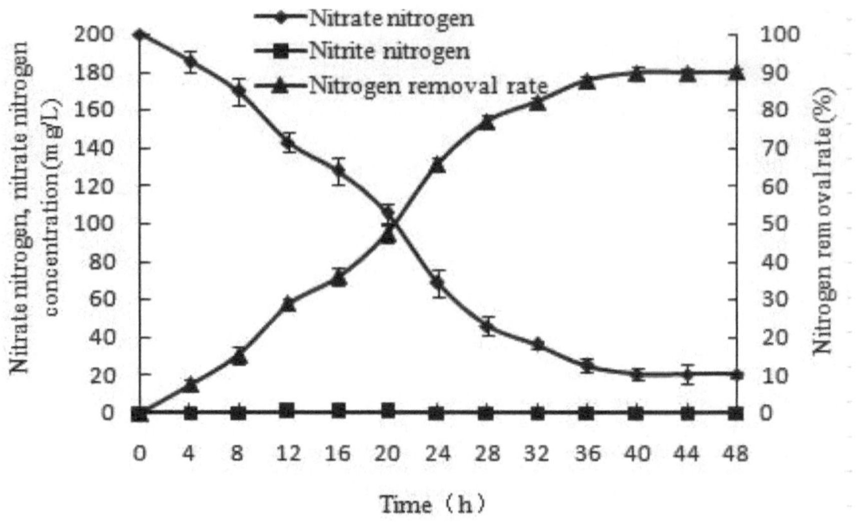

Figure 4. QX1 denitrification performance.

4 DISCUSSION

Denitrifying bacteria are mostly gram-negative bacteria; most of the denitrifying bacteria identified to date belong to *Pseudomonas* sp., *Acinetobacter* sp., *Paracoccus* sp., *Alcaligenes* sp., and Bacillus sp. Here, we screened denitrifying bacteria and identified strain QX1 as belonging to the *Arthrobacter* genus through morphology observation, physiological-biochemical identification, and 16S rDNA sequence alignment. A gas bath shaking table speed of 160 rpm is thought to ensure dissolved oxygen (DO) concentrations of 79–95% (Li 2014). In this study, DO remained above 2 mg/L by the actual measurement, and strains in BTB-denitrification medium showed blue halos, indicating that the bacteria had aerobic denitrification ability. Thus, these bacteria were identified as aerobic denitrifying bacteria.

By selection and optimization of pH, temperature, nitrogen source, and carbon source, we obtained optimal growth conditions for strain QX1: LB medium plus 1% acetic acid sodium (mass-to-volume ratio), 30°C, and pH 8. QX1 aerobic denitrification occurred mainly during the logarithmic phase, consistent with the results of Wang (Wang et al 2007) and Fan (Fan et al 2008). There was no nitrite nitrogen accumulation during the process of QX1 denitrification; the denitrification rate was 89.92% within 48 h. Aerobic denitrifying bacteria can also produce energy through aerobic denitrification, and oxygen acts as an electron acceptor in the respiratory chain (Chen 2006). The production efficiency of the former is lower than that of the latter (Wehrfritz et al 1993), and in the presence of sufficient amounts of DO, bacteria use their oxidizing capacity rather than their aerobic denitrification capacity. At the same time, a sufficient carbon source can facilitate the production of more carbon skeleton, conducive to cell growth. In this study, we found that strain QX1 could grow better in a composite organic nitrogen source and carbon source. Analyses showed that the QX1 degradation rate differed according to the type of culture medium. For example, when bacteria were grown in medium in which degradation and absorption rate were relatively quick, the cells were in a state of "feeding" and did not need to use their own materials or energy to undergo rapid growth and reproduction. In contrast, when the bacteria were grown in medium in which the matrix absorption rate was slower, cells were in the "hunger" state and needed to rely on their own materials and energy to maintain basal metabolism; therefore, there was no proliferation in the short term, only when the degradation of the substrate was sufficient to exit the "hunger" period could the cells grow and proliferate (Majone et al 1998). Thus, different culture medium could alter OD_{600} values.

Temperature and pH can affect the growth of microbial and enzymatic reactions, and each bacterium has an optimum growth temperature and pH range. In this study, temperature and pH significantly affected QX1 growth; the optimum temperature was 30°C, and the optimum pH was 8, similar to the standard optimum conditions for bacteria. At 40°C, the absorbance of QX1 cultures at OD_{600} was 1.28, whereas that at 15°C was 1.11, indicating that QX1 was resistant to low or high temperatures. These bacteria can be used as alternative bacteria showing alkali resistance, low temperature resistance, and high temperature resistance with efficient denitrification and may have broad applications in the enhancement of water quality at low or high temperatures.

5 CONCLUSIONS

In this study, we isolated an efficient aerobic denitrifying bacterium, QX1, from burdock fields. Strain QX1 was ultimately determined to be *Arthrobacter* sp. through the morphology observations, physiological-biochemical identification, and 16SrDNA sequence analysis. We further identified suitable growth conditions for this strain and found that the strain exhibited alkaline resistance and could grow at low and high temperatures. Moreover, QX1 exhibited efficient denitrification, reaching a nitrate nitrogen removal rate of 89.92%. Thus, this bacterium could be used as alternative bacterium in applications requiring alkaline resistance, low temperature resistance, and high temperature resistance in order to achieve efficient denitrification.

ACKNOWLEDGEMENTS

This study was supported by grants from the Young Project of Xuzhou Institute of Technology (grant no. XKY2016122), Jiangsu Province Six Talent Summit of China (grant no. JNHB-005), Projects of Ministry of Housing and Urban-Rural Development of China (grant nos. 2015-K6-014 and 2016-R2-013), Jiangsu province Natural Science Foundation of China (grant no. BK20161164) and Natural Science Foundation of China (grant no. 31600345).

REFERENCES

Chen F, Xia Q, Ju L K. Competition between oxygen and nitrate respirations in continuous culture of Pseudomonas aeruginosa performing aerobic denitrification [J]. Biotechnology and Bioengineering, 2006, 93(6):1069–1078.

Dong Xiuzhu and Cai Miaoying. Common bacteria manual system identification [M]. Beijing: Science Press, 2001.

Duan J M, Fang H D, Su B, et al. Characterization of a halophilic heterotrophic nitrification-aerobic denitrification bacterium and its application on treatment of saline wastewater [J]. Bioresource Technology, 2015,179: 421–428.

Fan Lirong, Huang Shaobin, Yang Jun, et al. Isolation and identification features of aerobic denitrifying bacteria biological filter tower system [J]. Microbiology Bulletin, 2008, 35 (5): 679–684.

Huang Tingling, Zhang Lina, Zhang Haihan, et al. Screening and denitrification characteristics of an oligotrophic heterotrophic nitrification—aerobic denitrifying bacteria [J]. Journal of Ecological Environment, 2015, 24 (1): 113–120.

Li Anfeng, Luo Jianping, Liu Yujuan, et al. The screening and analysis of denitrifying bacteria in SBR and MBR reactor aerobic [J]. Journal of Microbiology, 2014,34(5):65–69.

Li Jing, Wang Wenwen, Liang Lei, et al. Screening and denitrification characteristics of salt tolerance aerobic denitrifying bacteria [J]. Environmental Science & Technology, 2011,34(6):48–52.

Liu Yan, Gan Li, Huang Zheqiang, et al. Denitrification characteristics of deputy aureus YF1 [J]. Water Treatment and Technology, 2010, 36(10): 61–65.

Majone M, Massanisso P, Ramadori R. Comparison of carbon storage under aerobic and anoxic conditions [J]. Water Science and Technology, 1998, 38(8–9), 77–84.

Makoto S and Yoichi I. Heterotrophic nitrification and aerobic denitrification of high-strength ammonium in anaerobically digested sludge by Alcaligenes faecalis strain No. 4 [J]. Journal of Bioscience and Bioengineering, 2014, 117(6): 737–741.

Miyahara M, Kim S W, Zhou S, et al. Survival of the aerobic denitrifier Pseudomonas stuzeri strain TR2 during coculture with activated sludge under denitrifying conditions [J]. Bioscience, Biotechnology, and Biochemistry, 2012,76(3): 495–500.

Ren Y X, Lei Y, Liang X. The characteristics of a novel heterotrophic nitrifying and aerobic denitrifying bacterium, Acinetobacter junii YB [J]. 2014, Bioresource Technology, 171: 1–9.

Ren Y X, Yang L, Liang X. The characteristics of a novel heterotrophic nitrifying and aerobic denitrifying bacterium, Acinetobacter junii YB [J]. Bioresource Technology, 2014,171:1–9.

Shi Z, Zhang Y, Zhou J, et al. Biological removal of nitrate and ammonium under aerobic atmosphere by Paracoccus versutus LYM [J]. Bioresource Technology, 2013,148:144–148.

Sun Ling, Zhao Xinxin, Zhang Huifang, et al. Biological characteristics of a denitrifying phosphorus-accumulating bacterium [J]. Ecological Engineering, 2015, 81, 82–88.

The Chinese academic sciences institute of microbiology bacteria classification groups. General bacteria commonly used identification method [M]. Beijing: Science Press, 1978,114–115,125–127.

Wang Guoying, Cui Jie, Yue Xiuping, et al. Heterotrophic nitrification and aerobic denitrification bacteria denitrification phenol degradation characteristics at the same time [J]. China Environmental Science, 2015, 35 (9): 2644–2649.

Wang Hongyu, Ma Fangbu, Su Junfeng, et al. Identification and denitrification characteristics of aerobic denitrifying strain [J]. Environmental Science, 2007, 28 (7): 1548–1552.

Wang Jingfeng, Jin Min, Chen Zhiqiang, et al. Screening and denitrification characteristics of an aerobic denitrifiring bactaria [J]. Environmental Science, 2011,32(8): 2409–2413.

Wang X, Li T L, Dong M Y, et al. Reducing nitrate in groundwater by stabilized iron nanoparticles and denitrifying bacteria [J]. Journal of Agro-Environment Science, 2011,30(4): 739–745.

Wang Zhaoyang, Chen Guoyao, Jiang Ke, et al. Separation and identification and denitrifying properties of an plant resistance to cold and basophilic aerobic denitrifying bacteria [J]. Environmental Science, 2014, 35(6): 2341–2347.

Wehrfritz J M, Reilly A, Spiro S, et al. Purification of hydroxylamine oxidase from Thiosphaera pantotropha, identification of electron acceptors that couple heterotrophic nitrification to aerobic denitrification [J]. FEBS Letters, 1993, 335(2):246–250.

Xiao jibo, Jiang huixia, Chu shuzi. Denitrification characteristics of aerobic denitrifying bacteria Defluvibacter lusatiensis STR. DN7 under different nitrogen source [J]. Acta Ecologica Sinica, 2012,32(20):6463–6470.

Yang Y, Zhu H, Colvin V L, et al. Cellular and transcriptional response of Pseudomonas stutzeri to quantum dots under aerobic and denitrifying conditions [J]. Environmental Science & Technology, 2011, 45(11): 4988–4994.

Innovation and Application of Engineering Technology – Oscar, Chen & Wang (Eds)
© 2017 Taylor & Francis Group, London, ISBN 978-1-138-05057-0

Study on the treatment of upper corner gas based on the roof-drilling holes

Muguang Qin
State Key Laboratory of Gas Disaster Detecting, Preventing and Emergency Controlling, Chongqing, China
Chongqing Research Institute Co. Ltd., China Coal Technology and Engineering Group, Chongqing, China

ABSTRACT: With the development of drilling equipment and sealing material improvement, long-distance large aperture will gradually replace the special high pumping lane, tail lane, and other mining measures to become the development trend. Roof long drill mainly makes use of the mining face dynamic pressure formed by the roof crack channel to extract the adjacent layer and goaf gas. In view of the unfavorable present situation that the ordinary high-level drilling method is used to control the gas in the upper corner of 2201 working face in Lu Tai Mountain Mine, the method of roof drifting is proposed to solve the problem of gas overrun at the upper corner of the 2201 working face. According to the roof overlaying strata discipline and the pressure relief "O"-ring theory, in this paper, we calculated the height and horizontal distance of the long-drilled hole in the 2201 working face. According to the actual situation of working face in Lu Tai Mountain Mine, it also has determined the reasonable parameters of the top-drift long-drilled hole in the roof of the 2201 working face and drilling field, which provides a new idea and method for gas control at the upper corner of the working face of Lu Tai Mountain Mine.

1 INTRODUCTION

Lu Tai Mountain Mine belongs to Shanxi Coal Import and Export Group. It is approved to exploit No. 2 and No. 15 coal beds. The mining depth is from +940 m to +610 m. The east to west width of the well is 1.5–2.0 km. The length of the north to the south is 2.6 km. The area is 4.7 km^2. The production capacity of the coal mine is 600,000 t/a. At present, No. 2 coal bed is mining back to the mining face 2201. The cutting length is 160 m. The tending length is 1210 m. The average thickness is 2.2 m. The working face uses single "U" ventilation. And Lu Tai Mountain Mine mainly uses the high-level drill hole to extract the roof crack gas supplemented by the goaf area intubation to extract the goaf area gas to control the working angle upper corner gas, because the high-position drill field extraction effect is not ideal. In the process of mining face, goaf and surrounding rock massively pour out. The working face, upper corner, and return air show different degrees of gas overrun, restricting the normal production safety of the mine.

2 NECESSITY OF THE ROOF LONG DRILL TO SOLVE THE UPPER CORNER GAS

In China, many methods are available for the control of the upper corner gas: the ground drilling method, tail road method, plug (buried) pipe method, roof pumping method, and the ordinary high-bit drilling method. Although the first four kinds of mining methods have achieved good results in the country, most of them are highly expensive, maintenance costly, and difficult to manage with other shortcomings. As to the ordinary high-bit drilling method, although it has characteristics like strong targeting, small construction project, relatively simple operation, with the passage of the mining face, coal roof caving and the leakage resulted from the cutting of high drilling holes, the ordinary high-bit drilling method has shortcomings like short extraction time and low utility ratio. With the development of drilling equipment and sealing material improvement, long-distance large aperture will gradually replace the special high pumping lane, tail lane, and other mining measures to become the development trend. Roof long drill mainly makes use of the mining face dynamic pressure formed by the roof crack channel to extract the adjacent layer and goaf gas. There is a lot of gas in the goaf of the 2201 working face in Lu Tai Mountain Mine, which is the main source of gas flowing into the working face. In addition, the permeability of the direct top and basic roof of No. 2 coal seam is poor and rock is well closed, which is not beneficial to the release of gas. Enormous gas remains here. When the roof of the working face presses, the highly concentrated gas will gather here. It is a main reason for the occurrence of the upper corner gas overrun. Therefore, on the basis of analyzing the sources of gas in the upper corner and the cause of gas overrun, it proposes that the upper corner of the 2201 working face should adopt the roof drilling method to extract the upper corner area and the goaf gas, fundamentally solving the problem of the upper corner gas.

3 DESIGN OF THE ROOF-DRILLING METHOD FOR THE 2201 WORKING FACE

3.1 *Roof overlaying strata discipline and pressure relief "O"-ring theory*

1. Roof overlaying strata discipline

Figure 1 is the masonry beam theory proposed by Professor Qian Minggao, referring to "horizontal three areas" and "vertical three strips": along the working face, the overlying strata will pass through the coal wall bearing influence zone, the separation zone, and the recompression zone. The movement of the overlaying strata can be divided into collapse zone, fissure zone, and blending zone, where the first two are called the hydraulic fracture zone. The theoretical formulas for the collapse zone and fracture zone are derived in Tables 1 and 2, respectively.

2. Pressure relief "O"-ring theory

The theory proposes that the fractures in the overlying strata are formed in the overlying strata after mining. The fractal distribution of the separated strata is characterized by two stages: the first stage starts from the open-off cut, and the fractures continue to increase with the advance of the working face. In the middle of the goaf, the fractures are the most developed; the middle part of the goaf in the second stage tends to be compacted, while there is a connected mining area around the mined-out area, called the mining fracture "O" ring, as shown in Figure 2 (the numbers are separation layer ratio).

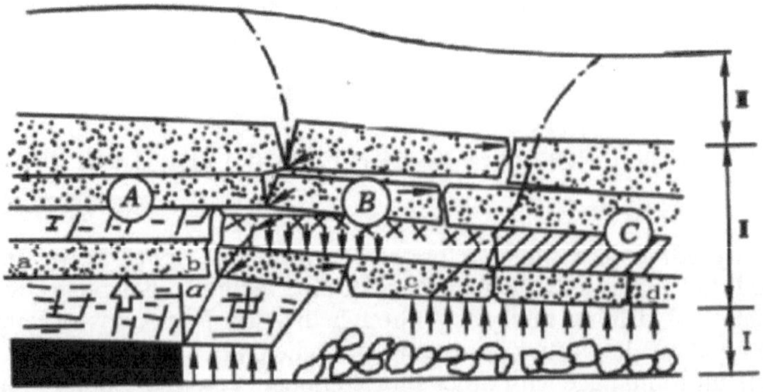

Figure 1. "Horizontal three areas" and "vertical three strips" destroyed by the overlaying strata.

Table 1. Formula for the height of the collapse zone.

Overlaying strata property (single compressive strength and main rock names)	Formula (m)
Hard (40–80 MPa, quartz sand, limestone, arenaceous shale and conglomerate)	$H_m = \dfrac{100\sum M}{2.1\sum M + 16} \pm 2.5$
Medium hard (20–40 MPa, sandstone, argillaceous limestone, arenaceous shale and shale)	$H_m = \dfrac{100\sum M}{4.7\sum M + 19} \pm 2.2$
Soft (10–20 MPa, mudstone and argillaceous sandstone)	$H_m = \dfrac{100\sum M}{6.2\sum M + 32} \pm 1.5$

Note: ΣM is the accumulated thickness, which cannot exceed 15 m; ± items are mean square error.

Table 2. Formula of the height of the fracture zone.

Rock property	Formula (m)
Hard (40–80 MPa, quartz sand, limestone, arenaceous shale and conglomerate)	$H_{li} = \dfrac{100\sum M}{1.2\sum M + 2.0} \pm 8.9$
Medium hard (20–40 MPa, sandstone, argillaceous limestone, arenaceous shale and shale)	$H_{li} = \dfrac{100\sum M}{1.6\sum M + 3.6} \pm 5.6$
Soft (10–20 MPa, mudstone and argillaceous sandstone)	$H_{li} = \dfrac{100\sum M}{3.1\sum M + 5.0} \pm 4.0$

Note: ΣM is the accumulated thickness; accumulated thickness cannot exceed 15 m; ± items are mean square error.

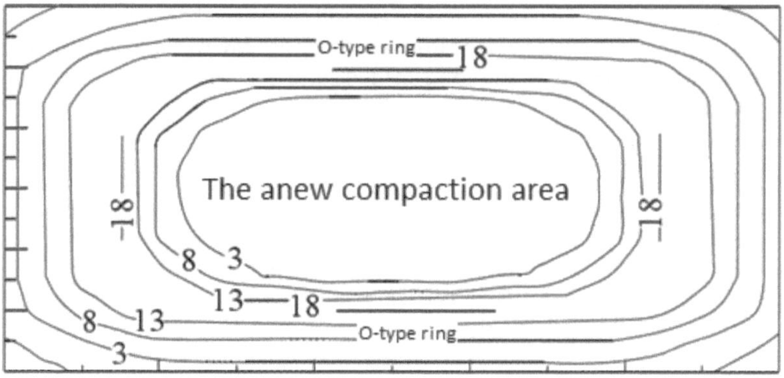

Figure 2. "O" ring distributed around the mined-out area.

The flow of the pressure relief gas is a continuous two-step process. The first step is to diffuse the gas from the nonfissured coal to the surrounding fissures. In the second step, in the form of seepage, the gas is discharged along the fracture flow to the drainage hole through the mining cracks. Obviously, the drainage borehole is arranged in the area where the fractures are developed and can be maintained for a long time, so that the pressure relief gas can flow into the drainage borehole. According to the "O"-ring theory of the distribution of overlying strata, the "O" ring is equivalent to a gas river and the gas in the surrounding coal and rock mass is seeped continuously into this gas river after resolution.

Therefore, the gas drainage drilling should be hit within the scope of the mining crack "O" ring to ensure that drilling has a longer drainage time, a larger drainage area, and a higher gas drainage rate.

3.2 *Research on the parameters of the roof long drill*

1. Vertical distance

One of the main factors influencing the control effect of the upper corner gas is the choice of the long drilled holes in the roof, that is, the vertical distance from the long drilling to the working face. It can be seen from the characteristics of the gas flow in goaf that a large number of highly concentrated gas remains in the top of "collapse zone" and "fracture zone", so the high hole is arranged in the fracture development area in the roof. The height of the ending hole of H_m conforms to formula (1):

$$H_1 < H_m < H_2 \tag{1}$$

where H_1 = the height of the collapse zone (m) and H_2 = the height of the fracture zone (m).

According to the comprehensive histogram of Lu Tai Mountain Mine, the top layer of the No. 2 coal seam is sandy mudstone with a compressive strength of 22.2–51.4 MPa, with an average of 36.9 MPa. It is medium hard rock. The height of the collapse zone (H_1) and the height of the fracture zone (H_2) can be calculated by using the empirical formula of the rock movement law:

$$H_1 = \frac{100\sum M}{4.7\sum M + 19} \pm 2.2 \tag{2}$$

$$H_2 = \frac{100\sum M}{1.6\sum M + 3.6} \pm 5.6 \tag{3}$$

where ΣM refers to the accumulated thickness. The 2201 working face can be taken with 2.2 m.

The average height of the caving zone is 5.3–9.7 m (average value is 7.5 m), and the height of the fissure zone is 25.3–36.5 m (average value is 30.9 m), so the rock between 7.5–30.9 m can be regarded as in the fissure zone. In order to give full play to the effect of the long drill, it is preliminarily determined that the long drilled hole can be located on the layer of the vertical distance of 30.0 m.

2. Horizontal distance

The horizontal distance between the roof long drill and the return airway in the top plate must be controlled within a reasonable range. According to the laws of overlying strata movement during coal mining, several studies have shown that the crack relief pressure angle of the roof is about 60°, and the boundary of the cracked "O" ring is about 60 m away from the roof crack relief pressure boundary. The vertical distance of the roof long drill is determined as 30.0 m, and the pressure relief layer in the boundary distance from the return airway is:

$$L = 30/\cot 60° = 17.3 \text{ m} \tag{4}$$

Therefore, through theoretical calculation, it is determined that the horizontal position of the long drilled hole in the roof direction should be 30.0 m vertical distance and 17.3–77.3 m horizontal distance from the return airway. Figure 3 shows the schematic diagram of the layout of the roof long drilled hole.

3.3 *Layout plan of the roof long drilled hole and the drilling field*

Design of the drilling field: the drilling field is bell-shaped, with a depth of 4 m and height of 3.8 m. The opening width of the drilling field is 8.4 m, and its internal width is 5 m. It is

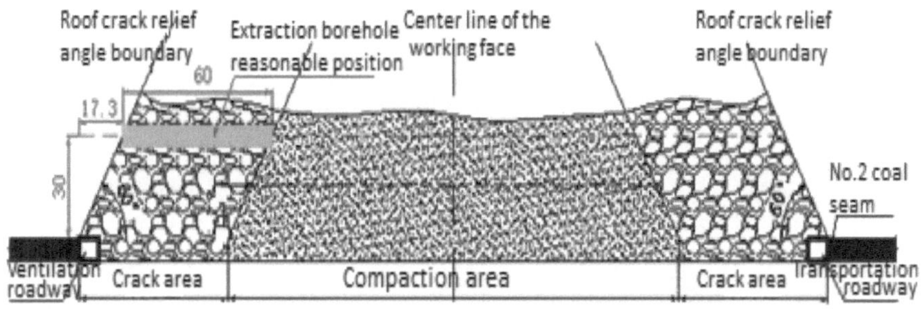

Figure 3. Schematic diagram of the layout of the roof long drilled hole (unit: m).

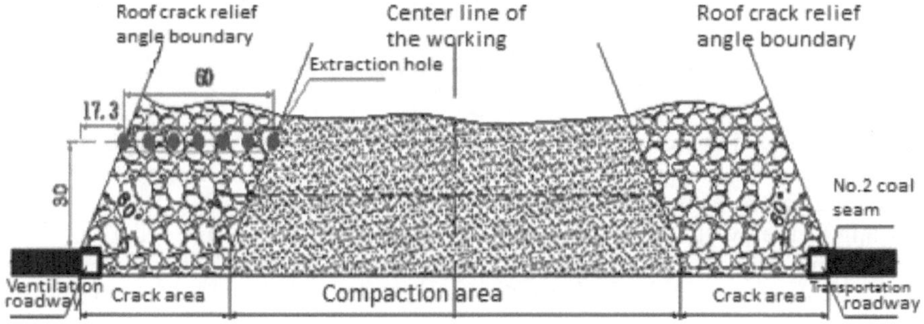

Figure 4. Profile positions of the roof long drilling holes (unit: m).

9 m between both sides of the pitch. The drilling field and lane shall be hardened by a cement mortar with a thickness no less than 300 mm. The ground is expected to be flat. Bricking (cement plaster) is adopted for the back part of the drilling field for the water silo 2 m in length, 1 m in width, and 2 m in height. The water silo should be as close to the alley as possible, and the water level should be as low as possible. Along the lane, do the ditch to the drill field. Warning is required to be completed after the completion of the water silo.

1. Space between the drilling fields: the space between the drilling fields is mainly determined by the gob roof collapse step and rig capacity and roof rock. At present, the drill hole in the underground drilling machine can reach 500 m. It preliminarily determines the distance between drilling and drilling site to be 200 m. Lay the first directional drilling rig near the working face in the return airway of the 2201 working face, 300 m away from the mining surface. Then, the horizontal space of the drilling field shall be laid by 200 m.
2. Length of the drilling hole: roof drilling length mainly depends on the distance between the drilling field, the overlap length between the previous and latter drill holes and the drilling capacity. According to the experience, it is appropriate for the overlap length to be about one-third of the drilling field spacing. In consideration of the initial test, increase the overlap part to be 100 m. In this way, the horizontal projection length is initially determined to be 300 m.
3. Space between the drilling holes: to ensure a longer drainage time, a larger drainage area, a higher gas drainage rate, combined with 2201 coal mining conditions, on the basis of the theoretical calculation, determine the roof long drilling in Lu Tai Mountain Mine to be 30.0 m in straight distance and 17.3–77.3 m in the horizontal distance of the return airway. Regarding the drilling spacing of 10 m, it is best suitable for five roof long drilling holes within the best drainage area in the upper corner. The length of the main hole is 300 m, and the diameter of the final hole is 96 mm. In order to improve the gas extraction effect in the upper corner of the borehole, two main holes in each drilling field are opened with a branch every 50 m, and the length of the branch hole is about 50 m. The roof long drilling holes are shown in Figures 4 and 5.

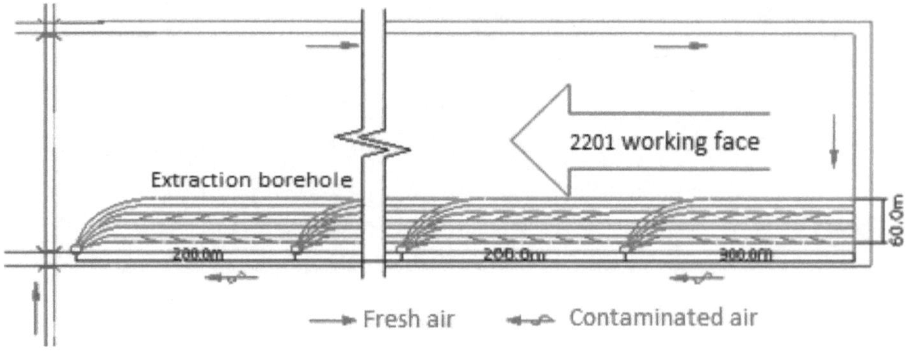

Figure 5. Plane layout of the roof long drilling holes (unit: m).

4 CONCLUSION

1. This paper analyzed the following main reasons for the overrun of the upper corner gas in the working face in Lu Tai Mountain Mine: residual coal in goaf and over surging of the gas from the nearby layer and surrounding rock.
2. According to the overlying strata movement discipline and pressure relief "O"-ring theory, it preliminarily determines that the layout of the roof long drilling holes suitable for Lu Tai Mountain Mine shall be 30.0 m for vertical distance and 17.3–77.3 m for horizontal distance from the return airway.
3. According to the practical situation in Lu Tai Mountain Mine, it determines the layout parameters of the roof long drilling holes and the drilling field: the space between the drilling fields is 200 m, the length of the drilling hole is 300 m, the overlap distance is 100 m, and the space between drilling holes is 10 m.

ACKNOWLEDGMENTS

This study was financially supported by the State Key Research Development Program of China (Grant No. 2016YFC0801404) and Chongqing Technological Innovation Leading Talent Support Program (CSTCKJCXLJRC14).

REFERENCES

Cai W.P., Liu J. & Sun D.S. 2013. Research and application of gas drainage technique with high-located drilling method along roof strike. *Journal of Safety Science and Technology* 9(12): 35–38.

Cui Y.J., Yang S.Q. & Zhang X.S. 2009. Application of gas drainage from long holes along roof strike in the working face upper corner. *Zhongzhou Coal* 2: 69–71.

Dong Z.J. 2012. Study on technology of solving as overrun with highly-located drilling method at upper corner. *Coal Technology* 31(11): 21–23.

Huang X.F. 2012. Research on technology of highly-located drilling method. *Mining Safety & Environmental Protection* 39(S1): 58–59.

Wu G.H. 2009. Applying technology of high drilling drainage to preventing gas accumulation at upper corner of full mechanized mining face. *Coal Mining Technology* 14(1): 94–95.

Xin X.Y., Wang L. & Yang S.M. 2008. Effect analysis on gas drainage to gob with long bore-holes along roof strike. *Safety in Coal Mines* 39(12): 83–85.

Zhang H.Q, Wang H.D. & Wang X.D. 2012. Study on gas drainage with large diameter high level bore-hole to replace high level gas drainage gateway. *Coal Science and Technology* 40(6): 51–53.

Zhu G.W., Fang E.C. & Lu P. 2003. Trial study of gas drainage by roof bore-holes on strike under different geological conditions. *Mining Safety & Environmental Protection* 30(S1): 1–4.

Innovation and Application of Engineering Technology – Oscar, Chen & Wang (Eds)
© 2017 Taylor & Francis Group, London, ISBN 978-1-138-05057-0

Pseudo-seismic imaging system in watery goaf detection using transient electromagnetic method

Hao Qiu, Yi Mu, Hui Xu & Zongyun Shu
Mine Safety Technology Branch, China Coal Research Institute, Beijing, China

ABSTRACT: In order to accurately detect watery goaf distribution and impact area in coal mine and on the basis of transient electromagnetic detection for the Gushuyuan coal mine, simulation of seismic processing experimental scheme based on wave-field conversion was established. Meanwhile, the comparison between traditional transient electromagnetic data processing method and pseudo-seismic imaging system was analyzed. The results showed an obvious relationship between the boundary of watery goaf and the geological electrical interface. Pseudo-seismic imaging system based on wave-field conversion can show more efficiency to distinguish depth and boundary range of watery goaf than traditional transient electromagnetic data processing method. This system can also overcome adverse effects on volume, improve vertical resolution, and provide reference for goaf disaster prevention and treatment.

1 INTRODUCTION

Mine water disaster is a great security threat to coal mine production. Because of the large amount of water and uncertain depth in the watery goaf, water inrush disaster probably occurs during mining. Water disaster prevention and control is always the focus of different point for mine water control work. The transient electromagnetic method is one of the most effective methods to detect the area of watery goaf. Because of the volume effect in the transient electromagnetic method, it appears that the low-resistance anomaly boundary does not represent the actual boundary of the water-rich region in traditional transient electromagnetic apparent resistivity map. It is difficult to determine the boundary of the watery goaf.

The transient electromagnetic imaging technique based on wave-field conversion theory provides a new idea for the detection of watery goaf by solving physical properties and geometrical parameters of the anomalous body, which traditional transient electromagnetic field cannot provide, an issue that has been investigated by numerous researchers (Levy *et al.*, 1988; Lee *et al.*, 1993; Chen *et al.*, 1999). It is generally accepted that there is a correspondence between the diffusion field and wave field of the transient electromagnetic method. Chen *et al.* (1999) analyzed simple wave-field conversion model under half-space condition based on the wave-field conversion theory. Li *et al.* (2005) designed a normalized method to make the converted wave field stable and reliable. Xue *et al.* (2006) constructed an equivalent conversion relationship between time and frequency domains after theoretical model forward calculation, curve comparison, error analysis, and the theoretical discussion about the propagation of diffusion field and wave field. Li *et al.* (2005), Xue *et al.* (2011), and Guo *et al.* (2012) put forward an idea of using the deconvolution technology to eliminate the waveform broadening effect in the wave-field conversion. Cheng *et al* (2013) studied the method of mine transient electromagnetic synthetic aperture in full space based on the idea of synthetic aperture in radar. Cheng *et al.* (2013) discussed the theory of transient electromagnetic field conversion, proposed an algorithm of mine transient electromagnetic field conversion, and calculated the wave field with actual surveying data. Li *et al.* (2015) studied Kirchhoff migration of the transient electromagnetic wave field of the air source device. Cheng *et al.*

(2016) designed an algorithm to convert the transient electromagnetic field to the pseudo-seismic wave field for advanced detection based on the function relationship between spreading electromagnetic field and seismic wave field in whole space.

Previous works have comprehensively studied the conversion algorithm of transient electromagnetic field, but the method of processing the converted wave field data based on the wave field conversion theory has not been considered completely. As a common transient electromagnetic detection target, specific detection and analysis method of watery goaf is still to be studied. In this study, a numerical calculation methodology was provided to operate this system based on the wave field conversion theory. A new data processing system was designed to deal with transient electromagnetic data in the wave field. Process comparison between traditional data processing and pseudo-seismic imaging was analyzed. A comparison experiment between traditional transient electromagnetic data processing method and simulated seismic processing method was set up for an effect analysis.

2 METHODOLOGY

2.1 *Conversion equation*

Mathematical integral form (Levy *et al.*, 1988; Lee *et al.*, 1993; Chen *et al.*, 1999) between transient electromagnetic field component and its corresponding wave field component is:

$$P(r,t) = \frac{1}{2\sqrt{\pi t^3}} \int_0^\infty q e^{-q^2/4t} U(r,q) dq \tag{1}$$

where $P(r, t)$ is the TEM component, $U(r, q)$ is the wave field component; t is the TEM time; q is the wave field travel time; r is the field value; the class time variable q corresponding to time t is quantity in $\sqrt{s}$ pending wave field $U(r, q)$ is the quantity in $1/\sqrt{\mu\sigma}$ σ is conductivity, and μ is permeability.

The significance of this conversion can convert the transient electromagnetic field component attenuation curve with time t into the propagation curve of the wave field with time q. The propagation velocity of this wave field is inversely proportional to the square root of the conductivity, which satisfies wave equation. Physical and geometric parameters can be solved by using data processing methods in seismology.

The conversion equation can be converted to discretization by using the variable step size trapezoidal, whose matrix form is:

$$AU = P \tag{2}$$

$$U = (U_1, \dots U_i, \dots, U_n)^T, \; P = (P_1, \dots P_i, \dots P_n)^T \tag{3}$$

where A is the $m \times n$ matrix in Eq. (2). The wave field conversion process can be realized by solving Eq. (2). Eq. (2) is a misconvergence integral equation, which can be derived by the method of matrix decomposition, regularization, and so on.

2.2 *Pseudo-seismic data process imaging*

Known from wave field conversion theory, transient electromagnetic field variables are needed to be completely converted from diffuse field to wave field. Induced electromotive force can be used to calculate the magnetic induction intensity change rate with time, which can be used to calculate vertical magnetic field components. Parameters t and P in Eq. (1) based on transient electromagnetic diffusion field can be obtained. The wave field value U of the transient electromagnetic field at time q can be solved by Eq. (2). The data of transient electromagnetic wave field can be processed and imaged by various methods in seismic data processing.

When dealing with pseudo-seismic data, the method of cubic spline interpolation is mainly used to interpolate high-density wave field conversion data, to make the wave field curve smooth and continuous. And cubic spline interpolation can also reduce the effect of the following process. The high-density interpolation method is defined by cubic spline interpolation. Function $S(x) \in C[a, b]$ is a cubic polynomial in each node $[x_j, x_j + 1]$, where $a < x_0 < x_1 < \cdots x_n = b$ are given nodes, if $y_j = f(x_j), (j = 0,1, \cdots, n), S(x_j) = y_j, (j = 0,1, \cdots, N)$.

As data collected by the transient electromagnetic method are mostly low-frequency components in the middle and late stages, velocity and wavelength of the converted wave field are much higher than those of the seismic wave field and hence vertical resolution is reduced. Therefore, the deconvolution method is used to deal with wave field conversion data to compress the seismic wavelet and improve the resolution. The deconvolution algorithm is defined by the least-squares-inverse filtering algorithm. The interpolated wave field data are used as an input signal:

$$b(t) = (b(0), b(1), \cdots, b(n)) \tag{4}$$

$$a(t) = (a(0), a(1), \cdots, a(m)) \tag{5}$$

$$y(t) = a(t) * b(t) = \Sigma\tau a(\tau)b(t - \tau) = (y(0), y(1), \cdots, y(M)), \; M = m + n \tag{6}$$

where $b(t)$ is input signal in Eq. (4), $a(t)$ is inverse filter factor in Eq. (5), $y(t)$ is output signal in Eq. (6), and $y(t)$ is the output wave field data after deconvolution processing.

Time–depth conversion is carried out to solve the depth of wave field corresponding to different times according to the velocity of the equivalent eddy current loop of the transient electromagnetic field. The wave field conversion curve corresponding to depth is obtained. The time–depth conversion algorithm is defined by the smoke ring inversion algorithm:

$$v = \frac{\Delta d}{\Delta t} = \frac{d_j - d_i}{t_j - t_i} = \frac{4}{\sqrt{\pi\mu}}\left[\frac{\sqrt{t_j\rho_j} - \sqrt{t_i\rho_i}}{t_j - t_i}\right] \tag{7}$$

where v is the propagation velocity of TEM field, Δt is TEM diffusion-field period, t is TEM time, ρ is the apparent resistivity, and μ is permeability. The propagation velocity of TEM field in each diffusion field period can be determined by Eq. (7):

$$D(t) = \int_0^t v(\rho,t)dt \tag{8}$$

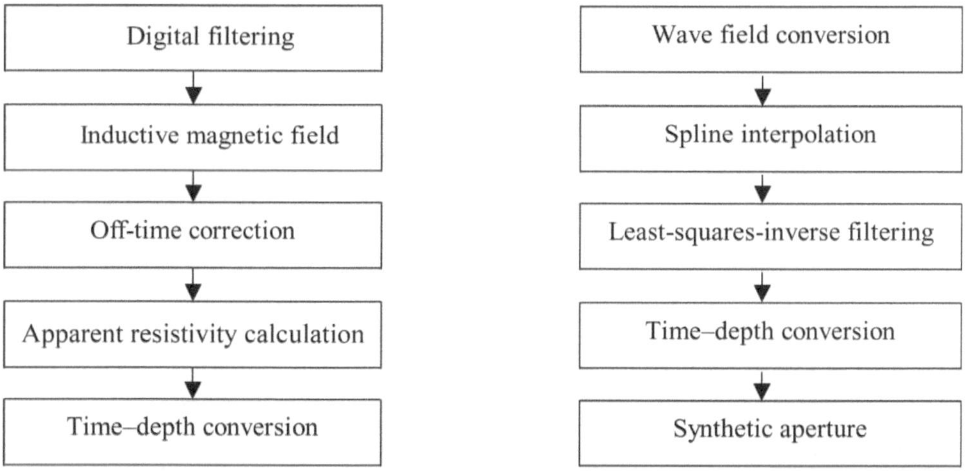

(a) Traditional transient electromagnetic data processing

(b) Pseudo-seismic imaging data processing

Figure 1. Processing flows of traditional TEM processing and pseudo seismic imaging data processing.

where ρ is the apparent resistivity, t is TEM time, and $D(t)$ is the propagating depth of TEM field. According to the apparent resistivity ρ and the wave field travel time q of time-series extracted from the transient electromagnetic original data, the depth corresponding to the sampling point of the wave field conversion data at different times can be calculated by Eq. (8).

Finally, correlation analysis is carried out to synthesize the conversion curves of multiple measuring points by using synthetic aperture technique to improve the resolution of wave field curve. Figure 1(a) and 1(b) show the process contrast between traditional data processing and pseudo-seismic imaging. All the processes of transient electromagnetic pseudo-seismic method are carried out under seismic wave field conditions. Seismic wave field is strong in resolution of geometric interface. The method can solve problem of transient electromagnetic method not being sufficient for resolving the geometrical interface. The pseudo-seismic imaging system can be used to determine the distribution and boundary of watery goaf.

3 EXPERIMENTAL METHOD

3.1 *Experimental background*

Working face designed in the Gushuyuan coal mine 91304 is No. 9, working elevation is 630–660 m, average thickness is 1.2 m, working face length is 240 m, inclined length is 150 m, and average dip angle is 3°. Figure 2 shows the parameters of the 91304 working face. There is No. 3 coal seam watery goaf in the upper part of the working face. Water content of the goaf is unknown. Average spacing between No. 3 coal seam and No. 9 coal seam is 45 m. In order to determine water yield property and the effect of No. 3 coal seam watery goaf, the transient electromagnetic method has been used to detect the area of the No. 3 coal seam watery goaf.

3.2 *Experimental design*

Instrument parameters were determined by data acquisition experiment. A parallel line to the working face was arranged in the corresponding surface area. The length of TEM survey line is 280 m. The dot-to-survey line distance is 20 m, comprising 15 test points. Figure 3 shows the construction layout and detection direction. On the basis of processing and interpretation of traditional transient electromagnetic data, distribution and boundary of the watery goaf in No. 3 coal seam were determined by the pseudo-seismic imaging system of the transient electromagnetic method. By comparing with the traditional apparent resistivity contour map, the effect of two kinds of data processing methods is analyzed.

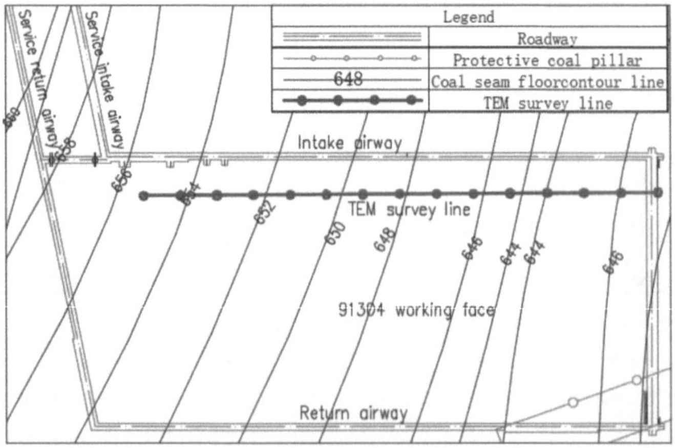

Figure 2. Mining engineering ichnography of 91304 working face.

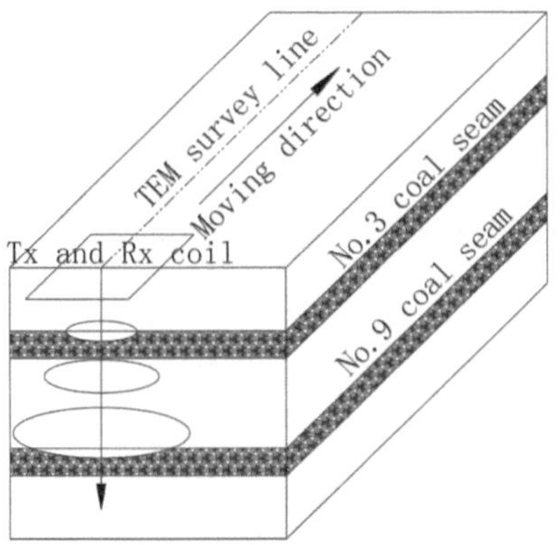

Figure 3. Schematic diagram of transient electromagnetic exploration.

4 EXPERIMENTAL RESULTS AND ANALYSIS

4.1 *Geophysical characteristics and explanation*

Regional geophysical characteristic is an important evidence to distinguish abnormal objective and watery goaf. Furthermore, the abnormal body has characteristics of low resistance and layered distribution. If relative apparent resistivity value is no more than 25 Ω·m, then this region can be defined as the watery anomaly region, whose water content is rich, with obvious low resistance characteristics. In terms of converted wave field characteristics, the shape of the wave field curve reflects the sharp change of resistivity. According to the wave field curve, it is feasible to analyze the electrical interface position, as there is an obvious amplitude jump curve near it. The more violent the amplitude curve, the more obvious the electrical interface. Therefore, the position of the electrical interface is determined by wave amplitude of the wave field curve.

4.2 *Results explanation*

Figure 4 shows the result of traditional transient electromagnetic method. It can be seen from the figure that the transient electromagnetic method results in a high lateral resolution and layered distribution character is obvious. There are two obvious low-resistivity anomaly areas in the No. 3 coal seam. The No. 1 anomaly area is distributed in the 0–80 m section of surveying point, and the depth range is 30–53 m. The No. 2 anomaly area is distributed in the 140–280 m section of surveying point, and the depth range is 53–70 m. Because of volume effect of transient electromagnetic method, the actual boundary of watery goaf area could not be reflected accurately by a relative low-resistivity anomaly range boundary. Therefore, an apparent resistivity section diagram can only determine the draft spatial orientation of the low-resistivity anomaly, and it is difficult to determine actual boundary and influence range of watery goaf in No. 3 coal seam.

4.3 *Results analysis*

Figure 5 shows the result of the transient electromagnetic pseudo-seismic imaging system. The curve link to each measurement point is the converted wave field curve, resulting from the electromagnetic pseudo-seismic process. The two black solid lines are the coal seam. The wave field curve shows a relatively strong fluctuation in the depth range of 60–85 m, as seen from the figure. The two blue dashed lines are electrical geometry interfaces, which are named A and B. The most obvious vibration amplitude curve is electrical interface position, with a strong water-rich region boundary. Determination of the boundaries of watery goaf improves the vertical resolution of the transient electromagnetic method. Lines A and B denote the actual vertical boundaries of the water-rich goaf.

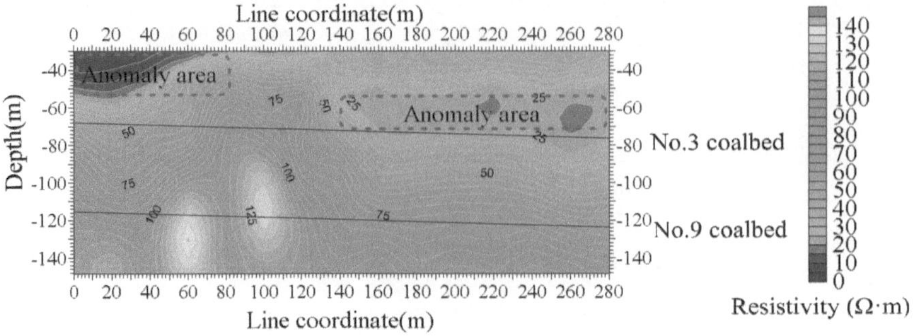

Figure 4. Transient electromagnetic apparent resistivity section diagram.

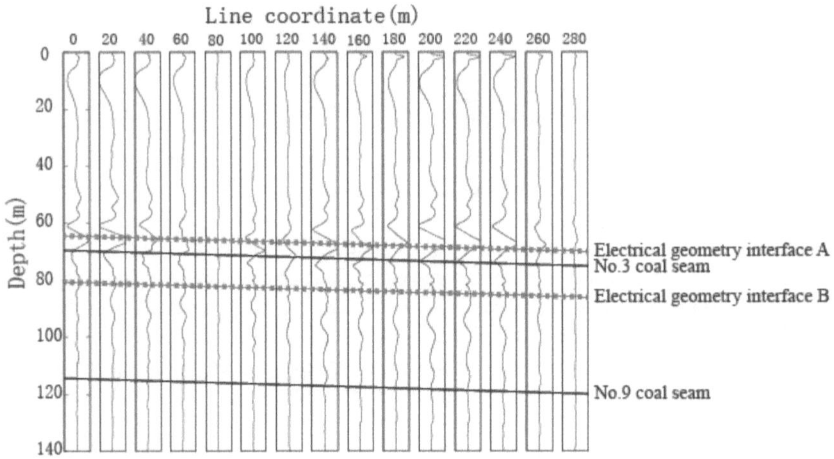

Figure 5. Transient electromagnetic pseudo-seismic imaging section diagram.

Low-resistance anomaly depth ranges in the two marked points are from 30 to 70 m in the transient electromagnetic apparent resistivity section diagram (Figure 4), both on top of the No. 3 coal seam. It is impossible to determine the actual boundaries and influence range of the watery goaf by traditional transient electromagnetic data processing, so actual boundaries and impact range of the watery goaf are determined by electrical geometry interfaces A and B (Figure 5). Most of the anomalous areas are located between No. 3 coal seam and No. 9 coal seam, and the depth range is 64–85 m. Water in goaf enters from the fracture, actual watery goaf boundary is lower than depth of the No. 3 coal seam. Pseudo-seismic processing is more accurate than the traditional transient electromagnetic data processing, validated by actual drilling experiment. Vertical drilling experiment was arranged from the transportation tunnel to the roof. There is a water point at 38 m. The water point is at a distance of 7 m from No. 3 coal seam. The instantaneous water flow is 23 m^3/h. The drilling results match pseudo-seismic imaging results.

5 CONCLUSIONS

This paper contributes to the numerical analysis of transient electromagnetic using pseudo-seismic imaging system. It is concluded that the pseudo-seismic imaging system is based on wave field condition. Results of this study show that:

1. Pseudo-seismic imaging system process is carried out on wave field conversion, and seismic wave field is strong in electrical geometry interfaces.
2. The experiment results show that adverse impact of volume effect in transient electromagnetic method is reduced by wave field data processing system.
3. Watery goaf boundary can be accurately detected as two geometric interfaces, which are important reference for water hazard prevention. The resolution of the transient electromagnetic method can be improved by using pseudo-seismic imaging system.

REFERENCES

Chen B.C., Li J.M. & Zhou F.T. (1999) Wave-field conversion method for transient electromagnetic field. *OGP(in Chinese)*, 34(5), 539–545.

Cheng J.L., Chen D., Xue G.Q., et al. (2016) Synthetic aperture imaging in advanced detection of roadway using the mine transient electromagnetic method. *Chinese J. Geophys. (in Chinese)*, 59(2), 731–738.

Cheng J.L., Qiu H., Ye Y.T., et al. (2013) Research on wave-field transformation and data processing of the mine transient electromagnetic method. *Journal of China coal society*, 38(9), 1646–1650.

Guo W.B., Xue G.Q., Li X., et al. (2012) Correlation analysis and imaging technique of TEM data. *Exploration Geophysics*, 43(3): 137–148.

Lee K.H., & Xie G.Q. (1993) A new approach to imaging with low-frequency electromagnetic fields. *Geophysics*, 58(1), 780–796.

Levy S., Oldenburg D. & Wang J. (1988) Subsurface imaging using magnetotelluric data. *Geophysics*, 53(3), 104–117.

Li X, Qi Z.P., Liu Y.A., et al. (2010) A research of synthetic aperture rapid imaging of ATEM. *In 26rd CGS, Chinese Geophysics 2010(in Chinese)*. 639.

Li X., Xue G.Q., Song J.P., et al. (2005) An optimize method for transient electromagnetic field-wave field conversion. *Chinese J. Geophys. (in Chinese)*, 48(5), 1185–1190.

Li X., Zhang Y.Y., Lu X.S., et al. (2015) Inverse Synthetic Aperture Imaging of Ground-Airborne transient electromagnetic method with a galvanic source. *Chinese J. Geophys. (in Chinese)*, 58(1), 277–288.

Xue G.Q., Li X., Guo W.B., et al. (2006) Equivalent transformation from TEM field sounding data to plane-wave electromagnetic sounding data. *Chinese J. Geophys. (in Chinese)*, 49(5), 1539–1545.

Xue G.Q., Li X., Qi Z.P., et al. (2011) Study of sharpening the TEM pseudo-seismic wave-form. *Chinese J. Geophys. (in Chinese)*, 54(5), 1384–1390.

Innovation and Application of Engineering Technology – Oscar, Chen & Wang (Eds)
© 2017 Taylor & Francis Group, London, ISBN 978-1-138-05057-0

Integrated prevention and control technologies of coal mine goaf water disaster in Linfen mining area

Wen Li, Zongyun Shu, Yuguang Lian & Jianghua Li
Mine Safety Technology Branch of China Coal Research Institute, Beijing, China
State Key Laboratory of Coal Mining and Clean Utilization (China Coal Research Institute), Beijing, China

ABSTRACT: In order to study the integrated prevention and control technologies of coal mine goaf water disaster in the Linfen mining area, China, based on the characters and classification division of the water disaster, technologies such as surface survey, surface comprehensive geophysical detection, integrated prevention and control were proposed. The results showed that, goaf water disaster was the most serious water disaster in the Linfen mining area; the surface survey was the basis method to determine the range of subsequent geophysical detection; and, in order to reach the aim of fine detection, comprehensive geophysical detection should be adopted with the Transient Electromagnetic Method (TEM) and the Controlled Source Audio Magneto-Telluric Method (CSAMT), to detect goaf water disaster. In addition, methods such as boreholes drilling certification, geophysical detection under coalmine, and exploration could be referred in the explanation of geological results. Last but not least, the mainly integrated prevention and control technologies involved TEM detection on the heading face and the mining working face, boreholes drilling drainage, and chemical prospecting of goaf water. These technologies and methods have been applied in the Linfen mining area and achieved good results.

1 INTRODUCTION

The goaf water disaster is one of the most serious coal mine disasters in China. According to statistics, the accidents induced by goaf water accounted for more than 80% of the total coal mine water accidents since 2000. The goaf water disaster has seriously threatened the safety production of coal mines. In view of the prevention and control of goaf water disaster, some experts and scholars have proposed measures from the perspective of type classification (Wu *et al.*, 2013), geophysical detection (Cheng *et al.*, 2008, Li *et al.*, 2011, Fu 2014, Li 2017), boreholes drilling (Liu *et al.*, 2009, Yang & Wu 2010, Zhang *et al.*, 2014), and geochemical prospecting of goaf water disaster and achieved good results (Li Wen & Li Jian 2015, Li 2016). Unfortunately, the geological conditions are very different in the same mining area because of extremely complex geological and mining conditions in China. Therefore, it is extremely important and necessary to carry out the prevention and control of goaf water disaster in different mining areas.

Linfen city is located in the south of Shanxi Province. From October 2014 to December 2016, the authors' research team carried out the projects of surface survey, geological detection, integrated control, and plan reports preparation; the goaf water disaster distribution conditions were investigated in the Linfen mining area, the water disaster type and damage degree were analyzed and evaluated, and the integrated control measures of the goaf water disaster were put forward. On the basis of the above practice, the position and water volume of goaf water was clear.

Table 1. Water source classification of 42 coal mines in Linfen mining area.

Number	County and District (coal mine number)	Coalfield	Major water source	Secondary water source	Remarks
1	Xiangning (14)	Hedong	ABCEF	DG	16 coal mines involved
2	Jixian (1)		AC	EF	
3	Yaodu (1)		A	CEF	
4	Yaodu (3)	Huoxi	A	CDEF	15 coal mines involved
5	Puxian (8)		AF	C	
6	Hongtong (3)		AC	BDF	
7	Huozhou (1)		ABDE	C	
8	Guxian (6)	Qinshui	ABDE	CF	11 coal mines involved
9	Anze (3)		AB	CDF	
10	Fushan (1)		ABD	CE	
11	Yicheng (1)		BD	ACF	

Note: A-Goaf water disaster; B-Ordovician limestone water disaster; C-Surface water disaster; D-Taiyuan group limestone water disaster; E-Geological structures water disaster; F-Sandstone water disaster of roof and floor; G-Borehole drilling water disaster.

2 FEATURES AND MAJOR TYPES OF WATER DISASTER IN THE LINFEN MINING AREA

Linfen mining area has major coal reserves in southern Shanxi. The coal-bearing area reaches 15,400 km, accounting for 76.2% of the city land area. It is mainly composed of southern Hedong Coalfield, southern Huoxi Coalfield, and western Qinshui Coalfield. The total coal reserves reached 41.67 billion tons.

The coalfield distribution and water disaster type of 42 coal mines in the Linfen mining area are shown in Table 1. Analysis showed that the major types of water disaster in the Linfen mining area involved goaf water disaster, Ordovician limestone water disaster, surface water disaster, Taiyuan group limestone water disaster, geological structures water disaster, sandstone water disaster of roof and floor, and borehole drilling water disaster. Among the 42 coal mines, 40 would have faced goaf water disaster from 2016 to 2020, accounting for 95.24%. Therefore, the goaf water disaster is the most serious type of water disasters in the Linfen mining area.

3 SURFACE SURVEY TECHNOLOGY OF GOAF WATER DISASTER

The surface survey technology of goaf water disaster was first adapted to the 42 coal mines of the Linfen mining area. The survey device was S86 RTK GPS with a precision of 2 cm + 1 ppm and a scope of about 452.38 km. Results showed that in the 42 coal mines, there were totally 718 closed shafts, with 142 bad closed ones. In addition, there were totally 333 water reservoirs or wells, 207 springs, and 158 surface cracks or subsidence areas. The surface water could conduct into goafs through these bad closed shafts or cracks and cause a sudden increase of goaf water, thereby harming the safety production of mines. Typical survey results are shown in Figure 1.

Through surface survey, such conditions could be whole grasped as the goaf distribution, water bearing situation, unknown goaf general location, and so on. Surface survey could provide TARGET or scopes for subsequent geophysical detection to goaf water disaster.

4 COMPREHENSIVE GEOPHYSICAL DETECTION TECHNOLOGIES OF GOAF WATER DISASTER

In the Linfen mining area, there are many experiences in the aspect of geophysical detection. Experiences showed that the appropriate geophysical detection technologies mainly involved

(1) Bad closed kiln (2) Surface cracks

Figure 1. Typical survey results.

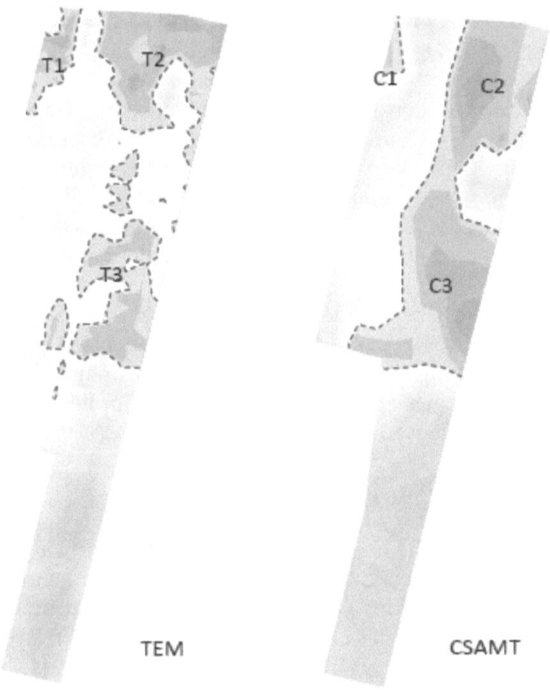

Figure 2. Abnormal distribution comparison between TEM and CSAMT near Coal 2 and Coal 3.

transient electromagnetic method (TEM) and controlled source audio magneto telluric method (CSAMT) to detect goaf water disaster. TEM is generally applicable to detect water-bearing goafs with buried depth of less than 600 m with high efficiency, even in complex terrain regions. Compared with TEM, CSAMT has larger detecting depth, higher resolution, and lower efficiency.

In 2016, among 42 coal mines, 15 coal mines carried out comprehensive geophysical detection with TEM and CSAMT, to detect goaf water. The detection area of TEM was about 29.042 km, whereas the detection area of CSAMT was about 12.550 km. In addition, in the process of geophysical exploration, results of processes such as borehole-drilling verification and geophysical detection under coal mines and mining conditions were referred, and the multiple solutions were reduced and the detection precision was greatly improved.

Case analysis was as follows. HMG Coalmine in the Linfen mining area is mainly extracting Coal 2 and Coal 3. The average thickness of Coal 2 is 2.20 m and Coal 3 is 2.33 m with an average distance of 7.24 m between them. The coal seam inclined by 3–9°. Because the goaf water distribution range was unknown, comprehensive geophysical detection methods were carried out on the surface with TEM and CSAMT. TEM adopted V8 multifunction electromagnetic detectors, with a measuring density of 20 m, whereas CSAMT adopted V8 multifunction workstations, with a measuring density of 40 m.

The abnormal distribution comparison between TEM and CSAMT near Coal 2 and Coal 3 is shown in Figure 2. The abnormal area TI of TEM had good corresponding relationship with the abnormal area Cl of CSAMT. In addition, the abnormal area T2 and T3 of TEM had not so good corresponding relationship but only similar forms with the abnormal area C2 and C3 of CSAMT, due to the different arrangement and density between two geophysical detection methods. When the two methods could not determine which one was near physical truth, these methods could make their respective advantages complementary to each other and this was just the superiority of comprehensive geophysical detection methods.

5 INTEGRATED PREVENTION AND CONTROL TECHNOLOGIES AND MEASURES OF GOAF WATER DISASTER

The integrated prevention and control of goaf water disaster should insisted basic principles such as *forecast firstly, detection where problem existing, detection before extraction,* and *mining after disaster control.* The keywords of water disaster-integrated prevention and control are *prevention, blocking, evacuation, drainage, and cutting*. Through analysis, there were 806 goaf water blocks in 42 coal mines, and the total water area was 37,967,469 m^2 with water yield estimated to be about 22,392,047 m^3. Among the 806 water blocks, the water yield of 455 blocks was more than 5000 m^3.

According to the characteristics of goaf water disaster in the Linfen mining area, four main technologies and measures were put forward: TEM detection on the heading face, TEM detection on the mining working face, boreholes drilling drainage, and chemical prospecting of goaf water. These technologies and methods have been applied in the Linfen mining area and good results have been achieved.

5.1 *TEM detection on the heading face*

The aims of TEM detection on the heading face are goaf water disaster detection of heading face and water-bearing body detection of roof and floor. The heading detection generally involves eight horizontal orientations (see Figure 3; seven horizontal orientations (0°; roof 15°, 30°, 45°; floor 15°, 30°, 45°)) and one vertical orientation (90°). Every horizontal orientation and the vertical orientation generally set 13 detection angles, as shown in Figure 4. The detection orientations and angles can be adjusted according to the actual geological conditions, hydrogeology conditions, and detection aims.

5.2 *TEM detection on the mining working face*

The aims of TEM detection at the mining working face are goaf and water bodiey detection overlying the wording face. The detection generally involves one horizontal orientation and 10 vertical orientations. The detection of orientations can be adjusted according to the actual geological conditions, hydrogeology conditions, and detection aims.

5.3 *Boreholes drilling drainage of goaf water*

Boreholes drilling drainage of goaf water includes overlying goaf water and the same coal seam goaf water. The boreholes drilling drainage sites of overlying goaf water are generally arranged in some place of two roadways, and the number of each drilling site is not less than three. The boreholes drilling drainage of the same coal seam goaf water is usually designed by the combination of deep holes and medium deep holes to ensure a heading distance of at least 30 m.

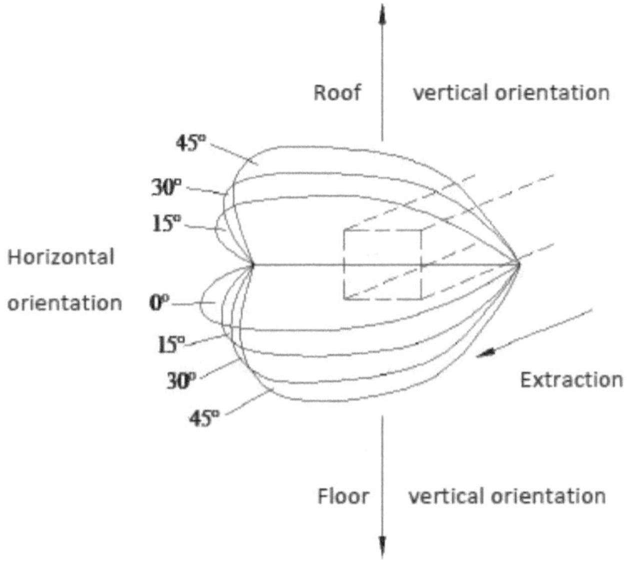

Figure 3. Eight orientations of TEM.

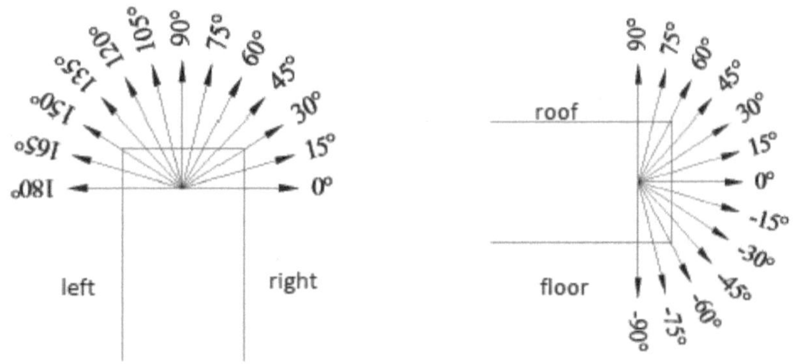

Figure 4. Thirteen detection angles of TEM.

5.4 *Chemical prospecting of goaf water*

Goaf water samples should be obtained during the boreholes drilling drainage. Then, water quality testing is carried out and the water quality characteristic database is established. Before and after planning working face mining, the water quantity and quality of each effluent site should be comprehensive and monitored for a long time. After abnormal water gushing, the position and source can be determined through the changes of water quantity and comparison of water quality database.

5.5 *Other technologies and measures*

In addition to the above four technologies and measures, the prevention and control of goaf water disaster also involve reserved waterproof coal and rock pillars, waterproof gates and walls, reasonable arrangement of mining methods, boreholes sealing quality investigation, hidden water disaster investigation in rainy season, timely filling of surface subsidence, cracks, small kilns, and so on.

6 CONCLUSIONS

Analysis showed that the major types of water disaster in the Linfen mining area involved goaf water disaster, Ordovician limestone water disaster, surface water disaster, Taiyuan group limestone water disaster, geological structures water disaster, sandstone water disaster of roof and floor, and borehole drilling water disaster. The goaf water disaster is the most serious type of water disasters in the Linfen mining area.

Surface survey technology of goaf water disaster is the basic method, which can provide TARGET or scopes for subsequent geophysical detection to goaf water disaster. Results showed that in the 42 coal mines, there were totally 718 closed shafts, with 142 bad closed ones. In addition, there were totally 333 water reservoirs or wells, 207 springs, and 158 surface cracks or subsidence areas. The surface water could conduct into goafs through these bad closed shafts or cracks and suddenly increase goaf water, thereby harming the safety production of the mine.

Experiences showed that the appropriate geophysical detection technologies mainly involved TEM and CSAMT to detect goaf water disaster. In the process of geophysical exploration, results of processes such as borehole drilling verification and geophysical detection under coal mines and mining conditions were referred. Therefore, multiple solutions were reduced and the detection precision was greatly improved. The detection area of TEM was about 29.042 km, whereas the detection area of CSAMT was about 12.550 km in the Linfen mining area. Practice showed that the comprehensive geophysical detection methods were effective.

In this paper, four main technologies and measures were put forward: TEM detection on the heading face, TEM detection on the mining working face, boreholes drilling drainage, and chemical prospecting of goaf water. These technologies and methods have been applied in the Chinese Linfen mining area and good results were achieved.

ACKNOWLEDGMENTS

The authors gratefully acknowledge the funding by the National Natural Science Foundation of China (51674142, 51404139) and the National Science and Technology Major Project of China (2016ZX05045001-004).

REFERENCES

Cheng Jianyuan, Sun Hongxing, Zhao Qingbiao, et al. (2008) The detection technology of excavated region in coal mine and case study. *Journal of China Coal Society*, 33(3): 251–255.

Fu Tianguang. (2014) Study on technology of comprehensive geophysical method exploration of mine goaf and water accumulated area. *Coal Science and Technology*, 42(8): 90–94.

Li Wen, Li Jian. (2015) Coafs disaster characteristics of resource integration coal mines and their control countermeasures. *Safety in Coal Mines*, 46(7): 179–181,185.

Li Wen, Mu Yi, Zhang Junying, et al. (2011) Optimization of surface detection technology and method of mine goaf. *Coal Science and Technology*, 39(1): 102–106.

Li Wen. (2016) Research on internet plus comprehensive prevention and control technology of hidden disaster in abandoned coal mine. *Coal Science and Technology*, 44(7): 86–91.

Li Wen. (2017) Optimization study of surface comprehensive geophysical detection methods of coal-mine goafs. *Coal Science and Technology,* 45(1): 194–199.

Liu Guolin, Pan Mao, Yin Shangxian. (2009) Coal mine water disasters: characteristics, prevention and control technology. *Safety in Coal Mines*, 35(1): 78–80.

Wu Qiang, Cui Fangpeng & Zhao Suqi, et al. (2013) Type classification and main characteristics of mine water disasters. *Journal of China Coal Society*, 38(4): 561–565.

Yang Chunlin, Wu Shuquan. (2010) Technology of comprehensive and integrated water detection and drainage for gob seeper. *Coal Mining Technology*, 15(2): 90–94.

Zhang Junying, Li Wen, Yang Junzhe, et al. (2014) Research on Hazard Assessment and Control Technology of Room and Pillar Mining Method Goaf in Shendong Mining Area. *Coal Science and Technology*, 42(10): 14–19.

Innovation and Application of Engineering Technology – Oscar, Chen & Wang (Eds)
© 2017 Taylor & Francis Group, London, ISBN 978-1-138-05057-0

An experimental study of loading rate effects on mechanical characteristics of cemented paste backfill

Y.G. Li
School of Civil and Resources Engineering, University of Science and Technology Beijing, Beijing, China

L.Z. Jin
School of Civil and Resources Engineering, University of Science and Technology Beijing, Beijing, China
Mine Emergency Technology Research Center, University of Science and Technology Beijing, Beijing, China

H. Tan & Y.D. Li
School of Civil and Resources Engineering, University of Science and Technology Beijing, Beijing, China

ABSTRACT: In order to investigate the mechanical parameter variation patterns and failure modes of Cemented Paste Backfill (CPB) under different loading rates, uniaxial compressive tests were conducted at various loading rates, using a digital-display press machine. Experimental results demonstrate that: a. in the range of 2.0–8.0 mm/min, the loading rate has a significant strengthening effect on both compressive strength and secant modulus of the CPB; Ultimate Compressive Strength (UCS) and secant modulus both increase with the increase of the loading rate; b. ultimate compressive strength manifests itself as a polynomial function of the loading rate, whereas the growth of secant modulus upon the loading rate is revealed to be exponential; and c. with the increase of the loading rate, failure mode of CPB specimens exhibits a gradual transformation from comprehensive tensile-shear damage to simplex shear damage. Research results in this work have demonstrated the effects of the loading rate on mechanical characteristics of CPB, which are expected to serve as a reference for future research on backfill failure mode.

1 INTRODUCTION

Underground mining is a major method of mine exploitation in China. Goaf filling can not only prevent the ground surface from subsidence, but also reduce the amount of tailing stocks, which has a great significance for environmental protection (Ghirian *et al*, 2013). Researchers around the world have conducted a large number of studies on backfill under different force conditions using various methods such as fundamental test, numerical simulation, and theoretical prediction. However, only a few investigations on strength variations of cemented tailings backfill under different loading rates have been reported (e.g., see Cao *et al*, 2015, p. 923–928, and Li *et al*, 2006, p. 2116–2120). Loading rate, as one of the primary parameters in dynamics research, usually exhibits a fairly wide range of variations. Mechanical properties of rock and coal-rock mass under various loading velocities are comprehensively studied internationally (Rosa *et al*, 2012). Generally, a strain rate less than 10^{-4} s^{-1} is considered as a low strain rate and that greater than 10^{-2} s^{-1} as a high strain rate, i.e., dynamic loading rate. Brazilian splitting tension tests on granite and marble under high temperature conditions at different loading rates (Yin *et al*, 2010) have been implemented to analyze the influences of temperature and loading rate on fracture characteristics of granite. Compressive tests on coal under various loading rates have been carried out, where the growth-to-decay trend of coal strength over the increasing loading rate was observed (Li *et al*, 2015). The loading rate corresponding to the turning point was defined as the critical loading rate. The influences of the strain rate effect of micro-components can be investigated using numerical simulation methods by discussing the impact of the loading rate as well as its sudden change on the compressive failure mode and macro-mechanical properties of concrete (Jin *et al*, 2014). In China, researchers also conducted a certain number of associated studies on different rock specimens. Uniaxial compressive tests on these specimens were carried out under different loading rates to investigate their mechanical parameter variation patterns and failure modes (e.g., see Zhou *et al*, 2013, p. 1868–1875, and Ji *et al*, 2011, p. 2507–2513).

2 SPECIMEN PREPARATION AND LOADING DEVICE

2.1 *Experimental materials and device*

Classified tailings of a gold mine were adopted as the raw materials for the tests, and 42.5R Portland cement was used as the aggregate material. Subsequent to the drying process of classified tailings, distribution characteristics of particle sizes were studied using a LS-909 laser particle size analyzer. Figure 1 illustrates the particle size distribution curve of tailings obtained from this mine.

It can be obtained from experimental data analysis that the cumulative distribution results of tailing particle size are $d_{10} = 30.74$ μm, $d_{50} = 100.54$ μm, $d_{90} = 190.04$ μm, and $d_{av=}$ 107.43 μm. With about 50% of the particles having a diameter greater than 100.54 μm, these classified tailings should belong to the coarse category.

Stress loading experiments were conducted by adopting a SANS digital-display fixed-displacement press machine, where desired load was applied on the specimens by exerting fixed displacement. This machine can also record information of axial load and deformation in real time and output experimental data by exporting Excel files.

2.2 *Experimental design and hypothesis*

The specimens were designed to be prepared using backfill slurry with a cement-sand ratio of 1/6 and concentration of 75%. Curing period, curing temperature, and relative humidity were designed to be 60 d, (20 ± 5)°C, and 95% ± 5%, respectively. Both ends of the specimens were meticulously polished using abrasive papers in order to control the smoothness under 0.02 mm. Acquiring of the CPB specimens was carried out as shown in Figure 2.

The loading rate was taken as the primary study object in this research. Axial compressive tests were conducted on the CPB specimens of classified tailings under four loading rates, 2.0 mm/min, 4.0 mm/min, 6.0 mm/min, and 8.0 mm/min. From the tests, the stress-strain curve of the CPB specimens in the whole process was obtained, and failure modes and crack development patterns under different loading velocities were analyzed.

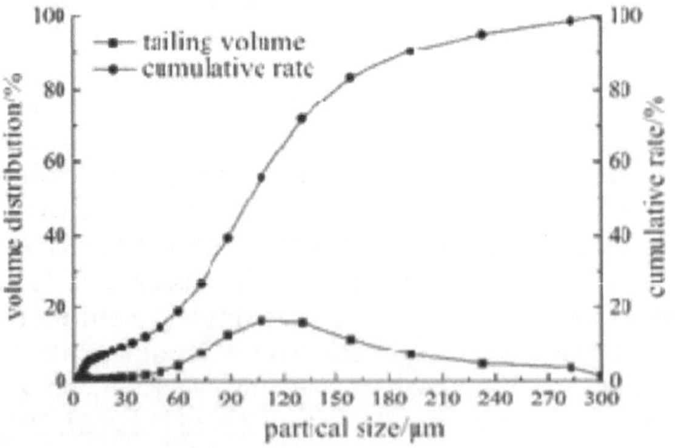

Figure 1. Curve of tailings' particle size distribution.

Figure 2. Cemented tailing specimens.

3 EXPERIMENTAL RESULTS ANALYSIS

3.1 *Analysis on the quantitative correlation between the loading rate and ultimate compressive strength*

According to the design plan, axial compressive tests were implemented on 36 CPB specimens by controlling the loading rate to vary from low to high. Ultimate compressive strengths of CPB specimens obtained at different loading rates are shown in Table 1. We can see that the ultimate compressive strengths corresponding to the loading rates of 2.0 mm/min, 4.0 mm/min, 6.0 mm/min, and 8.0 mm/min are 2.98 MPa, 3.01 MPa, 3.06 MPa, and 3.12 MPa, respectively.

It should be noticed that due to the narrow initially set loading range of Specimen 6–4, when the axial stress exceeded 6 kN, the loading equipment automatically halted. Therefore, the abnormal data needed to be eliminated. Experimental data gained from Specimen 8–1 were also removed due to the significant discrepancies with those of the other five sets of experiments.

It can be seen from the statistical results shown in Table 1 that within the loading rate range of 2.0–8.0 mm/min, the ultimate compressive strength of the CPB specimen presents a rising trend over the increasing loading rate, indicating that the loading rate has a strengthening effect on the ultimate compressive strength. Fitting was conducted on the ultimate compressive strength and loading rate using linear, exponential, logarithmic, and polynomial methods, respectively. Figure 3 illustrates the calculated results of fitting, where σ_p and v represent the ultimate compressive strength and the loading rate with units MPa and mm/min, respectively.

It can be learned by analyzing the fitting results in Figure 3 that all the three methods have a comparatively high fitting degree. However, polynomial fitting has an even higher multiple correlation coefficient than any other method, reaching 99.96%. This demonstrates that the polynomial fitting, $\sigma_p = 0.002v^2 + 0.005v + 2.963$, can better reflect the quantitative correlation between the ultimate compressive strength of cemented paste backfill and the loading rate.

3.2 *Quantitative characterization of the loading rate and secant modulus*

In order to explore the quantitative correlation between the secant modulus of the CPB and the loading rate, linear, exponential, and logarithmic fittings were conducted on the computed results in Table 2, and multiple correlation coefficients R2 of the computed results were taken as the evaluation indicator. Figure 4 displays the fitting curves of the three fitting methods.

It can be discovered from the three fitting methods shown in Figure 4 that, multiple correlation coefficients of linear, exponential, and logarithmic fittings all demonstrate

Table 1. Uniaxial compression strength values under different loading rates (MPa).

Specimen number	Loading rate/ (mm/min)	Ultimate compressive strengths/MPa
2–1	2.0	2.56
2–2	2.0	3.21
2–3	2.0	2.66
2–4	2.0	3.02
2–5	2.0	3.26
2–6	2.0	3.17
4–1	4.0	3.28
4–2	4.0	3.00
4–3	4.0	2.83
4–4	4.0	—
4–5	4.0	3.03
4–6	4.0	2.91
6–1	6.0	2.77
6–2	6.0	3.26
6–3	6.0	2.70
6–4	6.0	—
6–5	6.0	—
6–6	6.0	3.49
8–1	8.0	—
8–2	8.0	2.54
8–3	8.0	2.35
8–4	8.0	3.25
8–5	8.0	3.55
8–6	8.0	3.91

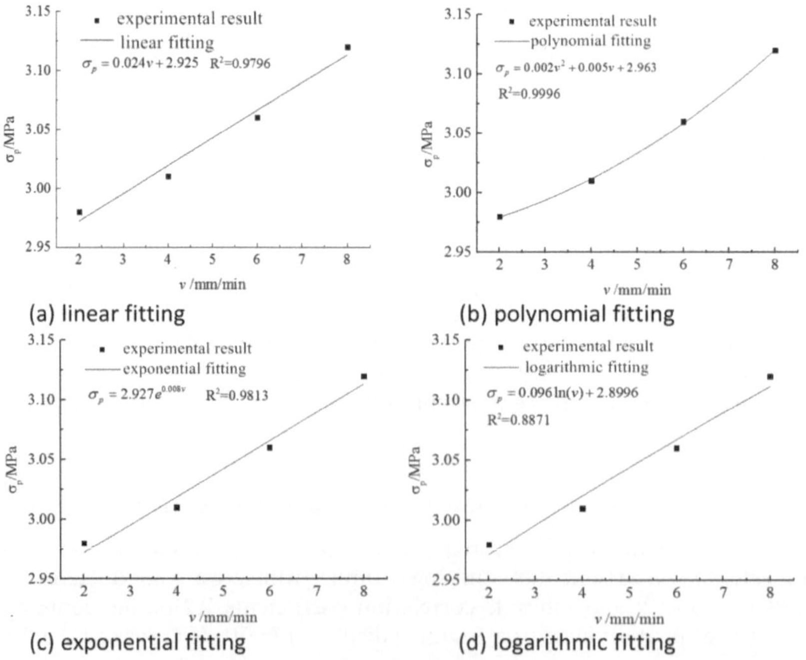

Figure 3. Fitting results between peak uniaxial compression strength and loading rates.

Table 2. The calculation of secant modulus value of the cemented filling body.

Specimen number	Loading rate/ (mm·min^{-1})	Secant modulus/ GPa
2–1	2.0	0.945
2–2	2.0	1.083
2–3	2.0	1.434
2–4	2.0	1.045
2–5	2.0	—
2–6	2.0	1.686
4–1	4.0	1.739
4–2	4.0	0.937
4–3	4.0	1.061
4–4	4.0	—
4–5	4.0	2.161
4–6	4.0	1.645
6–1	6.0	1.537
6–2	6.0	1.648
6–3	6.0	2.175
6–4	6.0	—
6–5	6.0	—
6–6	6.0	2.362
8–1	8.0	—
8–2	8.0	2.228
8–3	8.0	—
8–4	8.0	2.274
8–5	8.0	2.053
8–6	8.0	2.736

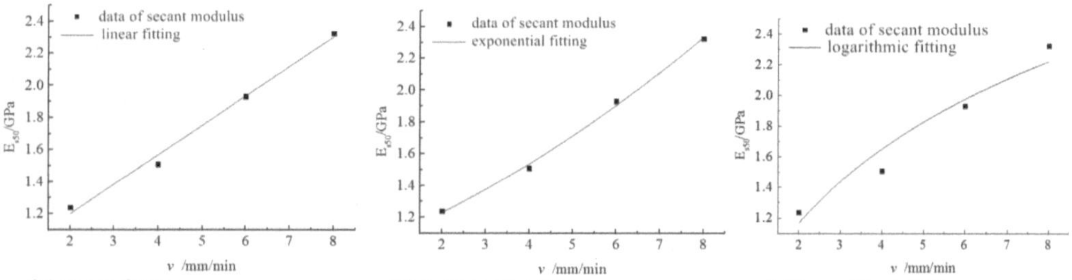

(a) Linear fitting (b) Exponential fitting (c) Logarithmic fitting

Figure 4. Fitting curves of the three fitting methods.

superior fitting characteristics. Among the three, exponential fitting possesses the highest fitting degree, which indicates that quantitative relationship between the secant modulus of the cemented paste backfill and the loading rate can be best characterized using exponential fitting. Moreover, an increasing loading velocity can result in the growth of secant modulus, which shows that the loading rate has a strengthening effect on the average stiffness of the cemented paste backfill.

4 ANALYSIS ON THE FAILURE MODE OF THE CEMENTED PASTE BACKFILL

Damaged CPB specimens and their failure modes at four loading rates are illustrated in Figure 5.

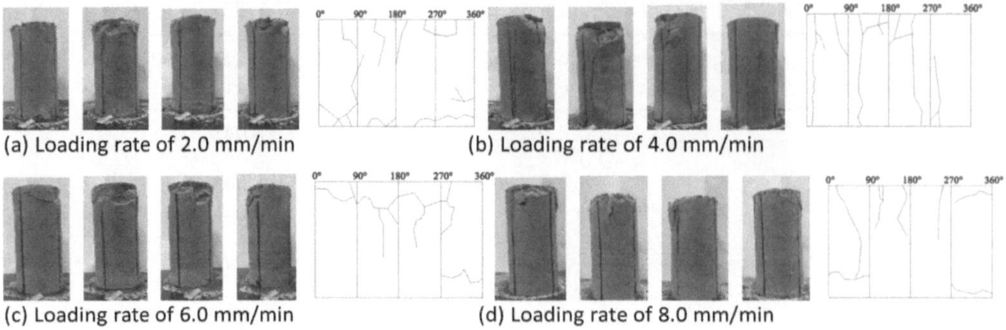

Figure 5. Photograph and failure sketch of specimens at different loading rates.

It can be observed from Figure 5 that, at the loading rate of 2.0 mm/min, specimens primarily exhibit a comprehensive failure mode of a tensile damage that is parallel to the loading direction and a shear damage; At the loading rates of 4.0 mm/min and 6.0 mm/min, the failure mode presents an interlaced distribution of longitudinal fracture and oblique shear; when the loading rate reaches 8.0 mm/min, the failure mode begins to transform to the simplex shear damage, even forming two dramatic fastigiated shear surfaces; with the increase of the loading rate, the specimen failure mode shows a gradual transformation from the comprehensive tensile-shear damage to the simplex shear damage.

5 SUMMARY

The loading rate has an obvious strengthening effect on compressive strength and secant modulus of cemented paste backfill. Ultimate compressive strength and secant modulus both rise with the increase of the loading rate. Ultimate compressive strength variation versus loading rate exhibits a polynomial pattern, while the secant modulus variation against the loading rate demonstrates an exponential distribution. Analysis of the CPB failure mode indicates that, with the increase of the loading rate, the specimen failure mode can be gradually transformed from comprehensive tensile-shear damage to simplex shear damage. A higher loading rate could generate a shorter time for CPB failure occurrence. In this paper, experimental results are acquired from the tests only on the specimens with a cement-sand ratio of 1/6 and concentration of 75%. Therefore, in order to determine the applicability of the obtained conclusions on other backfill materials, subsequent investigations are required to be conducted.

ACKNOWLEDGMENT

This work was supported in part by the National Natural Science Foundation of China (Grant no. 51574017).

REFERENCES

Cao Anye, Jing Guangcheng & Dou Linming, et al. (2015) Damage evolution law based on acoustic emission of sandy mudstone under different uniaxial loading rate. *Journal of Mining & Safety Engineering*, 32(6): 923–928.

Ghirian, A. & Fall, M. (2013) Coupled thermos-hydro-mechanical-chemical behaviour of cemented paste backfill in column experiments part I: physical, hydraulic and thermal processes and characteristics. *Engineering Geology*, 164: 195–207.

Ji Wendong, Yang Chunhe & Yao Yuanfeng, et al. (2011) Effects of loading strain rate on mechanical performances of salt rock. *Chinese Journal of Rock Mechanics and Engineering*, 30(12): 2507–2513.

Jin Liu & Du Xiuli. (2014) Effects of loading rate and its sudden change on concrete compressive failure. *Journal of Vibration and Shock*, 33(19): 187–193.

Li Haitao, Jiang Chunxiang & Jiang Yaodong, et al. (2015) Mechanical behavior and mechanism analysis of coal samples based on loading rate effect. *Journal of China University of Mining & Technology*, 44(3): 430–436.

Nasir, O. & Fall, M. (2010) Coupling binder hydration, temperature and compressive strength development of underground cemented paste backfill at early ages. *Tunnelling and Underground Space Technology*, 25: 9–20.

Nistico, N., Pallini, F. & Rousakis, T., et al. (2014) Peak strength and ultimate strain prediction for FRP confined square and circular concrete sections. *Composites: Part B*, 67: 543–554.

Rosa, L., Yu, R.C. & Ruiz, G., et al. (2012) A loading rate dependent cohesive model for concrete fracture. *Engineering Fracture Mechanics*, 82(3): 195–208.

Wei Wei. & Gao Qian. (2013) Strength prediction of backfilling body based on modified BP neural network. *Journal of Harbin Institute of Technology*, 45(6): 90–95.

Xu Jinyu & Liu Shi. (2013) Effect of impact velocity on dynamic mechanical behaviors of marble after high temperatures. *Chinese Journal of Geotechnical Engineering*, 35(5): 879–883.

Yin Xiaotao, Ge Xiurun & Li Chunguang, et al. (2010) Influences of loading rates on mechanical behaviors of rock materials. *Chinese Journal of Rock Mechanics and Engineering*, 29(suppl): 2610–2615.

Zhou Hui, Yang Yanshuang & Xiao Haibing, et al. (2013) Research on loading rate effect of tensile strength property of hard brittle marble-test characteristics and mechanism. *Chinese Journal of Rock Mechanics and Engineering*, 32(9): 1868–1875.

Innovation and Application of Engineering Technology – Oscar, Chen & Wang (Eds)
© 2017 Taylor & Francis Group, London, ISBN 978-1-138-05057-0

Research on basic characteristics of soft coal geological mechanics simulation of solid-gas coupling similar materials

S.Q. He[1]
School of Civil and Environmental Engineering, University of Science and Technology Beijing, Beijing, China

L.Z. Jin
School of Civil and Environmental Engineering, University of Science and Technology Beijing, Beijing, China
Mine Emergency Technology Research Center, Beijing, China

Y.D. Li, Y.G. Li, Y. Lu & X.Q. He
School of Civil and Environmental Engineering, University of Science and Technology Beijing, Beijing, China

ABSTRACT: To study the basic characteristics of soft coal similar materials of solid-gas coupling, in this paper, the molding moisture, humic acid sodium solution concentration, and aggregate size ratio to set orthogonal test with three factors and five levels are used. In the experiment, we adopted range analysis, linear regression, electromagnetic radiation, and other methods to study the influence of molding moisture, humic acid sodium solution concentration, and aggregate size ratio on the density, compressive strength, gas radiation initial velocity, permeability and elastic modulus of similar materials. The results show that the order of a significant effect on the comprehensive effect of materials is the concentration of humic acid sodium solution, aggregate size ratio, and molding moisture. The order of a significant effect on the density and permeability is molding moisture, aggregate size ratio, and humic acid sodium solution concentration. The order of a significant effect on the compressive strength and elastic modulus was humic acid sodium solution concentration, aggregate size ratio, and molding moisture. The order of a significant effect on the gas radiation initial velocity was molding moisture, humic acid sodium solution concentration, and aggregate size ratio. Permeability and humic acid sodium solution concentration presented an exponential and logarithmical relationship when the confining pressure and gas pressure are fixed, respectively. The adsorption/desorption and electromagnetic radiation characteristics of similar materials were similar to raw coal, which lays a foundation for the solid-gas coupling physically simulating experiment of soft coal.

1 INTRODUCTION

Soft coal mines have become high incidence area of major disasters in China; because of the low permeability and poor drilling stability of soft coal seams, gas control is difficult (Yuan, 2010). Gas drainage[1] is a fundamental measure to prevent gas accidents (Qu *et al.*, 2011). Previous research on this aspect was conducted through the theoretical and numerical simulation method (Yao *et al.*, 2010), those often cannot accurately reflect the actual situation of drilling around the borehole. A similar simulation experiment has the characteristics of speed, intuitive image, and so on. It has become an important method of experimental research in recent years. Domestic scholars have carried out a lot of research on similar material simulation techniques. Zhang *et al.* (2015) used cement as the binder; shredded coal as the aggregate; and water, sand and activated carbon as accessories to develop experimental materials of coal and gas outburst. Kang *et al.* (2015) used coal powder or sand as aggregates, gypsum and cement as binders for comparative experiments, and concluded that coal powder is a better aggregate than sand. Cheng *et al.* (2016) used sand, activated carbon, and coal as aggregates and cement as the binder to develop similar materials. Wei *et al.* (2015) used paraffin as the binder, hydraulic oil as a regulator, sand and talc as aggregates, developing submarine tunnel solid-liquid coupling similar materials. Above-mentioned research focused on solid and solid-liquid coupling similar materials, which cannot solve the problem of solid-gas coupling physical simulation. There are relatively few studies on the coupling of similar materials of solid-gas (Li *et al.*, 2015). In view of this, the authors, on the basis of predecessors, based on the principles of the similarity theory, through the orthogonal ratio test and the poor analysis, developed a material similar to soft coal. And through adsorption and desorption, gas permeability, electromagnetic radiation experiments of similar material specimens, the authors built a physical model to intuitively reveal the mechanism of major disasters in soft coal mines, with a view to provide a basis for similar soft coal solid-gas physical coupling simulations.

2 DEVELOPMENT OF SIMILAR MATERIALS FOR SOFT COAL SEAMS

2.1 *Material selection*

The similar materials required for the soft coal geomechanical model experiment not only to meet the requirements of the similarity theory, but also the adsorption and other characteristics should be similar to that of raw coal (Wang *et al.*, 2006). Soft coal has adsorption/desorption and low permeability properties; therefore, coal powder with different particle sizes was chosen as the aggregate. The different ratios of coal particles with different particle sizes were used to study the effect of particle size distribution on its performance and found the appropriate ratio. According to previous studies, in order to better suppress the molding and improve the strength, the binder should be added to the aggregate. In this paper, the preparation of similar materials of soft coal was studied to solve the problems of soft coal gas outburst and drainage. Therefore, the adsorption/desorption and low permeability of materials should be similar to that of raw coal. Because of the adsorption characteristics of sodium humate, it was selected as the binder. The experiments were carried out using molding moisture, humic acid sodium solution concentration, quality of (1–3 mm coal power) / (0–1 mm coal power) as three factors (A, B, C) for orthogonal design, and each factor set five levels; these are given in Table 1.

2.2 *Specimen production*

According to the experimental scheme given in Table 1, at room temperature (at 25°C), press the developed similar materials into the mold, retain the pressure for about 15 minutes and then remove the specimen from the mold. Paste the label and place the specimen in a vacuum oven (105°C) for drying for 2 h. Afterwards, place it in a dry room. According to the Code

Table 1. Design of similar material orthogonal experiments.

	Factor		
Number of level	A	B	C
1	6%	2.5%	1:5
2	7%	5%	1:4
3	8%	10%	1:3
4	9%	15%	2:3
5	10%	20%	1:1

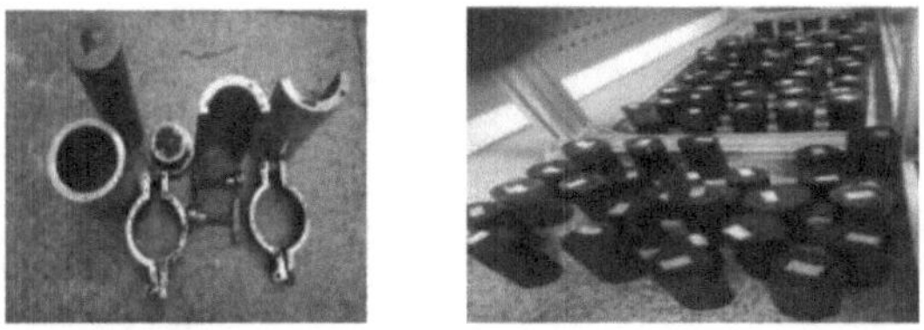

Figure 1. Pictures of the mold and specimens.

for rock test of water conservancy and hydropower project (SL264-2001) and Engineering rock mass test method standard (GB/T50266-2013) requirements, the specimen size is ø50 mm × 100 mm, as shown in Figure 1.

3 EXPERIMENTAL RESULTS

3.1 *Each parameter test results*

In order to study the density (γ), compressive strength (σ_c), elastic modulus (E), gas radiation initial velocity (Δp), gas permeability (k) and electromagnetic radiation characteristics of the materials, the different test programs were designed according to Table 1. From the experiment, we studied the effects of different factors on similar materials characteristics. We concluded from the experiment that the density distribution of similar materials was 10.95~12.21 kN/m^3. The compressive strength was 0.53~1.92 MPa, the gas radiation initial velocity was 19.20~23.70 mmHg, which was similar to raw coal (21.1 mmHg). It can simulate the soft coal outburst seam. The elasticity modulus was 28.4~159.22 MPa, the gas permeability was (0.15~5.61) × 10^{-16} m^2, which was conformed to the low permeability characteristics of soft coal. Each parameter distribution range was wide, and we can, according to the parameters of soft coal, select a reasonable similarity constant, when constructing the geomechanical model.

3.2 *Sensitivity analysis*

The density, compressive strength, gas radiation initial velocity, elastic modulus and gas permeability of similar materials was studied by using the comprehensive grading method and range analysis (Huang *et al.,* 2007) to find out the impact factors of each parameter, and the results are given in Table 2.

As it is given in Table 2, the five indicators which were selected have the same weight. The weight of each indicator was 0.2 and the sum of the five indicators was the comprehensive quantitative indicators. For density, the order of significant effect is as follows: $R_A > R_C > R_B$. For compressive strength, the order of significant effect is as follows: $R_B > R_C > R_A$. For gas radiation initial velocity, the order of significant effect is as follows: $R_A > R_B > R_C$. For elasticity modulus, the order of significant effect is as follows: $R_B > R_C > R_A$. For gas permeability, the order of significant effect is as follows: $R_A > R_C > R_B$. For comprehensive index, the order of significant effect is as follows: $R_B > R_C > R_A$. The order of significant effect on the similar material was humic acid sodium solution concentration, aggregate size ratio, and molding moisture. The humic acid sodium solution concentration and aggregate size ratio had a great influence on the mechanical index. The reason was that the adhesion of the humic acid sodium solution to the aggregate enables the particles of different particle sizes to bind together, and the rational arrangement of different size particles can reduce the number of breakable pores. The great effect of the molding moisture and humic acid sodium solution concentration on the gas radiation initial velocity was that the humic acid sodium solution was a multi-functional polymer compound, and contains more active groups such as hydroxyl groups and carboxyl groups. It has a large internal surface area and a strong adsorption, affecting the gas dispersion. The great effect of the molding moisture and aggregate size ratio on gas permeability was due to its effect on the pore structure.

Table 2. Comprehensive score results of similar materials.

	Γ			σ_c			Δp			E			K			Overall rating		
Factor	A	B	C	A	B	C	A	B	C	A	B	C	A	B	C	A	B	C
K_1	56.09	56.72	56.71	4.16	5.25	5.26	104.40	108.50	108.30	402.4	428.49	387.21	3.27	9.02	6.60	114.06	121.59	112.82
K_2	56.99	56.79	56.42	5.42	3.88	4.98	108.70	106.10	105.90	442.63	316.88	399.47	5.16	6.93	6.93	123.78	98.12	114.56
K_3	58.51	57.12	57.86	5.91	5.45	4.80	101.40	103.60	108.20	450.48	493.48	435.53	16.88	10.59	9.55	126.64	135.25	123.19
K_4	57.38	56.79	57.43	6.37	5.70	5.18	113.50	101.00	101.70	543.74	461.66	443.74	7.94	5.39	10.61	145.79	126.11	123.73
K_5	56.27	57.82	56.83	5.60	7.18	7.24	101.50	104.30	105.40	439.73	578.47	613.03	3.90	5.22	4.36	121.40	150.60	157.37
R	2.42	1.10	1.44	2.21	3.30	2.44	12.10	7.50	6.60	141.34	261.59	225.82	13.61	5.37	6.25	31.72	52.48	44.56

*K_i represents the sum of the results of each test with the same number of bits in the ith factor. R indicates the significance of the corresponding factors in the test.

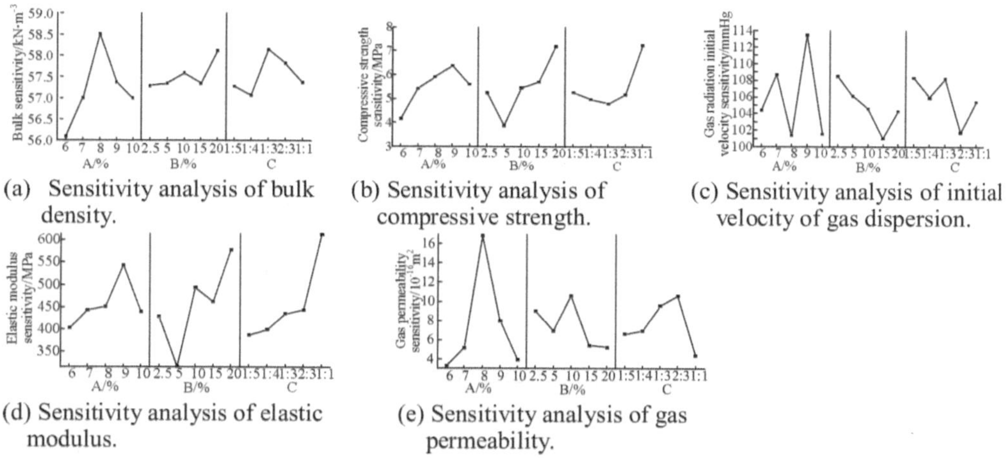

(a) Sensitivity analysis of bulk density.

(b) Sensitivity analysis of compressive strength.

(c) Sensitivity analysis of initial velocity of gas dispersion.

(d) Sensitivity analysis of elastic modulus.

(e) Sensitivity analysis of gas permeability.

Figure 2. Sensitivity analysis of each parameter.

In order to analyze the effect of each factor on the properties of similar materials, according to Table 2, OriginPro9.0 was used to make the intuitive analysis chart, as shown in Figure 2.

As shown in Figure 2, the physical parameters increase with the increase of molding moisture and humic acid sodium solution concentration on the whole, the optimum molding moisture was 8%. This was because the moisture content was too large to be pressed and dried. The gas radiation initial velocity decreased with the increase of the humic acid sodium solution concentration and the gas permeability increased with the increase of aggregate size ratio.

4 VARIATION LAW OF SIMILAR MATERIAL PROPERTIES

4.1 *Effect of humic acid sodium solution concentration on gas permeability*

The molding moisture is 8%, aggregate size ratio is 1: 1, and confining pressure and gas pressure are fixed, respectively. The impact of humic acid sodium solution concentration on permeability was measured under different confining pressures, and it is shown in Figure 3.

As shown in Figure 3, when other factors are fixed, the permeability of the specimen decreased with the increase of the humic acid sodium solution concentration. The permeability increased with the increase of gas pressure when the confining pressure fixed. The permeability decreased with the increase of confining pressure when the gas pressure fixed. Because the permeability of similar materials was low, it was suitable for simulating soft and low permeability coal seams. However, from the influence of confining pressure and gas pressure, the solid-gas coupling geomechanical model simulation with similar materials should be combined with the actual situation of the coal seam and confining pressure and gas pressure should be imposed to reduce the experimental error. The linear regression method was used to fit some experimental data. The relationships are given in Tables 3 and 4.

4.2 *Effect of humic acid sodium solution concentration on the desorption characteristics*

The desorption of raw coal and similar materials is shown in Figure 4.

Figure 4 shows that the desorption of the original soft coal gas reached 4.5 ml/g in 500 s, indicating that the initial desorption resolved quickly and it could release large amounts of gas in a short time. The release of the initial velocity is an important factor that affected coal and gas outburst accidents. It can be seen that the study of the desorption characteristics of similar materials is important to improve the effect of gas extraction (M *et al.*, 2011).

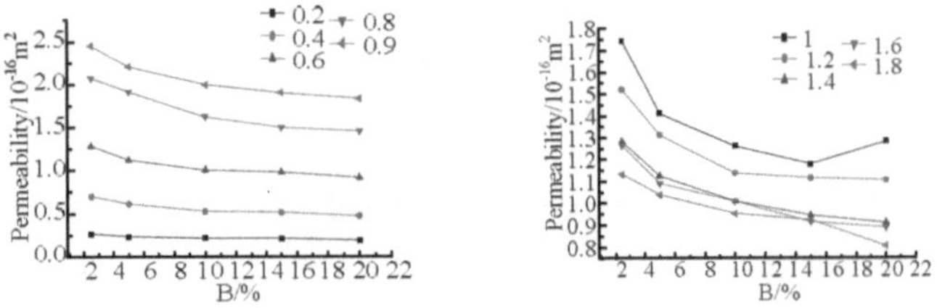

(a) Confining pressure 1.4 MP. (b) Gas pressure 0.6 MPa.

Figure 3. Relationship curves of permeability and cementing agents of test specimens.

Table 3. Relationship between permeability and cementing agents under confining pressure 1.4 MPa.

Gas pressure/ MPa	Confining pressure/MPa	Factor	Fitting relationship	Fitting degree/R^2
0.2	1.4	B	$k=0.104e^{-\frac{x}{10.049}}+0.178$	0.8878
0.4			$k=0.352e^{-\frac{x}{5.422}}+0.476$	0.9792
0.6			$k=0.578e^{-\frac{x}{5.117}}+0.923$	0.9669
0.8			$k=0.994e^{-\frac{x}{7.526}}+1.377$	0.9899
0.9			$k=0.974e^{-\frac{x}{5.651}}+1.821$	0.9926

Table 4. Fitting relation between permeability and cementing agents under gas pressure 0.6 MPa.

Confining pressure/MPa	Gas pressure/ MPa	Factor	Fitting relationship	Fitting degree/R^2
1	0.6	B	$k=-0.441\ln(-0.023\ln x)$	0.8827
1.2			$k=-0.367\ln(-0.018\ln x)$	0.9770
1.4			$k=-0.314\ln(-0.018\ln x)$	0.9959
1.6			$k=-0.314\ln(-0.019\ln x)$	0.9853
1.8			$k=-0.240\ln(-0.009\ln x)$	0.8632

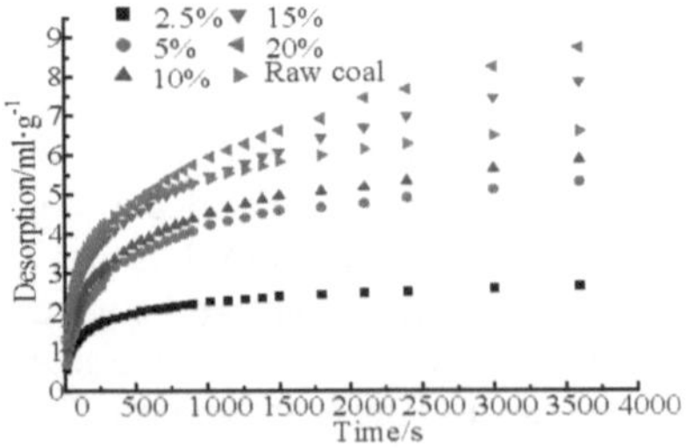

Figure 4. Relationship curve of desorption of similar materials with cementing agents.

The amount of gas desorption increased with the increase of the concentration of humic acid solution, which was attributed to the humic acid sodium that belongs to the polymer compounds group and has a strong adsorption. Therefore, for the desorption characteristics of different soft coal seams, the material can be prepared by adjusting the concentration of humic acid solution.

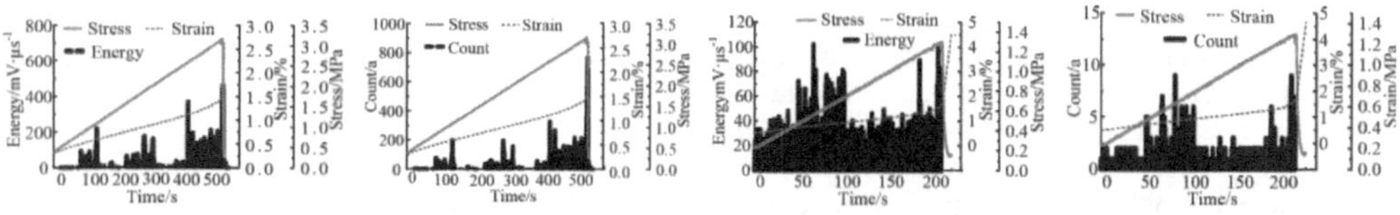

(a) Relation between the electromagnetic radiation phenomenon, stress and strain of the raw coal 0–500.

(b) Relation between the electromagnetic radiation phenomenon, stress and strain of the raw coal 0–200.

Figure 5. Relation between the electromagnetic radiation phenomenon, stress and strain.

4.3 *Electromagnetic radiation performance research*

In this paper, the electromagnetic radiation characteristics of raw coal and similar material specimens under uniaxial compression were studied, as shown in Figure 5.

Figure 5 shows that the energy and counts generated by the different stages of the specimen loading damage were different. The type of electromagnetic radiation change between raw coal and similar materials is consistent with one of the studies of Nie et al. (2002). At the beginning of loading, the energy and the counts are high and the medium loading is calm. The energy and counts gradually increased before the main damage occurred, the strength was higher when damaged and it weakened after the damage. The electromagnetic radiation characteristics of coal and similar materials are similar, and so the geomechanical model can be constructed by using soft coal similar materials, and the electromagnetic radiation detection technology can be used to predict the drilling deformation, gas outburst and impact pressure of soft coal seams.

5 SUMMARY

The density of similar materials was 10.95~12.21 kN/m^3, the compressive strength was 0.53~1.92 MPa, the gas radiation initial velocity was 19.20~23.70 mmHg, elastic modulus was 28.4~159.22 MPa, and gas permeability was (0.15~5.61) × 10^{-16} m^2. Each parameter distribution range was wide, which could simulate different soft coal seams. The order of significant effect on the comprehensive effect was concentration of sodium humate solution, aggregate size ratio, and molding moisture. The experimental data and the fitting formula showed that the permeability and humic acid sodium solution concentration presented an exponential relationship at different gas pressures when the confining pressure was fixed at 1.4 MPa. The permeability and humic acid sodium solution concentration presented a logarithmic relationship at different confining pressures when the gas pressure was fixed at 0.6 MPa. By using the electromagnetic radiation experiment, we concluded that the energy and counts were different at different stages of specimen loading damage. The adsorption and electromagnetic radiation characteristics of similar materials were similar to raw coal, which lays a foundation for the solid-gas coupling physically simulating experiment of soft coal.

ACKNOWLEDGMENT

This study was supported by the National Natural Science Foundation of China (no. 51574017).

REFERENCES

Cheng, Weimin., Sun, Lulu. & Wang, Gang., et al. 2016. Similar material simulation test of steep-inclined extra-thick coal seam. *Journal of Mining & Safety Engineering* 33(3): 387–392.

Kang, Xiangtao., Huang, Gun. & Deng, Bozhi., et al. 2015. Experimental study on similar material for simulating raw coal. *Journal of Northeastern University (Natural Science)* 36(1): 138–142.

Li, Shugang., Zhao, Pengxiang. & Lin, Haifei., et al. 2015. Study on character of physical simulation similar material of coal-rock and gas solid-gas coupling. *Journal of China Coal Society* 40(1): 80–86.
M, P., S, H. & Liu, S. 2011. Gas diffusion behavior of coal and its impact on production from coalbed methane reservoirs. *International Journal of Coal Geology* 86(4): 342–348.
Nie, Baisheng., He, Xueqiu. & Wang, Enyuan., et al. 2002. Study on electromagnetic radiation and acoustic emission in shearing process of coal. *Journal of China University of Mining & Technology* 31(6): 609–611.
Qu, P., Shen, R. & Fu L., et al. 2011. Time delay effect due to pore pressure changes and existence of cleats on borehole stability in coal seam. *International Journal of Coal Geology* 85(2): 212–218.
Wang, Hanpeng., Li, Shucai. & Zhang, Qiangyong., et al. 2006. Development of a new geomechanical similar material. *Chinese Journal of Rock Mechanics and Engineering* 25(9): 1842–1847.
Wei, Liyuan., Jing, Hongwen. & Xu, Bangshu., et al. 2015. Solid-fluid coupling analogous material test for subsea tunnel. *Journal of Central South University (Science and Technology)* 46(3): 983–990.
Yao, Xiangrong., Cheng, Gonglin. & Shi, Biming. 2010. Analysis on gas extraction drilling instability and control method of pore-forming in deep surrounding-rock with weak structure. *Journal of China Coal Society* 35(12): 2073–2081.
Yuan, Liang. 2010. Concept of gas control and simultaneous extraction of coal and gas. *China Coal* 36(6): 5–12.
Zhang, Shutong., Dai, Linchao. & Wang, Bo., et al. 2015. Experiment study on mixture ratio of similar material for simulation of coal and gas outburst. *Coal Science and Technology* 43(6): 76–80.
Zhou, Junping., Xian, Xuefu. & Li, Xiaohong., et al. 2010. Effect of different adsorptional gases on permeability of coal. *Chinese Journal of Rock Mechanics and Engineering* 29(11): 2256–2262.

Innovation and Application of Engineering Technology – Oscar, Chen & Wang (Eds)
© 2017 Taylor & Francis Group, London, ISBN 978-1-138-05057-0

Discussion on the construction technology of the shield machine across Yong River

Long Ma & Guangliang Xiao
Ningbo Rail Transit Group Co. Ltd., Ningbo, Zhejiang Province, China

Xiuheng Ding
Shanghai Civil Engineering Bureau Group, Co. Ltd., CREC China, Shanghai, China

Yanbin Ren
Ningbo Administration of Work Safety, Ningbo, Zhejiang Province, China

ABSTRACT: According to the project case that the shield machine crosses Yong River in the interval from Datong Bridge Station to Zhongxing Bridge South Station of TJ3101 Bid for Line 3 of Ningbo Rail Transit, in this article, the selection of shield machines based on engineering hydrographic, geologic, and environmental risks, and comprehensive management of the shield machine with regard to earth pressure, discharge volume, thrust force, tunneling speed, construction gap filling and other parameter settings are mainly summarized. And also, it develops measures including detailed monitoring, spewing prevention, roof caving prevention and seepage prevention in advance, and finally achieves controllable quality and safety as well as remarkable economic benefit, and works out the construction method for shield machines crossing rivers, which provides plentiful experience for the design and construction.

Keywords: Shield; Yong River; Selection; Parameters; Segment; Tunnel; Management

1 INTRODUCTION

1.1 *Project profile*

In the interval from Datong Bridge Station to Zhongxing Bridge South Station of TJ3101 Bid for Phase 1 of Line 3 of Ningbo Rail Transit, the left line is 1546.697 m long (1291 rings), and the right line is 1543.426 m long (1288 rings), which sums up to 2579 rings. The buried depth of the tunnel top is about 18.1~34.9 m, the maximum longitudinal slope is 28‰, and the minimum curve radius of the line plane is 449.8 m. The tunnel segment has an inner diameter of 5500 mm, outer diameter of 6200 mm, measures 350 mm in thickness, 1200 mm in width, C50 in concrete strength and P10 in impermeability grade.

The interval from Datong Bridge Station to Zhongxing Bridge South Station Datong Bridge station under has to cross Yong River and the cross-section is 330 m wide, and the maximum water pressure of the cross-river section is 0.4 Mpa, where the formation has good water permeability. Yong River is 6.4 m deep. The cross-river tunnel plane is straight and the cut plane has an ascending grade of 28‰, and the earth backing on the top of the cross-river section is 20.6 m~32.8 m deep. Because the cross-river section has high water pressure and the formation has good water permeability, long distance operation under such environmental conditions is prone to risk. The chart for crossing Yong River is shown below (see details in Fig. 1).

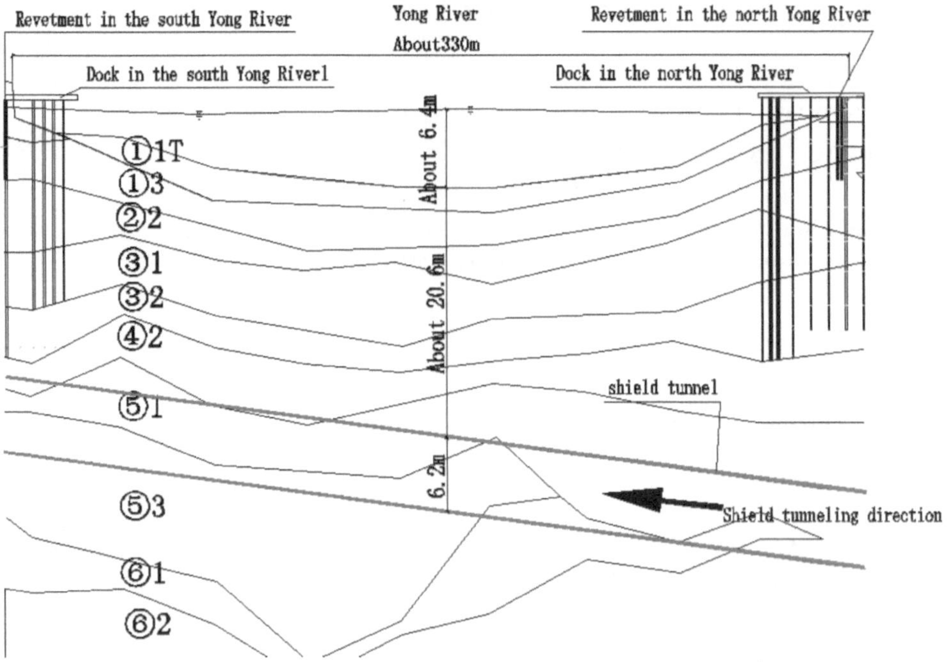

Figure 1. Chart for crossing Yong River.

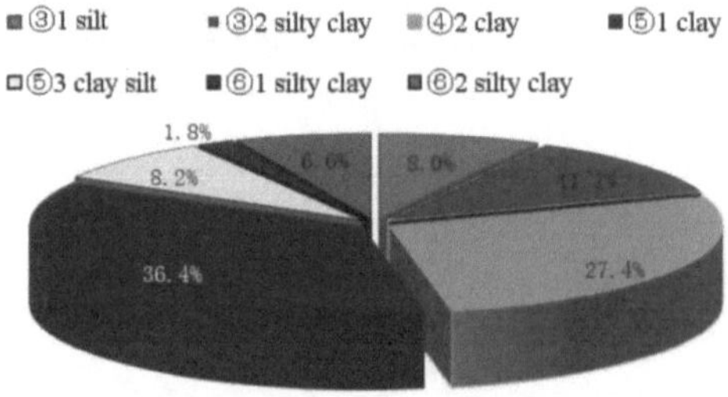

Figure 2. Distributing graph of the soil layer crossing the Yong River section.

1.2 *Engineering hydrology and geology*

The strata for tunnel crossing in the interval from Datong Bridge station to Zhongxing Bridge South Station is mainly composed of ②2 silty clay, ③1 silt, ③2 silty clay, ④2 clay, ⑤1 clay, ⑤3 clay silt, ⑥1 silty clay, and ⑥2 silty clay. The distributing graph of the soil layer crossing the Yong River section is shown below (see details in Fig. 2).

The soil layers including ②2 and ④2 have high water content, large porosity ratio, and poor permeability, presenting flow to soft plastic state, high compressibility, and low strength. Under the action of an external force, it is easy to be disturbed and the strength is reduced. It is very difficult to keep the balanced earth pressure in the shield tunneling and often early settlement may occur, and the settlement sees no shrinkage after the shield passes after a long time. The soil in ③1 and ⑤3 layers contains artesian water and it is prone to drift sand, water inflow and soil inflow during shield construction, and therefore water seal work is required. The ⑤1, ⑥1, and ⑥2 layers are silty clay of soft plasticity, having high viscosity, and are easily stuck to the shield equipment or cause pipeline

Table 1. Statistics of the physical and mechanical indexes of each stratum while crossing Yong River.

Rock no.	Rock name	Natural water content ω %	Gravity density γ kN/m³	Natural porosity ratio e	Degree of saturation Sr %	Specific gravity of solid matter Gs	Liquid limit ωL %	Plastic limit ωp %	Index of plasticity IP	Index of liquidity IL	Standard penetration blow count N (blow/30 cm)
②2	Silty clay	47.22	16.97	1.338	96.98	2.75	42.41	23.2	19.21	1.26	13.0
③1	Silt	26.24	19.1	0.756	94.59	2.71	28.11	20.32	7.79	0.73	12.6
③2	Silty clay	28.96	18.73	0.837	94.45	2.72	29.79	17.78	12.01	0.94	11.0
④2	Clay	40.61	17.5	1.168	95.7	2.75	42.92	22.98	19.94	0.89	
⑤1	Clay	29.31	19.07	0.825	97.14	2.75	44.22	23.39	20.83	0.3	
⑤3	Clay silt	28.85	18.94	0.807	96.92	2.71	29.54	21.51	8.02	0.91	24.2
⑥1	Silty clay	30.95	18.88	0.862	98.28	2.72	33.57	19.44	14.13	0.81	13.0
⑥2	Silty clay	29.75	18.79	0.842	96.1	2.72	32.81	19.75	13.06	0.75	

block, leading to idling of the cutter head and blocked rabbet and discharge pipe, resulting in upheaval or subsidence of the strata. The interval shield tunneling and cut surface is located on the stratum alternated between softness and hardness, possibly due to excessive soil removed from the soft stratum, the surface subsidence exceeds the expected and the shield axial deviation is too large. Therefore, during construction, it is essential to control the stability of the cut surface and to always keep the shield under good conditions. Statistics of the physical and mechanical indexes of each stratum while crossing Yong River is given in Table 1.

2 SELECTION OF THE SHIELD MACHINE

2.1 *Considerations for selection of the shield machine and risk analysis*

The selection is based on the information concerning important and difficult points in the interval as well as field hydrogeological conditions, with reference to existing shield project cases at home and abroad as well as shield-related specifications, experience accumulated in the construction of the Ningbo metro project, according to the principles of applicability, reliability, advancement and economic efficiency.

The risk analysis is conducted mainly because (1) shield damage is caused by lack of strength and stiffness of the shield shell; (2) it is prone to water inflow and mud and sand inflow, which caused the instability of the cut surface, if not controlled in time, it may cause serious catastrophic accidents; (3) high pressure may lead to seepage during use that affects tunnel operation.

2.2 *Selection of the shield machine*

In this project, two slurry-articulated earth pressure balance shield machines having a diameter of 6,340 mm are employed, which are composed of the shield shell, thrust unit, segment erector, cutter head device (at an opening rate of 44%), screw conveyor (2 hydraulic gates + 1 manual gate), power set, electrical equipment, piping and wiring device, frame, etc. The main performance parameters of the shield machine are given in Table 2.

Table 2. Main performance parameters of the shield machine.

Main component description	Parameter
Total length of the entire machine (including backup system)	67.005 m
Length of the main engine	9.055 m
Total weight	About 410t
Maximum tunneling speed	80 mm/min
Maximum thrust force	40018 kN
Digging diameter	Φ6360 mm
Form of articulation	Active articulation
Shield outer diameter, shield shell thickness	Φ6340 mm, 40 mm
Tail void	30 mm
Tail seal	3-run wire brush

3 COMPREHENSIVE CONSTRUCTION MANAGEMENT OF SHIELD MACHINE

3.1 *Parameter setting of the shield machine*

3.1.1 *Earth pressure*

$$\text{Earth pressure } P = k{*}k_0{*}\gamma{*}h$$

where P is the balanced pressure; γ is the average gravity of the soil; h is the buried depth of the tunnel; k_0 is the lateral static balanced pressure coefficient, the generally measured value is 0.6; k is the correction factor of 1.1~1.3.

During construction, the set earth pressure is 0.01~0.02 MPa, which is larger than the theoretical value and is real-time adjusted depending on the monitored data. The settling volume of the ground to control the notch position of the shield machine rises about 1 mm, reducing the settling volume in the later stage.

3.1.2 *Discharge volume*

The theoretical discharge volume per ring is $\pi D^2 L/4 = \pi \times 6.34^2 \times 1.2/4 = 37.883$ m³/ring. The shield intake and discharge volume is controlled within 98%~100%, namely 37.126 m³/ring~37.883 m³/ring. In the process of construction, the ground is allowed to produce micro-uplift, to prevent excessive settlement, and the discharge volume is controlled at 98% of the theoretical value; and the discharge volume per ring is strictly controlled in the construction process to ensure that there is no phenomenon of over excavation. The smaller the thickness of the earth backing, the more sensitive are the construction settlement responses to the construction; the smaller the proportion of the later settlement in the accumulative settlement, the shorter is the duration of the later settlement. In the case of the basically same geological conditions, the accumulative settlement volume of the shallowly buried buildings after a disturbance is about two times that of the deeply buried buildings.

3.1.3 *Tunneling speed and thrust force*

For uniform speed construction, the tunneling speed is real-time adjusted from 20 to 40 mm/min, to ensure constant speed of shield tunneling. To reduce the cutter head torque and shield thrust and to reduce the disturbance to the surrounding soil, the total thrust must be determined based on the buried depth of the tunnel, chamber earth pressure, friction between the shield shell and soil, friction between the shield tail and segment as well as necessary traction for backup systems, in order to ensure continuous and balanced construction and avoid long-term postponement.

3.1.4 *Synchronous grouting*

The synchronous grouting in shield tunneling is not only the main means to fill the construction gap between the soil and segment ring and reduce late deformation, but also a key process in the shield tunneling construction. The grout should be rapidly, uniformly and sufficiently poured to ensure that its construction gap is timely and sufficiently filled, so that the construction disturbance over the soil is minimized. The surface deformation and segment offset are controlled to a minimum and the segment joints are protected from water leakage.

The construction gap per ring is given by the following relation: $1.2\pi(6.36^2-6.2^2)/4 = 1.89$ m^3 (the outer diameter of the excavated shield cutter head is 6.34 m; the outer diameter of the segment is 6.2 m). The grouting volume per ring is generally 120%~150% of the construction gap; that is to say, the synchronous grouting volume for each ring advanced is (2.3~2.8)m^3.

According to the buried depth of the tunnel, the grouting pressure should be controlled at 0.35~0.45 MPa, and not more than 0.45 MPa. Synchronous grouting directly affects the control effect of ground settlement and is fundamental to control ground settlement, and a uniform aquifuge is formed around the segment after four points are grouted and cured. Therefore, the thick grout is used, with the grout proportion detailed in Table 3. Such a grout has a high yield at early pouring, and also, small compressibility and bleeding rate, which can effectively control the ground subsidence and tunnel uplift. The grouting volume and grouting position are determined under subjection to the pressure and monitoring data of stratum deformation at the time of grouting.

A person is appointed to record the grouting position, volume and pressure and to adjust them on time according to the monitoring information. The pouring system at the working surface is periodically cleaned to prevent pipe clogging. The representation for measuring basic parameters of the synchronous grout is slump, density and bleeding rate.

For shield uplift, under the premise of land subsidence, the grouting volume at the upper part is properly increased to control the proportion of upper and lower grouting at 3:1~4:1. The shield height is controlled at about −30 mm, to offset partial tunnel uplift.

3.1.5 *Secondary grouting*

When the shield passes through the riverbed, continuous grouting is required. Continuous grouting includes three parts, namely synchronous grouting, intermittent grouting, and secondary grouting. Intermittent grouting is for the last 6 to 8 rings at shield tail and the supplemental grouting before secondary grouting is difficult to be realized, with the purpose to solve the saltative subsidence of the segment after the shield tail. However, secondary disturbance over the same section is avoided possibly, for the settling volume that resulted from secondary disturbance is at least two times of that of the first disturbance. For the backfill secondary and supplementary grouting, the grout proportion is temporarily determined as given below (see details in Table 4).

Technically, the secondary grouting is the main means to control late ground settlement, so that the secondary grouting is arranged in time when the segment is dislocated for 6~8 rings. During secondary grouting, a grouting pump is used to inject the cement slurry and water glass between the segment and surrounding soil via segment lifting holes, and the grouting

Table 3. Grout proportion at each 1 m^3 synchronous grouting (weight ratio).

Anhydrous lime Kg	Fly ash kg	Sand kg	Bentonite kg	SK-6 (admixture) Kg	Water kg
80	300	1180	50	3	320

Table 4. Temporarily determined grout proportion for backfill secondary and supplementary grouting (weight ratio).

Liquid A		Liquid B
32.5 grade cement Kg	Water L	Water glass L
1000	1000	250

is controlled by pressure, with the pressure being adjusted from 0.4~0.6 MPa. According to the monitoring of land subsidence, when the ground daily settling volume exceeds 0.6 mm or the accumulative settlement reaches 60% of the set value, lose no time to have supplementary grouting, in multiple and repeated manner. Secondary grouting pressure, amount of slurry and grouting interval: the secondary grouting pressure is generally controlled at the groundwater soil pressure +0.05 MPa at the buried depth of the tunnel; the amount of each grouting is generally controlled below 0.7 m^3, and the grouting should be conducted at an interval after the segment dislocates from the tail for 6~8 rings, having 5~10 rings as a group.

3.2 *Shield construction under supervision*

According to the settlement value at various monitoring points, an analysis is conducted to check whether the chamber earth pressure, grouting pressure and grouting amount are reasonable; according to the results of the analysis, the point is accurately selected and the grout is reasonably poured to achieve the purpose of controlling the riverbed sedimentation rate and accumulative settlement volume.

3.2.1 *Riverbed deformation monitoring*

In the scope subject to settlement along the shield tunnel, the permitted surface subsidence value is +10~–30 mm and the special plot is controlled at +5~–10 mm. Ship borne ultrasound is used for monitoring. In the axial direction along the tunnel, the monitoring length is measured as 80 m (30 m before the shield notch to 50 m after the notch), each 2 m water area at the left and right-hand side above the axis. Each monitoring provides the measurement data for a total of 80 m that is 30 m before the notch along the axial direction and 50 m after the notch, and after comparison with the background curve or previous monitoring curve, it also provides a set of 80 m long comparison curves along the tunnel axis to be used by the construction unit. The data of the 15~20 m before the notch are saved as the background data, for comparison in the next monitoring. The accumulative settlement alarm value is ±20 cm and the single settlement alarm value is ±3 cm.

3.2.2 *Shield tunneling direction and shield attitude control*

The tunnel generally adopts the automatic measuring system, namely the equipment to real-time detect shield position (3D coordinates) and attitude (cross and vertical inclination, drop in the level). Measure and move the station backward, that is to say, with the shield advancing, reference point and temporary bench mark are successively set in the shield hole. Due to segment deformation and displacement subject to the shield thrust, the reference point set in the shield hole also has slight displacement. Therefore, regular inspection is required, especially at the curve segment and riverbed. The change of the shield attitude cannot be too large or too frequent, in order to reduce the disturbance to the soil. The maximum allowable value of lateral displacement is 50 mm and the maximum allowable value of vertical displacement is 25 mm. The allowable value of the rotation angle is –5°~5°.

4 TECHNICAL MEASURES FOR SHIELD MACHINE CONSTRUCTION

4.1 *Preparation before tunnel boring*

① Before the shield machine crosses the flood wall and pier, a detailed investigation should be conducted on its structure, form of pile foundation, and relationship with the tunnel, to disclose to construction personnel;
② Review the riverbed elevation and accurately calculate the thickness of the tunnel earth backing. Check the exact relative relationship between the shield machine and surface river course;
③ Test section tunneling: 30 rings before the shield is advanced to the river course, start trial boring construction parameter control and enhance surface monitoring. When the shield crosses the flood wall and pier, lose no time to adjust the setting earth pressure, in order

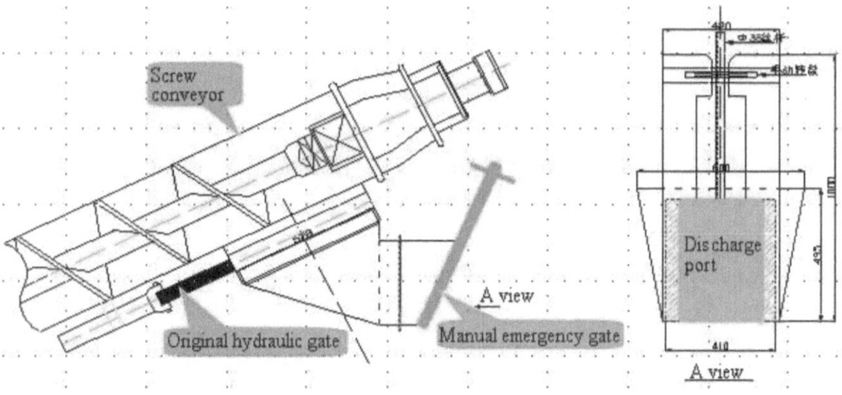

Figure 3. Manual gate diagram.

to reduce disturbance to the soil and protect its safety. In addition, when the shield crosses the river, the change of earth backing thickness and water level should be considered to set earth pressure.

④ When the screw conveyor is configured with two gates and starts construction across the river, once spewing occurs, lose no time to close the gates. The manual gate diagram is illustrated in Fig. 3.

⑤ In order to minimize the dwell time of the shield machine in the riverbed, before boring, a complete inspection, maintenance, and repair should be conducted over the main engine and backup system of the shield machine, and also sufficient wearing parts should be supplied.

4.2 *Measures against roof caving*

Due to sudden change of earth backing in the river-crossing section, the shield machine should have the erection pressure adjusted before arriving at Yong River, preventing blinded advancing due to excessive erection pressure leading to water penetration.

① In the boring process of the river-crossing section, the change of earth backing leads to pressure change, and so before boring, it is required to calculate the earth pressure of each ring of the river-crossing section, and gradually decrease the earth pressure subject to the monitoring data and discharge volume when the shield is advanced from the land area to water area, and gradually increase the earth pressure when the shield is advanced from the land area to water area.

② While crossing the river-crossing section, adjust the shield attitude and minimize deviation correction. The rule of frequent shield correction and little correction is followed and the single-ring slope correction should not exceed 1‰ and the plane correction should not exceed 2.5 mm.

4.3 *Measures against tail seepage*

The shield machine is employed with three shield tail brushes. The shield tail seal is improved by the use of high-quality grease filled at a fixed and sufficient amount; the sponge strip adhered to the segment and installation of the leak-proof steel plate at the tail. The segment wedge is controlled in advance to ensure the concentricity between the segment erected and advancing axis, to achieve uniform tail gap. Where necessary, seal the tail with PU hoop. Before erection, each segment should be longitudinally applied with anti-abrasive agent, to prevent waterproof materials from falling and breaking due to extrusion.

A pneumatic wrench is used to fasten the segment bolts, and retighten the 5-ring segment that is dislocating from the tail. The segment is made of waterproof material and putty is

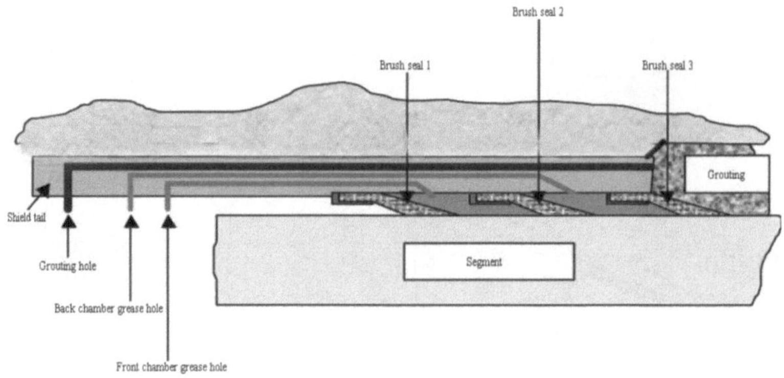

Figure 4. Schematic of the tail seal.

applied at the corner to seal. The waterproof material should be firmly pasted. The tail seal diagram is illustrated in Fig. 4.

4.4 *Tunnel quality control*

The quality of shield tunnel construction directly influences the service life of the tunnel. Great attention should be paid to the segment erection quality, shield attitude control, and other aspects. It is essential to make great efforts against following aspects including segment drop off, dislocation, water leakage, and large axial deviation:

4.4.1 *Segment quality control*

① Adjust shield condition, possibly control the plane and height difference between the notch and tail within 20 mm, neither fierce correction nor sharp correction is recommended, to ensure uniform gap around the tail. ② The segment should be selected to possibly avoid the capping block located within 90° of the arch bottom, ensure that the block L reaches such an aperture width so that it will not crack when the capping block is inserted. ③ Arrange an experienced operator to erect the segment and control the erection quality, especially the step among segments, and avoid any crack caused by uneven stress due to an excessive step among segments. ④ When the curve section is advanced, reasonably select wedge segments to ensure that the segment axis timely follows the change of the shield attitude. In case of any problem, the shield attitude should not be sharply corrected, but slowly corrected; follow relevant specifications to operate, including segment hoisting and erection, and control of shield parameters; it is required to take timely and effective measures to prevent the uplift of tunnel segments.

4.4.2 *Segment leakage*

The pipe making and erection quality should be controlled, because the failure itself, dislocation and water stop usually occur at the same time. For this reason, the quality requirements should be strictly controlled over manufacture, transport and erection of segments.

4.5 *Contingency control and countermeasures*

Information should be disclosed to construction personnel in advance to ensure careful construction, enhanced duty management and project monitoring; sufficient emergency materials should be supplied and used at the first time in case of any accident; designated personnel should be appointed for 24-h on-site monitoring; before advancing, the shield should be sufficiently tested to ensure the reliability of the shielding performance. And also, service personnel on duty for 24 hours should be equipped, for timely handling of any shield machine malfunction, so as to ensure the successful performance of the shield boring; when the shield machine crosses, if the ground deformation value reaches the alarm value, corresponding measures should be taken according to the specific situation.

5 CONCLUSION

The interval from Datong Bridge Station to Zhongxing Bridge South Station of TJ3101 Bid for Line 3 of Ningbo Rail Transit successfully crosses Yong River. Yong River is 330 m wide, having 6.4 m water depth, 20.6 m maximum buried depth, and a poor geologic environment of the soft soil. After risk analysis with regard to environmental geology, the Komatsu 6340 shield machine is selected. The parameters of the shield machine are managed in a comprehensive manner to select the optimum parameters. Lose no time to fill the construction gaps using synchronous grouting and secondary grouting, to minimize the subsidence. It develops detailed measures against spewing, roof caving, seepage, and uplift in advance. Finally, it safely crosses Yong River, resulting in good forming tunnel, controllable risk, and remarkable economic effect, which meets the policy of "no break", "no crack", "no uplift", "no subsidence", "no leakage", and "no seepage" developed by Ningbo Rail Transit, and works out the construction method for shield machines crossing rivers, which provides plentiful experience for similar work.

REFERENCES

GB 50157—2003 Code for design of metro [S]. Beijing: China Planning Press, 2003: 50–55.

Ningbo Rail Transit Co., Ltd., Certificate of conformity for tunnel shield mortar in the interval of TJ3101 bid of phase 1 project of Line 3 of Ningbo Rail Transit [R]. Ningbo 2016: 2–3.

Ningbo Rail Transit Co., Ltd., Construction scheme for interval of TJ3101 bid for Line 3 of Ningbo Rail Transit [G]. Ningbo 2015: 6–10.

Ningbo Rail Transit Co., Ltd., Construction technique for TJ3101 section of phase 1 project for Line 3 of Ningbo Rail Transit [G]. Ningbo 2014: 50–56.

Ningbo Rail Transit Co., Ltd., Design documentation for construction plan of phase 1 project of Line 3 of Ningbo Rail Transit [G], Ningbo 2014: 1–5.

Ningbo Rail Transit Co., Ltd., Reconnaissance report for the geotechnical engineering of KC302 section of phase 1 project of Line 3 of Ningbo Rail Transit [G]. Ningbo 2014: 17–26.

Innovation and Application of Engineering Technology – Oscar, Chen & Wang (Eds)

Research on the plane flange sealing structure with knife edge

H.X. An, Z.Y. Bi, X.G. Liu & X.Y. Wang
Institute of Systems Engineering, China Academy of Engineering Physics, Mianyang, China

ABSTRACT: The effectiveness of the sealing structure largely determines the reliability of the entire system. The knife edge sealing type has a simple structure and high sealing performance, which is widely used in the plane flange sealing structure of metal containers with high tightness requirements. The sealing structure is designed to be used at high temperatures and the sealing behavior of the knife edge under high-temperature conditions is investigated. The variations of these evaluation indexes of the sealing performance such as effective contact width and contact pressure are analyzed. This research provides references for the design of the knife edge sealing structure under high temperature conditions.

Keywords: Knife edge seal, high temperature condition, design parameters

1 INTRODUCTION

Sealing structures are typically used in many devices with high pressure, corrosive fluids, vacuum or in the presence of radiation. For some dangerous goods packaging, the pollution of the environment and humans would be caused if the packing cases leak. And so, the effectiveness of sealing is essential for accident-resistant packaging cases owing to the features of dangerous goods. Temperatures of some dangerous goods in packaging can reach hundreds of centigrade and therefore, the rubber sealing structure cannot be used in this condition. Knife edge sealing is a kind of a flat pad sealing structure with simple frame and high performance, which is widely used in metal packaging cases, with request of high vacuum and sealing, and the knife edge type metal seal flange and its optimum has been studied and found to be one of the most reliable metal seals (Matsuzaki 1992, Gu 2004, Zhang 1994, Xu 1993, Carpenter 1962).

Sealing performance and the design parameters of the knife edge sealing type have been studied in some accident-resistant packing cases (Li 2012, Li 2012). However, research on the sealing performance of the knife edge sealing at high temperatures is less. A model of the sealing structure and pressure is built under conditions of the packaging state. And then, the effective width of the metallic gasket is given to verify its ability at high temperatures.

2 KNIFE EDGE SEALING TECHNOLOGY

2.1 *Requirements of sealing*

For the packaging cases, the only port that passes through the outer containment vessel containment boundary is a vent port, through which air is evacuated form and argon is backfilled through the annulus between the outer containment vessel and the inner containment vessel and the interior of the inner containment vessel. The typical accident-resistant packaging cases typically adopts the double sealing structures of the rubber elastic ring and metal seal to meet the requirement of the leak rate of the whole packaging. However, temperature is essential for the sealing structure and its property for the packaging cases studied in this paper is at high temperature for the most of its transport time. Hence, the single metal seal type is chosen.

For gasket sealing, the fundamental cause of the leakage is the existence of an intermittent between the contact surfaces; the pressure difference and the concentration difference between sides of the contact surfaces are the driving forces of leakage (Gu 2004, Zhang 1994, Xu 1993). To reduce leakage, it is necessary to maximize the contact area. However, under the condition of high temperature, if materials of the gasket, the flange, the bolt, and the vessel are different, the difference of the linear expansion coefficient would be larger. When the temperature rises, an additional compression force would be exerted on the gasket, which makes it more compressed. When temperature drops, the bolt and the vessel are restored to the tightening state at room temperature, but the gasket cannot be restored and it will cause seal failure. Knife edge sealing can withstand strong stress; it can even be generated to perform diffusion welding and make the sealing more reliable.

2.2 *Knife edge sealing*

Knife edge sealing is a kind of a metal flat gasket sealing, shaping the contact surfaces with the blade and the metal sealing ring into a seal. It has advantages of a simple structure, high sealing degree, convenient operation, etc. Through the combination of a convex edge and a concave edge or combination of two convex edges, the knife edge sealing structure can generate a larger shear pressure and even generate the diffusion welding with large pressure, which makes the sealing more reliable (Gu 2004, Zhang 1994, Xu 1993).

According to the apex width of the knife edge, the contacting state of knife edge seals was divided into three types: penetration type, indentation type and intermediate type; and under pressure, nearly Meyer's hardness in the sealing surface layer, the optimum shape of the knife edge was the penetration type (Matsuzaki 1992, Feng 2008). And researchers found that the sharp blade shape can maintain a satisfactory sealing performance even under a smaller torque, but it is difficult to maintain its performance under high temperature conditions (Satoshi 2003). On the other hand, the gasket material of the knife edge sealing structure began to flow when the pressure exceeds the yield strength of the material during the pre-tensioning stage, a small clearance on the surface would be filled at the same time, so as to achieve a reliable sealing, and its application in an ultrahigh vacuum without baking and repeatedly polishing the flange (Van 1955).

3 PARAMETERS OF THE KNIFE EDGE SEALING STRUCTURE

Considering both the studies mentioned above and the packaging frame, the knife edge chosen in this study is a combination of two convex edges. The flange and knife edge structures are designed for their particular use. Figure 1 shows the structure of the flange and its knife edge.

Figure 2 shows the sketch of the whole structure studied in this paper. The whole structure consists of the head flange, the body flange, and the metal gasket. The upper and lower knife compress the metal gasket under the effect of a bolt pre-tightening force and this makes the material of the soft gasket surface embed the knife edge sealing surface and fill into the micro-leakage channel on the contact surface, so as to realize effective sealing. The notation D_{oh} is the outside diameter of the head flange and it is equal to the outside diameter of the body flange, D_{ih} is the inside diameter of the head flange, D_{ke} is the diameter of the knife edge, D_{ib} is the inside diameter of the body flange, t_{h1} is the thickness of the head flange, t_{h2} is the contact thickness of the head flange, and t_b is the thickness of the body flange.

Figure 3 shows the dimensions of the sealing structure, including the dimensions of the gasket, the knife edge, and parts of the flanges. The radius of the knife edge tip is 0.1 mm and the angle of the round apex is 70°.

Considering the fact that the packaging cases would be repeatedly used, a hard material was chosen for the flange and a soft material was chosen for the gasket. Silver can also be

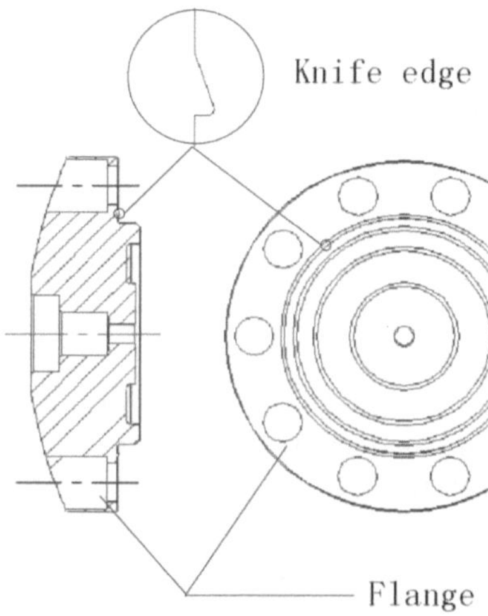

Figure 1. Structures of the flange and knife edge.

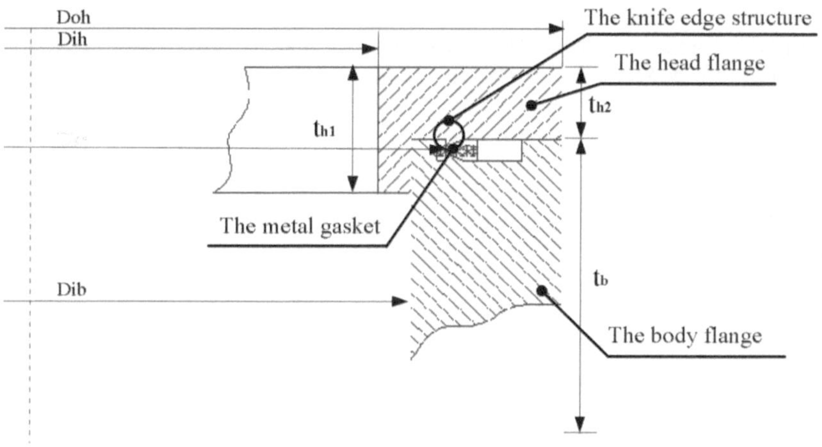

Figure 2. Sketch of the whole structure. (D_{oh} = 83.5 mm, D_{ih} = 44 mm, Dib = 52 mm, D_{ke} = 56 mm, t_{h1} = 28.9 mm, t_{h2} = 52.5 mm, and t_b = 52.5 mm).

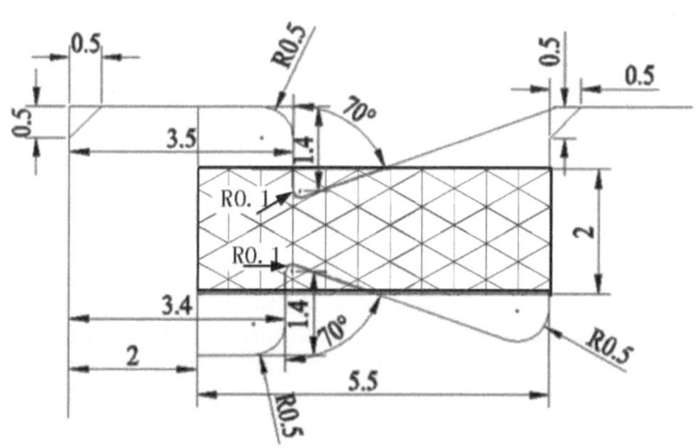

Figure 3. Dimensions of the sealing structure.

Table 1. The mechanical function parameters of the materials.

Material	E 105 MPa	Rm MPa	Re MPa	A %	ν	ρ g/cm^3	α $\times 10^{-5}$/°C
16Cr25 N	2	510	275	20	0.3	7.65	1.84
OFC	1.15	209	33.3	60	0.34	8.94	1.86

* E is the modulus of elasticity, Rm is the tensile strength, Re is the yield strength, A is the elongation after fracture, ν is Poission's ratio, ρ is the density and α is the coefficient of line-expansion.

used as plating on the edge tip in order to improve the life of the sealing ring. And then, the head flange and the body flange are made of 16Cr25 N heat-resistant stainless steel and the metal gasket is made of Oxygen-free Copper (OCF). To analyze the sealing performance, the mechanical function parameters of the materials are given in Table 1.

4 ANALYSES OF THE SEAL AREA AND KNIFE EDGE

4.1 *Request of the sealing performance*

To verify whether knife edge sealing structure works efficiently, the indexes used to evaluate the sealing performance should be given first. Research shows that there are many factors that affect the knife edge sealing performance and only the gasket coefficient and gasket pre seal pressure ratio values cannot completely describe the sealing performance of metal gaskets in this sealing structure. In this paper, the sealing surface is the contacting of metal to metal. To form an absolute seal in this structure, the contact surface should deform permanently by the contact pressure and the contact surface defects should be filled with the material generated from the plastic flow of the metal sealing contact surfaces. The requirements are as follows: the plastic deformation of the sealing surface and the contact pressure should be about two times the yield strength of the material; and the other is to form a closed seal ring, which meets the condition that the line contact width is greater than or equal to 0.5 mm at contact pressure[6, 9]. The criteria are given as expression 1 and expression 2:

$$CP \geq 2\,R_{p0.2} \quad (1)$$
$$CW \geq 0.5\text{ mm} \quad (2)$$

The notation CP is the minimum contact pressure of the sealing surface, which reflects the degree of compression, and the larger the CP's value, the compacter is the contact surface and also the higher is the sealing performance, Rp0.2 is the yield strength of the material of the gasket and CW is the line contact width of the sealing surface, which represents the ability of the sealing structure, CW's value is greater under the same load and this means that it is easier for the structure to achieve the sealing status and has a better sealing performance.

4.2 *Conditions*

The sealing structure is assembled at 20°C and the condition determines the initial status of the loading and location of the components. By calculating the whole packaging cases with goods, the highest temperature is 100°C. The sealing performance of the knife edge at high temperatures is concerned in the study, and so the working temperature of the sealing structure concerned ranges from 20°C to 100°C. The typical temperatures set in further calculation are as follows: the lowest temperature is 20°C and the highest temperature is 100°C. And so, the calculation is carried out under the two situations to confirm the performance of the sealing structure.

Figure 4 shows the schematic diagram of the loading model. In the structure, the sum of the bolts' pre-tightening press is 660 kN (Wen 2013), and the loading is displaced on the top

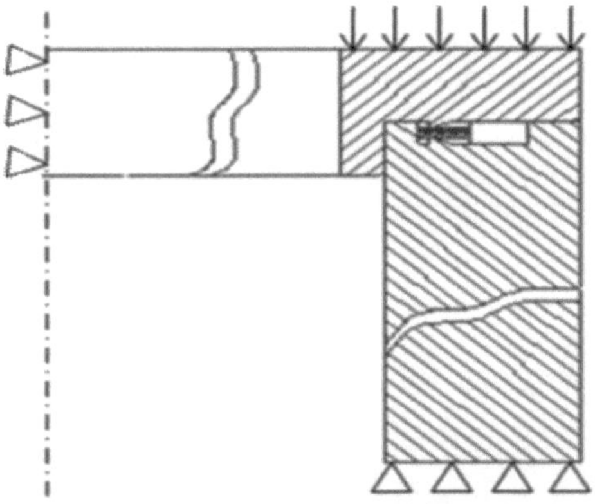

Figure 4. Schematic diagram of the loading model.

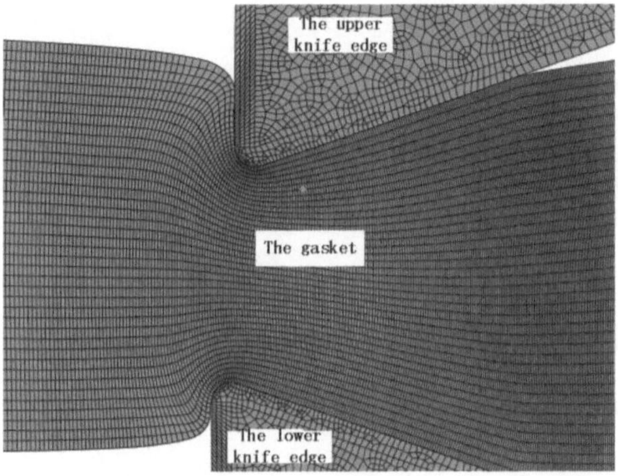

Figure 5. Deformation of the gasket.

of the head flange to exert a pre-tightening force in simulation. The radial motion of both flanges is constrained and the friction coefficient of each contact surface is 0.1.

4.3 *Results and Discussion*

The penetration depth of the flange into the gasket face is an important index of the sealing ability of the structure. And another important index is the contact press of the contact surface. The deformation of the gasket, contact press of its surface, and the line contact width are given below.

Figure 5 shows the deformation of the gasket after loading; the upper knife edge and the lower knife edge penetrated into the metal gasket and form an effective sealing. During the whole loading, the finite element meshes that are in contact with the edge are pressed heavily than others. Results confirm that the gasket is strong enough under such press at both 20°C and 100°C. Table 2 shows the efficient contact width of the gasket. The values are all above the criterion given before and the sealing performance at 100°C is better than that at 20°C.

Figure 6 shows the press on surfaces of the gasket. Although the contact width of the upper surface is larger than that of the lower surface, the contact press of the two surfaces is almost equal. The largest press emerges at the line that is pressed by the knife edge tip and the contact press also meets the request of the efficient sealing.

Table 2. Contact width of the gasket.

Temperature °C	CW of the upper surface Mm	CW of the lower surface Mm	Result
20	0.6	0.58	Efficiency
100	0.66	0.63	Efficiency

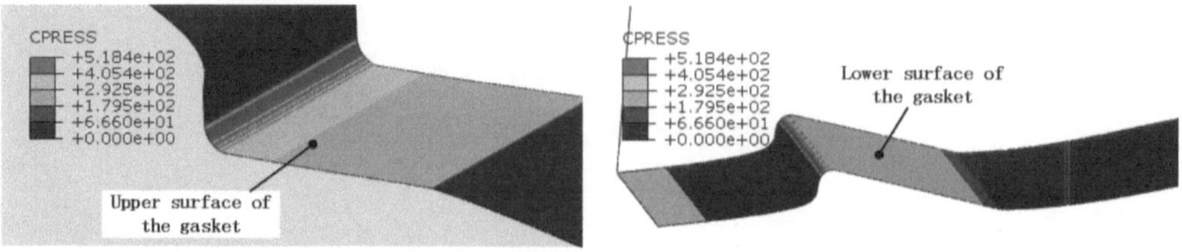

Figure 6. Contact press of the gasket.

5 CONCLUSION

Designing an effective sealing device has always been a main research focus of engineers in the sealing area. The effectiveness of the sealing structure largely determines the reliability of the entire system, and this seems especially important for dangerous goods packaging systems. The knife edge sealing structure consisting of the head flange and the body flange and a soft metal gasket is designed under high temperature conditions, and the contacting state of the knife edge sealing was simulated and calculated in this paper to verify its sealing performance. The sealing performance of the sealing structure was verified to meet the request, and it even shows a better performance under higher temperature. And the results provide references for the design of the knife edge sealing structure under high temperature conditions.

REFERENCES

Carpenter. R. (1962) A new version of the knife-edge vacuum seal. Measurement Science and Technology, 10(39), 532–533.

Feng Xiu & Wei Long. (2008) Finite Element Analysis for the Metallic Gasket Effective Width. Fluid Machinery, 09(36), 45–48.

Gu Bo-qin, Li Xin-hua & Tian Zheng, M. (2004) *Static Sealing Design Techology*. Beijing, China Standards Press. *(in Chinese)*

Li Jun-cheng, Zhang Si-cai, Zhang Fang-xiao. (2012) Optimization Design of Knife-edge Metal Sealing Structure Based on Response Surface Method. *Packing Engineering*, 5(33), 32–35. *(in Chinese)*

Li Jun-cheng. D. (2012) *Robust Design to the Plane Flange Sealing Structure with Knife-edge*. Institute of Structural Mechanics, China Academy of Engineering Physics.*(in Chinese)*

Matsuzaki. Y., Hosokawa. K. & Funabashi. K. (1992) Mechanism and optimum shape of knife edge for metal sealing. *Tribology International*, 6(25), 397–403.

Satoshi Kurokouchi, Satsuo Shinoda & Shinsaku Morita. (2003) Dependence of the Seal Property of ConFlat-type Flange on the Fine Dimensions of the Knife Edge. *J. Vac. Sci. Technol.*, A21(2), 438–448.

Van Heerden P.J. (1955) Metal Gaskets for Demountable Vacuum Systems. *Rev. Sci. Instrum.*, 26, 1130–1131.

Wen Bang-chun. etc. M. (2013) *Machine Design Handbook*. Beijing, China Machine Press. *(in Chinese)*

Xu Hao. M. (1993) *Sealing*. Beijing, China Metallurgical Press. *(in Chinese)*

Zhang Xiang-zhao & Shou Zhen-dong. M. (1994) *Sealing Gasket and Packing. Beijing*, China Machine Press. *(in Chinese)*

Innovation and Application of Engineering Technology – Oscar, Chen & Wang (Eds)
© 2017 Taylor & Francis Group, London, ISBN 978-1-138-05057-0

A top-down design method of a model-based definition and its application in an engineering system

Lei Wang, Xueliang Huang, Wenbin Tang & Lifan Wei
Institute of Systems Engineering, China Academy of Engineering Physics, Mianyang, China

ABSTRACT: There are a large number of iterative modifications in engineering systems design. The traditional serial design is accomplished by model accumulation and modifications, which may result in correlative modifications of both models and drawings, times without number, owing to the size change of some parts of the system. To achieve higher efficiency and quality of the design of engineering systems, the top-down method of the model-based definition was proposed, including the theoretical model of the top-down design method, design information transfer flow, control structure factors and the collaborative design method in a PDM system. An application of the rocket model design in this top-down method was given. The application validates that the top-down method can reduce unnecessary iterative modifications and can achieve greater efficiency in the design process.

Keywords: Top-down method; Collaborative design; Model-based definition; Control structure; Engineering system

1 INTRODUCTION

For complex engineering systems, there exist several rounds of structure optimization and model modifications in the design process. Yet, the traditional serial product design is completed by the continuous accumulation and modifications with the bottom-up modeling method. Mostly, the designers are involved in modifying plenty of the relevant models and drawings one-by-one, because of the change of a certain part's size parameters. Therefore, the repetitive labor of model modification accounts for more than 80% of the whole design work, which makes the development cycle long, the design changes difficult, and the design work repeated. As the engineering system's complexity increases and the function requirement enhances, it is necessary to develop a more effective and rapid method to improve the efficiency and quality of the design.

Now, the top-down design method is widely used in the mechanical modeling design, whose application can solve the problem of parallel design and collaboration work for a team. It is considered to accomplish the structure and function design from the level of integrity and the overall situation. In the structure design, it is first to consider the layout design. And then, divide the system into different sub-systems. And then, finish the design of each sub-system. It is a process from the abstract to the concrete, which is in conformity with the thinking mode, makes designers come closer to their potential, and reduces unnecessary and repetitive modification in the design phase. On account of the establishment of a reasonable top-down design planning, the upper and lower processes data and parallel data have established a relationship between part and part, parts and components, components and subsystems, subsystem and subsystem. They share characteristics by the publishing of control structures or skeleton models. When a certain parameter or characteristic changes, the corresponding update of the design can be automatically performed, thus shortening the design cycle and improving the accuracy.

In this paper, a top-down method of model-based definition and the related application are introduced, the theory model is studied, and research on the top-down method procedure is carried out based on NX software. And it is verified in a system engineering of the rocket model design.

2 RELATED WORK

For the large system engineering, it is important to realize the top-down design and parametric drive. The Boeing Company paved the new mode of digital design and manufacturing for aircrafts. It listed the top-down method as one of the top ten key technologies in the development of the Boeing 787 aircraft. They constructed the product's 3D models with both 3D product manufacture information and 3D design information. Early in the phase of concept design, it is available to control the interplay in relationships easily by making an associated design plan, dividing the aircraft into reasonable parts, and managing the hierarchy of the product's skeleton model. With the real-time online global collaboration environment, designers of different departments could achieve the explicit correlation information of the parts at the beginning of the design, to guarantee the consistency and integrity of the design. In the product design of the automobile industry, cooperative design technology has already been brought in. The product design is laid out by sharing information and related parameters in the top-down concept. For example, in the late 1980s, the concurrent engineering technology was carried out in three auto producers in the United States such as Ford Motor Company, General Motors, and Chrysler. They considered all the factors of the product life cycle at the beginning of the design and made the reliability of the downstream and technology as the constraint conditions of the design, in order to avoid or reduce the problems only found in the later stage of the design, which needed to back the early stage of design to do the modifications. In this way, they gained quantities of economic benefits.

3 THEORETICAL MODEL OF THE TOP-DOWN DESIGN METHOD

The theoretical model of the top-down design method is shown in Figure 1. We can consider it as a special form of the parametric design method, the driving parameters of which are the geometrical characteristics of the design at the above level, such as reference points, baselines,

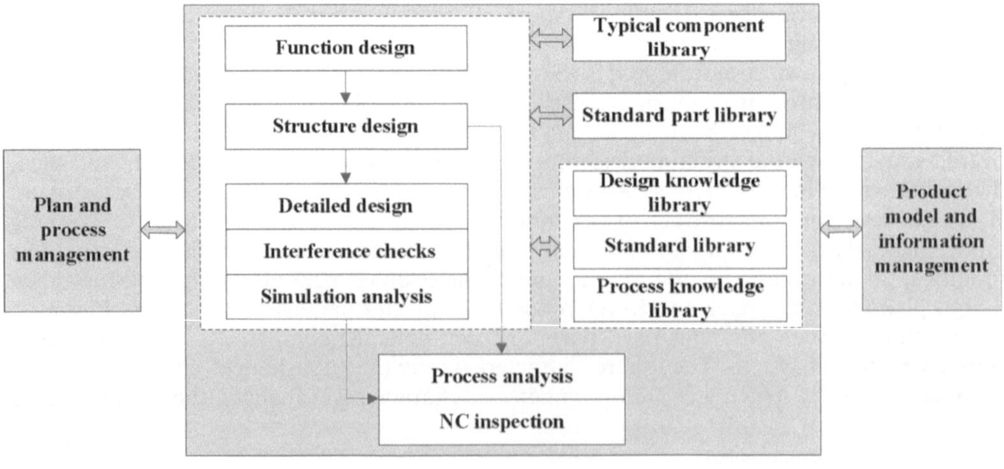

Figure 1. Theoretical model of the top-down design method.

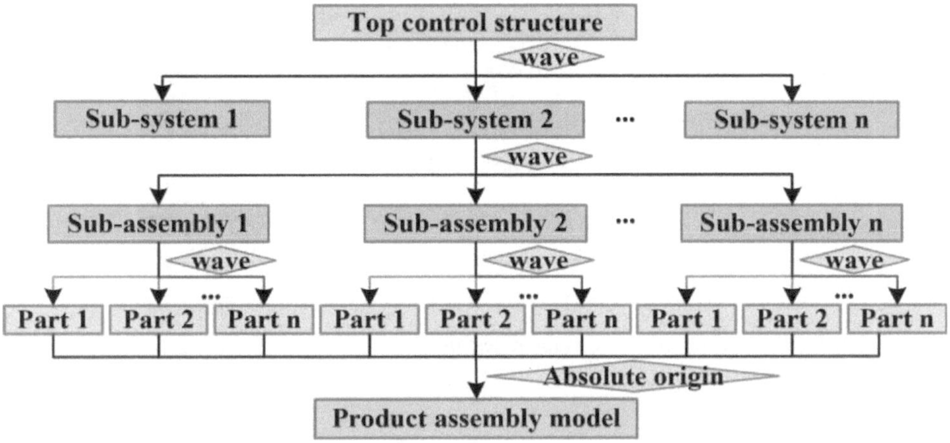

Figure 2. The top-down design method based on NX/WAVE.

datum, and coordinate systems. The relationship of the upstream design and the downstream design can be described by using Equation 1.

$$Y = \mathrm{f}\ (X) \tag{1}$$

where "X" stands for the input from the above level to the next level. "Y" stands for the output design at the next level based on the input at the above level. "f" stands for the relationship between the above level and the next level.

For large complex products, they are divided into "overall—subsystems—components—parts" levels decreasing step by step in NX software. Establish the control structure (also called as the skeleton model in some CAD software), and then establish the relationship of the system, sub-systems, components and parts by the WAVE technology, as shown in Figure 2.

The control structure of the system should have the following elements, including:

The datum. It consists of points, lines, surfaces, and the coordinate system. In the control structure, the datum plane and the datum line are used to describe the key size and geometry of products.

The geometric relation between components. It consists of an assembly geometric shape (the contour line of the product or the component) and the size parameters that can describe the geometry shape.

The driving parameters. It consists of indirect driving parameters (calculated by product features), direct driving parameters (determined by the assembly constraints of the components), and the interface parameters (the size of the components with assembly relationships.).

The associated expression of the size parameters.

4 WORKFLOW OF THE TOP-DOWN DESIGN METHOD

The workflow of top-down design method of the model-based design is shown in Figure 3. The application in engineering systems of this technology can make it possible to do collaboration teamwork of different units, to pass the design data in real time, and to synchronize the system and structure by update. The workflow includes the following steps:

- Build up the product structure tree in the Product Data Management (PDM) system.
- The chef designer creates the product's document folder, makes division of the system, and decides the team members and distributes the authority.
- Create the overall control structures in integration software of PDM and CAD.
- Function and layout design.

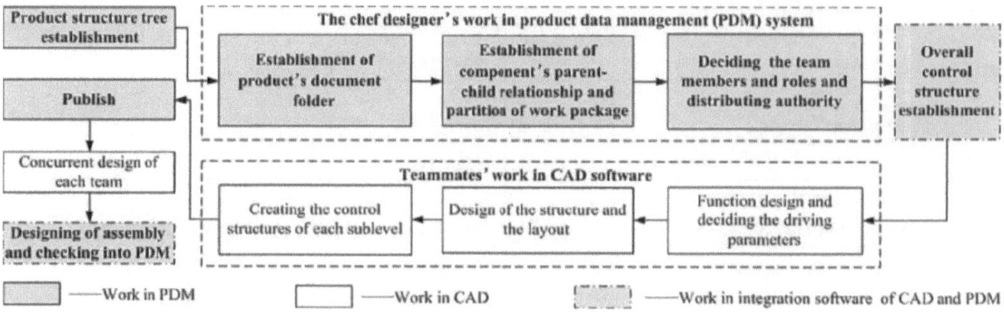

Figure 3. The basic workflow of the top-down design.

- Making publishment.
- Each work team does the detailed design in parallel.
- Make assembly design and check it into the PDM.

5 APPLICATION IN AN ENGINEERING SYSTEM

5.1 *Application object*

Take a small-scale model of the rocket as an example to validate the top-down design method based on the NX/WAVE. The rocket model can be divided into five parts including the shell body, the warhead, the instrumentation chamber, the engine, and the control surface according to the function of the system, as shown in Figure 4.

5.2 *Analysis of the key parameters and establishment of the product reference geometry*

According to the structure of the model, there are seven key parameters, including the length of the warhead, the length of the instrumentation chamber, the distance between the warhead and the instrumentation chamber, the length of the engine, the length of the control surface, the width of the control, and the length of the shell body (equal to the sum of the length of the warhead, the length of the instrumentation chamber, the distance between the warhead and the instrumentation chamber, and the length of the engine). Create the expressions of the parameters and use it to build the mode in order to make the sub-systems associated with each other and make the designers carry out modifications easily. Therefore, whatever kinds of the modifications are made, the model can gain the accurate parameters information after update. And then, create the reference geometry of the product, which consists of the absolute datum, the warhead top datum, the warhead base datum, the instrumentation chamber datum, the engine datum, the control surface's length datum, and the control surface's width datum, as shown in Figure 5.

5.3 *Creating the control structure*

Create the overall control structure, whose document name is "rocket_CS.prt" in NX. The detailed steps are as follows:

- Create the datum in an assembly environment according to the reference geometry relationship.
- Draw a contour line of the shell body in a drafting environment, which can be used to create the control structure of the sub-system.
- Mark the key dimensions on the overall control structure by using the PMI (Product Manufacturing Information) software module.

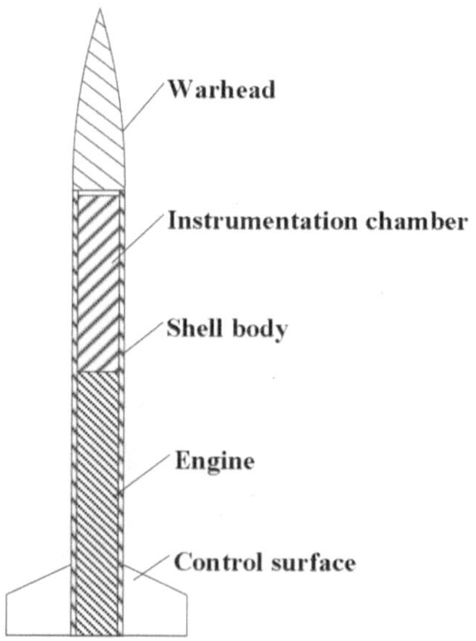

Figure 4. Diagram of the application object.

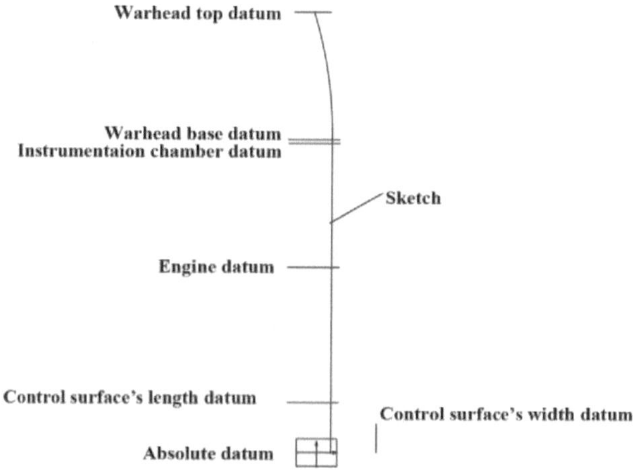

Figure 5. The reference geometry of the product.

Select the overall control structure in the assembly navigator, click the "new level" option of "WAVE", and create the warhead sub-system, the shell body sub-system, the engine sub-system, the instrumentation chamber sub-system, and the control surface sub-system by using the filter to select the reference information of each sub-system, as shown in Figure 6.

5.4 *Detailed design of sub-systems*

In the assembly navigator, choose the control structure of a certain sub-system as the "display component" and create the sub-system model with the key dimensions. Each sub-system is shown in Figure 7.

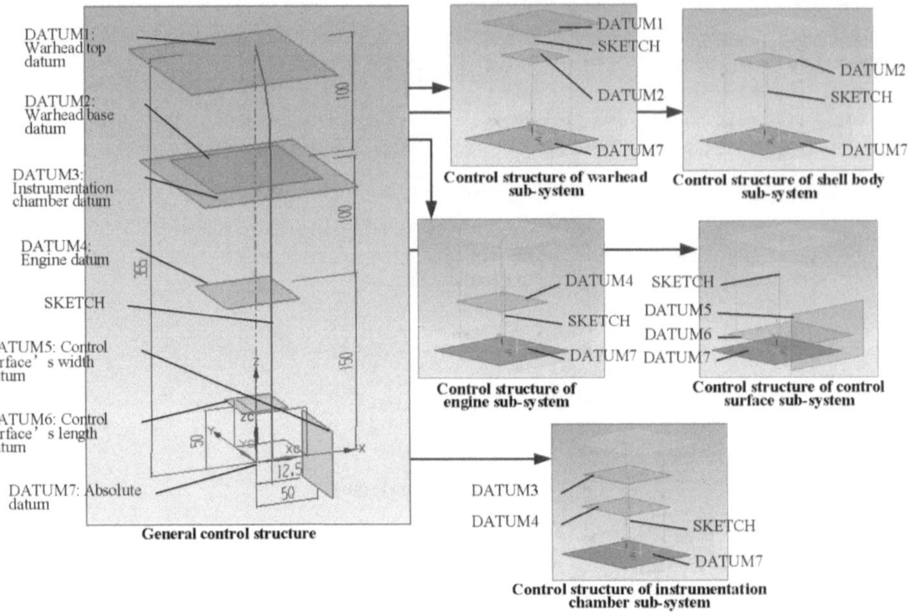

Figure 6. Control structures of all levels.

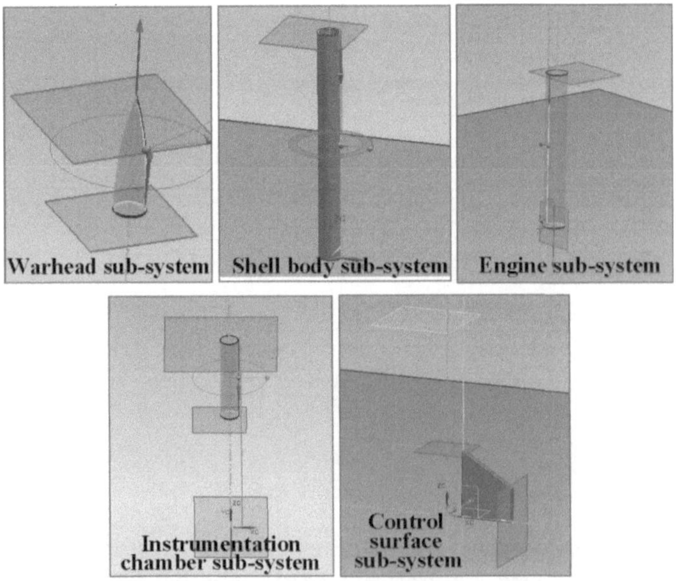

Figure 7. Detailed design of sub-systems.

5.5 *Creating of start parts and linked parts*

Create the start parts and linked parts. In the start parts, establish reference sets with the information passed from the parent start part to the child start part, as shown in Figure 8. And then, create different linked parts by setting the corresponding passing information. It is particularly applicable to the series of products development, such as the development of different grades of configuration for the same vehicles.

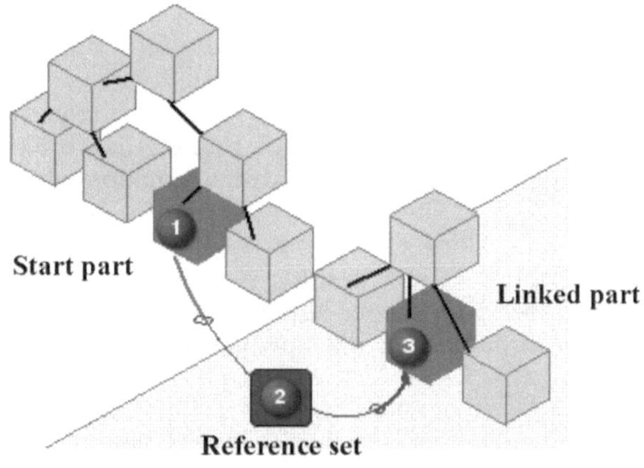

Figure 8. The relationship of start parts and linked parts.

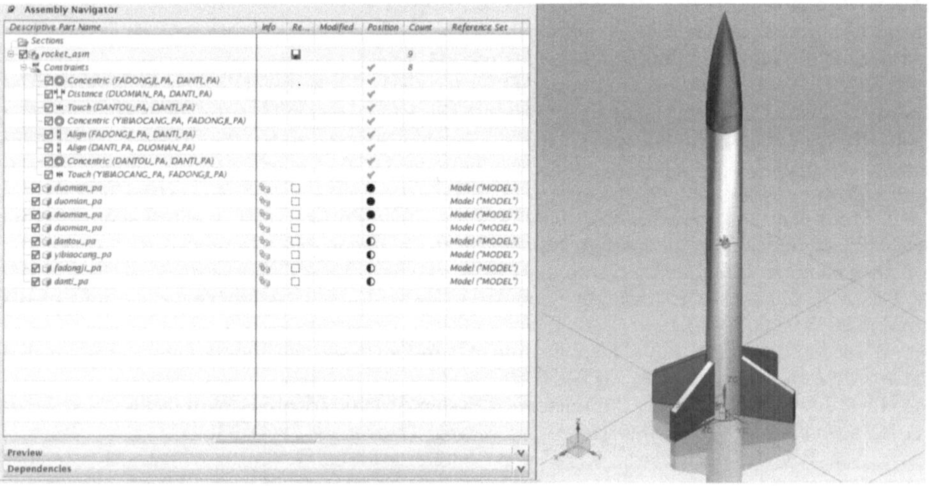

Figure 9. The assembly model and its information.

5.6 *Assembly design*

Create the assembly model of sub-systems with the constraint relationship, as shown in Figure 9. The assembly design will be modified automatically, as some important parameters in the overall control structure change.

6 CONCLUSIONS

In this paper, the recent status of the top-down design method of the model-based definition, theoretical model, and the implementation procedure is analyzed. WAVE technology based on NX software is introduced, including the steps of creating control structures, start parts, linked parts, and assembly designing. Take a small-scale model of the rocket as an applicational example, and it indicates that it is helpful to separate the layout design from the sub-system design premised on the guarantee of associations between them, to meet the need of overall and detail structure designs, and to provide a reference in the top-down collaborative method to the engineering systems.

For the top-down design method, however, there are still some problems in the actual collaborative modeling design, which needs to study in the future work. First, designers often consider more about the relationship between part's features that are usually established after modeling than the relationship between different parts and the relationship between parts and components, which will surely have a serious influence on the ultimate result of the product model. Second, according to the current application situation of the actual mechanical modeling in the top-down design method, a majority of the designers lack knowledge about the top-down method and skills. Third, for complex engineering systems, the CAD software can neither support the procedure of several sub-systems collaborative design, nor manage an enormous amount of model data, so that the deep integration and coordinated use of CAD and PDM that is good at publishing and managing data need to be further studied.

REFERENCES

Alemanni, M. & Destefanis, F. & Vezzetti, E. Dexin Chu, Xuening Chu, Yupeng Li, Guolin Lyu & Deyi Xue. (2016) A multi skeleton modelling approach based on top-down design and modular product design for development of complex product layouts, *Journal of Engineering Design*. [Online] 27 (10), 725–750, Available from: doi: 10.1080/09544828.2016.1227428.

Chunyan, Zhang & Chenyin, Che, & Qihua, MA. (2010) Application of NX WAVE in modeling of aircraft model. *Machinery Design & Manufacture,* (1), 91–93.

Demoly, F., X.T. Yan, B. Eynard, S. Gomes, and D. Kritsis. (2011) "Integrated Product Relationships Management: A Model to Enable Concurrent Product Design and Assembly Sequence Planning." *Journal of Engineering Design.* [Online] 23 (7): 544–561. Available from: doi: 10.1080/09544828.2011.629317.

Doran, D., and M.K. Starr. (2010) "Modular Production—a 45-Year-Old Concept." *International Journal of Operations & Production Management.* 30 (1): 7–19. doi: 10.1108/01443571011012352.

Duhwan, Muna & Jinsang, Hwang & Soonhung, Han. (2009) Protection of intellectual property based on a skeleton model in product design collaboration. *Computer-Aided Design*, 41(9), 641–648.

Gao, S., S. Zhang, X. Chen, and Y. Yang. (2013) "A Framework for Collaborative Top-Down Assembly Design". *Computers in Industry.* [Online] 64 (8): 967–983. Available from: doi: 10.1016/j.compind.2013.05.007.

Haipin, Liu & Jiangping, Rao. (2011) Application of Top-Down strategy to product serialization design. *Electro-Mechanical Engineering,* 27(5), 57–60.

Haitao, Li & Jianjun, Cai. (2011) Parameterized design of cylindrical in-volute helical gear by using UG-NX6. *Journal of Hebei University of Science and Technology*, 32(4), 330–332.

Quintana, V. & Rivest, L. & Pellerin, R. & Venne, F. & Kheddouci, F. (2010) Will model-based definition replace engineering drawings throughout the product lifecycle? A global perspective from aerospace industry. *Comput Ind,* 61(5), 497–508.

Xiang Chen, Shuming Gao, Youdong Yang, Shuting Zhang. (2012) Multi-level assembly model for top-down design of mechanical products. *Computer-Aided Design*. [Online] 44, 1033–1048, Available from: doi:10.1016/j.cad.2010.12.008.

Innovation and Application of Engineering Technology – Oscar, Chen & Wang (Eds)
© 2017 Taylor & Francis Group, London, ISBN 978-1-138-05057-0

Optimization analysis based on Ventsim of PG.4 ventilation in later reformation scheme

M.W. Xu
School of Civil and Environmental Engineering, University of Science and Technology Beijing, Beijing, China

L.Z. Jin
School of Civil and Environmental Engineering, University of Science and Technology Beijing, Beijing, China
Mine Emergency Technology Research Center, Beijing, China

ABSTRACT: On the basis of ventilation resistance determination PG.4, established 3D visualization model of PG.4 by Ventsim, the later stage of the mine's development scheme is analyzed one by one, which is presented in about the ventilation resistance changes, ventilation safety management feasibility analysis, economic rationality. Eventually, according to the results of the analysis to choose the best optimization scheme for late reconstruction scheme, which can save costs, improve mine safety management, accurate quantitative ventilation calculation, accelerate the digital mine construction.

1 INTRODUCTION

PG.4 began construction in 1955, and was put into operation in August 1958. As gas emissions and ground temperatures rise, the mine ventilation system becomes more and more complex. PG.4 is the transition from level two to level three, and forms a complex ventilation system with multiple intake shafts, air shafts, and levels. Due to the incomplete independent ventilation system, the ventilation network inevitably contains some diagonal branches, which greatly affects the airflow stability. Together with the mine exploit, this is bound to cause a huge waste of energy.

The mine roadway ventilation resistance is not reasonable. Because the coal mine contains a number of air intake and a return air shaft, the risk resistance capacity of the ventilation system is not strong. As a result, it is inevitable to produce some diagonal branches in the ventilation system, as the airflow stability is greatly threatened. In addition, with the downturn in the economic situation of coal, coal economy is experiencing difficulties, which brings severe challenges to the economic feasibility analysis of the mine exploit plans. In this paper, based on the collection of PG.4 ventilation parameters, the Ventsim software is used to establish a 3D visualization model of the entire mine ventilation system, and a simulation scheme is proposed.

2 DETERMINATION OF VENTILATION RESISTANCE

Combined with the location of production and the present situation of the ventilation system in PD.4, 4 longest airflow lines including large air volume, and the mining area or work face is chosen as the main line, which Can reflect the characteristics of the mine ventilation system, and other routes as auxiliary. The atmospheric pressure of the roadway branch is measured by the precision barometer method, and the wind speed, the section size and the weather condition of the roadway should be measured as well. The ventilation resistance between the two measuring points can be calculated using Equation 1.

$$h_r(i,i+1) = h_s(i,i+1) + h_z(i,i+1) + h_v(i,i+1) \tag{1}$$

where H_s represents static pressure difference, Pa; H_z represents the location pressure difference, Pa; H_v represents the dynamic pressure difference, Pa.

3 USING VENTSIM TO ESTABLISH THE 3D MODEL OF THE WHOLE MINE

3.1 *Theoretical analysis*

The software uses the Hardy-Cross iterative method to solve the ventilation network, which is based on the law of airflow balance, wind pressure law, wind resistance law and ventilation system attribute data. According to the initial air volume of each branch roadway in the ventilation network, the air volume increment ΔQk (correction value) of each airway is obtained, and the air volume of each branch roadway in the airway is corrected and iteratively calculated until the correction value ΔQk satisfies the given accuracy. For a ventilation network with node m and branch n, N = n − m + 1 residual tree branches and independent loop can be selected. Taking branch-piece volume as a variable, the tree branches volume can be expressed by the volume of the branch-piece. According to the law of wind pressure balance, each independent loop corresponds to an equation, which establishes a system of equations consisting of N variables and N equations. The root of the equations can be solved by finding the root of the equations, and then calculating the volume of the branches. After simplification, the general mathematical expression of the correction value of the volume is obtained in equation 2.

1. Fundamental theory of ventilation network calculation

The matrix equation of the energy balance equation and the matrix equation of the node air volume balance equation is as follows:

$$CR_{diag}\left|C^T Q_y\right|_{diag} C^T Q_y - CH_f - H_N = 0 \tag{2}$$

$$Q = C^T Q_y \tag{3}$$

2. Mathematical model of ventilation network solution

$$F\left(Q_y\right) = CR_{diag} \left| C^T Q_y \right|_{diag} C^T Q_y - CH_f - H_N = 0 \tag{4}$$

The iterative method can be used to solve the nonlinear equations. After K iterations, the approximate value of air volume is Q, and the (4) formula is expanded by Taylor series and a differential term of more than two times is obtained. It is assumed that the air volume is approximately Q_y after K iterations. Formula (4) is expanded by Taylor series and the differential term of more than two times is ignored.

$$F(Q_y^{k+1}) = F(Q_y^k) + \frac{\partial F}{\partial Q_y}\Big|_{Q_y=Q_y^k} \Delta Q_y^k = 0 \tag{5}$$

$$\Delta Q_y^k = -\left(\frac{\partial F}{\partial Q_y}\right)^{-1}_{Q_y=Q_y^k} * F(Q_y^k) \tag{6}$$

$\Delta Q_y^k = \Delta Q_y^{k+1} - Q_y^k$ is an independent loop air volume correction vector.

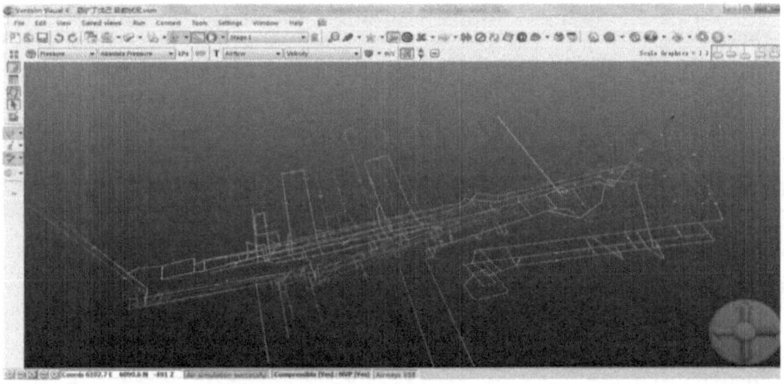

Figure 1. Mine ventilation model of PCD NO. 4.

$$\frac{\partial F}{\partial Q_y} = \begin{Bmatrix} \frac{\partial f_1}{\partial q_{y1}} & \frac{\partial f_1}{\partial q_{y2}} & \cdots & \frac{\partial f_1}{\partial q_{y_N}} \\ \frac{\partial f_2}{\partial q_{y1}} & \frac{\partial f_2}{\partial q_{y2}} & \cdots & \frac{\partial f_2}{\partial q_{yN}} \\ \vdots & \vdots & & \vdots \\ \frac{\partial f_N}{\partial q_{y1}} & \frac{\partial f_2}{\partial q_{y1}} & \cdots & \frac{\partial f_N}{\partial q_{yN}} \end{Bmatrix}_{N \bullet N} \tag{7}$$

$$\frac{\partial F}{\partial Q_y} = 2CR_{diag} \mid C^T Q_y \mid_{diag} C^T - C\frac{\partial H_f}{\partial Q_y} - \frac{\partial H_N}{\partial Q_y} \tag{8}$$

The air volume can be calculated by the following k + 1 times iteration.

$$Q_y^{k+1} = Q_y^k + \Delta Q_y^k \tag{9}$$

3. Check that the accuracy meets the requirements.

$$\Delta \mathrm{Q} = -\frac{\sum \mathrm{R}_i Q_i^2 \mp H_v \mp H_n}{2\sum |R_i Q_i|} \tag{10}$$

According to the formula (2), the air volume correction value of each loop is obtained, and the branch air quantity of each loop is corrected as follows:

$$\mathrm{Q}'_{\mathrm{ij}} = \mathrm{Q}_{\mathrm{ij}} \pm \Delta Q_i \tag{11}$$

3.2 *Error analysis of mine ventilation visualization simulation*

Whether the result of the mine ventilation system visualization is consistent with the predicted situation is key to mine optimization design. Only with the actual situation of the mine as much as possible, the later simulation of the mine optimization program is feasible. So it is necessary to verify the simulation results. In view of the PG.4 underground air volume data, the actual air volume and the simulated air volume are compared. Only if the simulation results are consistent with the actual situation of the mine, is mine optimization scheme feasible. Therefore, it is necessary to verify the simulation results. The research group compared the actual air volume and the simulated air volume in Table 1.

Table 1. Comparison of air quantity data for ventilation section.

Tunnel name	Measured air volume (m^3/min)	Simulated air volume (m^3/min)	Deviation rate (%)
East air shaft	1150	1146	8.35
Auxiliary shaft	5993	5910	1.38
1 level main shaft	1946	1892	2.77
3 level inclined shaft	4153	21.8	8.36
Ding-7 haulage-dip	964	957	0.73
Ji Wu connecting tune	825	813	1.45
Ding Wu cross drift	2568	2501	2.61
Nan cross drift	1883	1915	1.70
Wu910—19130 working face	1478	1517	2.64
Ding56—19170 working face	1546	1590	2.85
Ji15—23130 working face	1484	1520	2.43

The simulation results show that when the mine ventilation system is running stably, the deviation rate of the simulated air volume and the measured air volume is less than 10%. Therefore, the three-dimensional model can be used as a model of the PG.4.

4 OPTIMIZATION SCHEME AND ANALYSIS

4.1 *Put forward 3 sets of optimization schemes, and carried out simulation*

A: Exploit Geng-1 mining area and 2th level draft undertake the return airflow task of Geng-1

B: Exploit Geng-1 mining area, and add Geng-1 special return roadway. Exhaust dirty air to the 3th level, and stop 2th shaft fan, and let it be downcast.

C: Connect Wu-9 extension return airway with Ding-9 total return air roadway. Level 2th return draft independently undertake the dirty air task of Ji-3 mining area. Close the mining area of Ding-9 and Geng-1

Using Ventsim model to simulate the air pressure of tunnel is shown as Table 2.

After the simulation and analysis, the scheme of three haulage-dip airway speed is exceeded at Ji-3 area. The wind speed at the 2th level is exceeded in Scheme 2. The resistance can be reduced by 400 pa in solution 3. Through the comparative analysis of wind pressure, solution 3 has an advantage in reducing ventilation resistance. Taking PG.4's exploitation and mine planning at present into consideration, Ding-9 mining area would be finishing in few months. Intake tunnel of Ding Group and Ding nine return roadway would be closed. At this time, the air shaft of 1th level will be stopped, and it undertakes the task of return air of Ding-9 area.

Group G coal area is rich in sulfur coal gas; flammable. Group G area belongs to the high content of outburst coal seam. Considering the current situation of coal and mining cost, Geng coal group will be closed. Therefore, it is recommended to close the Geng group coal.

Wu-9 air return line is too long, and the ventilation resistance is too large. Let the Ding-9 extension and the total return of nine link up. It can greatly reduce ventilation efficiency. Considering the comprehensive factors, the reference solution 3 would be suggested.

4.2 *AHP analysis*

AHP is used to judge the relative importance of each factor in each level. The judgment matrix is used to quantify the scale of judgment. Quantitative relationship determination is used to express the degree of relative importance of the previous level factors. Making the

Figure 2. Solution B (This picture comes from Ventsim).

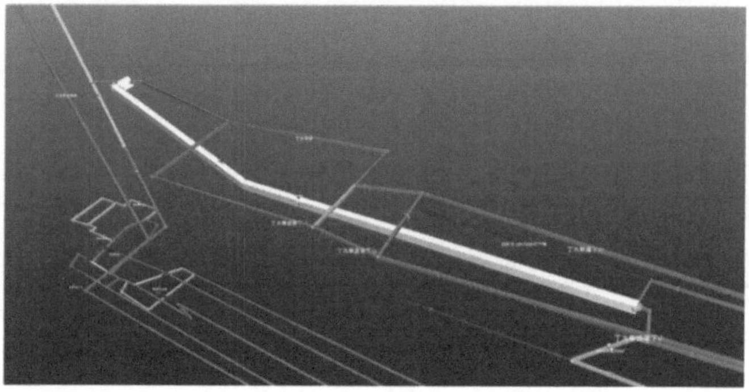

Figure 3. Solution C (This picture comes from Ventsim).

Table 2. Main roadway air volume.

Solution	Fan type	Air volume (m³/min)	Negative pressure	Main fan power (Kw/h)
1	AGF606-2.2-1.3-2	4920	2096.7	448
	YR1000-6/1180	3024	2922.2	405
	FBCDZ-10-№38	11870	4353.15	446
2	AGF606-2.2-1.3-2	8346	3678.8	534
	FBCDZ-10-№38	12600	4278.3	457
3	YR1000-6/1180	888	456	101
	FBCDZ-10-№38	21960	6207	376

hierarchy of the components in the system is the first step. This process requires a full analysis of the target. Thus, a hierarchy analysis model is made up of the elements according to the structure of the highest level, the middle level and the lowest level, which include security, technology, and economy. The AHP model is established as shown in the following figure.

According to the weight distribution of each single index to the total target and expert scoring of single factor, comprehensive evaluation is done:

$$S = 100\sum_{i=1}^{3} a_i f_i \tag{12}$$

Based on the results of the calculation, we can see that the comprehensive evaluation score of scheme three is the highest, which is the best solution.

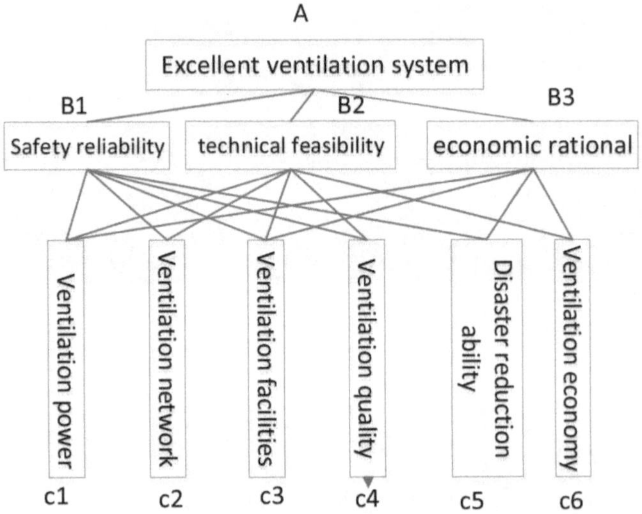

Figure 4. AHP structural analysis model of mine ventilation system.

Table 3. Computing hierarchical Sort weights.

C	B	B1, B2, ……, Bm a1, a2, ……, am	C Hierarchy sorting
C1		K11, K12, ……, K_{1m}	
C2		K21, K22, ……, K_{2m}	
……		…… ……… …	
Cn		Kn1, Kn2, ……, K_{nm}	

5 SUMMARY

According to the results of Ventsim simulation, it can be used to predict the ventilation in the scheme. This provides a high degree of credibility of the program based on the evaluation, but also to save cost. The scheme can be displayed in 3D visualization. At the same time, the air volume and the wind speed are clearly displayed. The performance of the fan can also be directly displayed, which greatly facilitates the optimization of the retrofit scheme. According to the simulation results, the scheme is in accord with mine ventilation safety management. The economic index is reasonable and the quantity of the project is small, which is in line with the long-term planning of the mine.

ACKNOWLEDGEMENTS

This study was supported by The National Key technologies Research & Development Program (No. 2016YFC0801706) and National Natural Science Foundation of China (No. 51574017).

REFERENCES

Bi Wenguang, Zhu Fang et al. Simulation technology of 3D ventilation in Coal Mine Based on Ventsim system. *Coal mine safety*, 2011 (2): 31–33.

Li Zijun, Chen Yanli. Metal mine parameter optimization of mine ventilation resistance, *Ventsim*, 2014 (3): 136–140.

Liu Zhaosheng, Zhang Hua, Zhou Yu, et al. Simulation of ventilation system based on Ventsim Zhang Zhuang iron mine. *Modern mining industry*, 2015 (11): 169–172.
Luo Zhenjiang, and other applications of Ventsim software in mine ventilation. *Nonferrous Metallurgy Design and research*, 2009 (3): 7–9.
Wang is outstanding. Reconstruction of the old mine ventilation system based on Ventsim—a case of. *Tongxing mining (Mining Department of nonferrous metals)* 2013 (1): 87–89.
Xu Xiangyu, Hou Zhenhai et al. Study on coal mine ventilation system based on Ventsim system. *Coal technology*, 2015 (9): 216–218.
Zhang Cambodia, Zhang Xiwei. Design of Ventsim software of mine ventilation system optimization based on non ferrous mining and metallurgy, 2013 (6): 7–9.
Zhang. Panjia Pingdingshan thirteen mine ventilation system optimization. Inner Mongolia, *Inner Mongolia University of Science and Technology*, 2014.
Zhu min. Research on Optimization of ventilation system in Wang Ping mine. *Liaoning, Liaoning Technical University*, 2011.

Innovation and Application of Engineering Technology – Oscar, Chen & Wang (Eds)
© 2017 Taylor & Francis Group, London, ISBN 978-1-138-05057-0

Microgrid network scheduling using a binary artificial bee colony algorithm

Yichuan Shao, Liwei Tian, Ji Zhang & Xin Li
College of Information Science and Engineering, Shenyang University, Shenyang, China

ABSTRACT: This paper aims to use a successful swarm intelligence technique called Artificial Bee Colony (ABC) algorithm to minimize both the Micro source-to-source interference and total system transaction time in Microgrid networks. As the Microgrid network scheduling model formulated in this work is a discrete problem, a Binary version of Artificial Bee Colony (BABC) algorithm is proposed in this study. Numerical results for four test cases with different scales, which ranging from 10 to 200 Micro sources, have been presented to demonstrate the performance of the proposed methodology.

1 INTRODUCTION

Mark Weiser in his famous article describes a vision of ubiquitous computing where a number of computers and sensors are deeply integrated into our daily lives and users can access information at any time and in any place (Weiser 1991). Microgrid is being developed as an essential elemental technology for realizing this vision (Bashan et al 2013 and Asztalos et al. 2014). However, there also some weaknesses associated with the deployment of the Space-based application (Brummitt et al. 2012 and Bashan et al. 2013). Particularly in a large-scale Microgrid network environment, where several Micro sources try to interrogate load at the same time in the same area (Asztalos 2012), the phenomenon of an unacceptable level of misreads occurs. In this paper, the main feature of our approach is that the Micro sources interference is solved by Microgrid network scheduling (Korkali et al. 2014).

Recently, by modeling the cooperative foraging behaviors of honey bee colony, the artificial bee colony (ABC) algorithm is developed by Karaboga16 ABC is a population-based SI optimization tool, which could be implemented and applied easily to solve multi-variable and multi-modal continuous functions. In this paper, a novel binary version of ABC (BABC) is proposed for solving the Micro sources interference problem in Microgrid networks. In BABC, we develop a differential expression, in which the relevant variables are interpreted in terms of changes of probabilities that changed per iteration. The main feature of this new operator is that it works in binary space, while still maintains the major characteristics of the original ABC's expression. In addition, in order to restrict the bees' positions within the range [0,1], the position clipping boundary condition (PCBC) strategy is applied in the proposed algorithm.

Experiments are performed on four Microgrid network scheduling cases with different scales. In the experiments, a comprehensive comparative study on the performances of the proposed BABC and two well-known evolutionary and swarm-based algorithms, namely the binary genetic algorithm (BGA) (Korkali et al. 2014) and particle swarm optimization (BPSO) (Kennedy and Eberhart 1997), is presented. Results show that the performance of the BABC is better than or similar to those of BGA and BPSO with the advantage of high convergence speed and precision.

The rest of this paper is organized as follows. Section 2 describes the problem of Microgrid networks scheduling. Section 3 describes the proposed BABC algorithm, including the discretized update formulas of employer, onlooker and scout, and the algorithmic framework. Section 4 provides the detailed design algorithm of optimal scheduling for Microgrid networks using BABC. Experiment results and discussion are presented in section 5. Finally, conclusions are drawn in Section 6.

2 MICROGRID NETWORK SCHEDULING PROBLEM

Given a Microgrid network laid out in some manner, we can construct the associated collision graph $G=(V, E)$ where each vertex $v\in V$ corresponds to a Microgrid Micro sources and each edge $e \in E$ indicates that those two Micro sources can be operated in parallel (i.e., there are no collisions between these two readers). It should be note that the graph partitioning problems (GPP)[5] is to partition a graph G into k subgraphs such that the number of edges connecting nodes in different subgraphs is minimized, and the number of edges connecting nodes of the same subgraph is maximized. The frequency allocation problem for networks of Microgrid Micro sources is to allocate frequencies to various readers. That is, when two Micro sources lie in each other's interference region, they are given different frequencies. Clearly this problem can be reduces to GPP model.

In our Microgrid network scheduling model, there are two objectives, namely finding the optimal structure of Micro sources and scheduling the Micro sources to reduce the total system processing time. The first objective is in order to avoid Micro sources interference or collision. At this point, this problem looks like the frequency allocation problem, except that the allocation is done along the time axis. Interfering Micro sources are allotted non-overlapping periods of time so as to avoid collision between them. Obviously, the second objective is in order to confirm the efficiency of the Microgrid system by minimizing the total processing time of the Microgrid networks. Then in each time step t, the solution variable is therefore given by $s^t = (s^t_1, s^t_2, \ldots, s^t_i)$ as a partition of the Microgrid network. Here $n(t)$ denotes the number of Micro sources that waiting for scheduling in this time step. Each element $s^t\in[0, 1]$, $i = 1, 2,\ldots, n(t)$, in the solution vector is corresponding to the presence or absence of the ith reader. That is, a bit "0" in a solution vector indicated the absence of the corresponding reader, while a bit "1" means the reader's presence. Then in each time step t, the Microgrid network scheduling model can be formulated as a discrete optimization problem that is given in what follows:

$$\begin{aligned} Max\, f\left(s^t\right) = {} & w_1 f_\alpha\left(s^t\right) + w_2 f_\beta\left(s^t\right) \\ & + w_3 f_\gamma\left(s^t\right) + w_4 f_\lambda\left(s^t\right) - \eta f_p\left(s^t\right) \end{aligned} \tag{1}$$

where w_1, w_2, w_3, w_4 are the weights given to each term of the fitness function and η is the punishment coefficient; f_a is the function to select the max transaction time of the solution partition; f_β is the function to calculate the number of Micro sources in the time step of this solution; f_r is the sum function of all the possible collision which the members of the partition have with the Micro sources still remaining to be partitioned in the Microgrid network. Intuitively, this makes sense since removal of the partition leaves a lot of scope for further formation of partitions in remaining nodes and would not cause much loss of edges in the graph (Microgrid network); f_λ is the weight function attached to this partition of Micro sources that is assigned to the average of the transaction times of the Microgrid forming the partition; f_p is the punish function that enable the searching algorithm exclude the illegitimate partitions that contain colliding readers.

3 BINARY VERSION OF THE ABC ALGORITHM

The implementation of novel binary ABC algorithm consists of employed bee, onlooker bee and scout bee phases.

3.1 *Employed bee phase*

After all the bees return to the hive with a certain amount of nectar, the first half (SN) that found the best food sources become "employed bees." At this stage, each employed bee xi generates a new food source vi in the neighborhood of its present position as follows.

$$v_{ij}(t) = x_{ij}(t-1) + \phi_{ij}\left(x_{kj}(t-1) - x_{ij}(t-1)\right) \\ for\ j = 1, 2, \ldots, D \tag{2}$$

where $k \in (1, 2, \ldots, SN)$ and $j \in (1, 2, \ldots, D)$ are randomly chosen indexes, and $k \neq i$. ψ_{ij} is a random number generated in the range (-1, 1). In sum, the employed bee formula remains unchanged in the proposed binary ABC, except that now x_{ij} and x_{kj} are integers in {0, 1} and each bit of xi will be updated simultaneously.

In order to restrict the bees' positions within the range [0,1], A normalization method and a threshold level has to be introduced to map all real valued numbers of vij to the range [0,1]. The round function and the position clipping boundary condition (PCBC) can be used to accomplish this last modification. The resulting change in position then is defined by the following rule:

$$\begin{aligned} & x_{ij}^{e}(t) = round\left(v_{ij}(t)\right) \\ \textbf{PCBC:}\quad & \text{if } x_{ij}^{e}(t) \geq UB, \text{then } x_{ij}^{e}(t) = UB \\ & \text{else if } x_{ij}^{e}(t) \leq LB, \text{then } x_{ij}^{e}(t) = LB \end{aligned} \tag{3}$$

3.2 *Onlooker bee phase*

The remainder of the bees ("onlooker bees") watches the waggle dance to decide which of the employed bees should be followed. An onlooker bee selects a food source found by the employed bee xi depending on its probability value pi calculated by the following expression:

$$p_i = {fit_i}/{SN} \tag{4}$$

where f_i is the nectar amount (i.e., the fitness value) of the ith food source.

3.3 *Scout bee phase*

If a position cannot be improved further through a predetermined number of cycles called "limit", then the food source is assumed to be abandoned, and the corresponding employed bee becomes a scout that randomly re-initializes in the fitness landscape.

4 MICROGRID NETWORK SCEDULING PROCEDURE BASED ON BABC

The detailed design of Microgrid network scheduling algorithm based on BABC is introduced in this section. The algorithm design reflects a multi-phase searching process that can be described as follows:

1. Initialization Phase

a. Micro sources Specification

This gives the details of the mobile Microgrid Micro sources that include the according interrogation range—the distance up to which a load can be read by the reader; the interference range—the distance within which if two Micro sources transmit simultaneously their signals would interfere; the process time; and the number of the mobile Micro sources to be used.

b. Topology Specification

This gives the details of the working area to be covered by Microgrid network according to the application scenario. It includes the shape and dimension of the region; the Microgrid network distribution (i.e., the Micro sources position) in the working area; and then the collision graph according to the readers' layout, interrogation and interference range.

c. Population Generation
N individuals forming the BABC population should be randomly generated. The ith artificial bee of the tth time step is defined as follows:

$$X_i^t = \left(x_{i1}^t, x_{i2}^t, \ldots, x_{in(t)}^t\right), x_{ij}^t \in [0,1] \tag{5}$$

In our work the direct encoding scheme is applied to encode the individuals. The possible solutions are represented as an artificial bee with dynamic dimension according to different time steps (i.e., solution vector dimension $n(t)$ = the number of Micro sources in t^{th} time step). Each element x^t_{ij} in the dimension is corresponding to the absence of the jth Micro sources that can only be "0" or "1". A bit "0" in an individual indicated the absence of the corresponding reader. Otherwise a bit "1" in an individual indicated the presence of the corresponding reader.

2. Optimization Phase
At the end of the initialization phase, all the information needed for the optimization phase is obtained for generating the optimal Microgrid network scheduling solution. The basic building blocks of this phase are:

a. Fitness Evaluation
At each iteration of every time step, for each individual x^t_i, evaluate its fitness using the objective optimization function as follows:

$$\begin{aligned} f\left(x_i^t\right) &= w_1 \max_{x_{ij}^t-1} P^t(j) + w_2 \sum_{j-1}^{n(t)} x_{ij}^t \\ &+ w_3 \sum_{x_{ij}^t-1} C^t(j) + w_4 \sum_{x_{ij}^t-1} P^t(j) / \sum_{j-1}^{n(t)} x_{ij}^t - \eta \sum_{j-1}^{n(t)} x_{ij}^t \end{aligned} \tag{6}$$

where $P^t(.)$ is the readers' processing time array of the time step t, $C^t(.)$ is the sum function of all the possible collision which the individual x^t_i has with the Micro sources still remaining to be partitioned in the Microgrid network, the punishment coefficient η is set to be 1000, and w_1 w_2 w_3 w_4 representing the importance of each term.

b. Population Evolution
Compare the evaluated fitness values and select employer, onlooker, and scout for each bee in the colony. Then update the food source and position of each bee according to Eq. (2–4), respectively.

c. Termination condition
In each time step, the BABC algorithm is performed until the fitness is small enough, or a pre-determined number of iterations are passed. Then obtained partition should be cut from the network, and the procedure will go back to the optimization phase. The computation is repeated a certain number of time steps, until all the Micro sources are grouped and the optimal scheduling can be obtained.

5 EXPERIMENTAL RESULTS AND DISCUSSION

In this section, four Microgrid network cases, which including 30, 60, 120 and 200 Micro sources respectively, are scheduled to validate the capability of the proposed method. In the simulation test, the proposed BABC, BGA and BPSO were tested on these four Microgrid network cases. The population size for all algorithms was set at 60. The max generation of each run is 300, 800, 3000, and 8000 for 30, 60, 120, and 200 Micro sources network respectively.

Table 1. Processing time of 30 Microgrid network.

Micro source	1	2	3	4	5	6	7	8	9	10
Time	13.52	17.19	4.62	5.61	1.22	8.04	3.64	6.41	10.84	8.07
Micro Source	11	12	13	14	15	16	17	18	19	20
Time	8.89	4.01	8.90	9.36	7.28	16.81	13.92	12.77	5.29	1.81
Micro source	21	22	23	24	25	26	27	28	29	30
Time	15.49	10.31	1.08	4.62	19.94	14.73	17.92	19.44	11.04	10.30

Table 2. Scheduling results on time steps of four Microgrid networks.

	BABC			BPSO			BGA		
Sensors	Best fitness	Mean fitness	Standard deviation	Best fitness	Mean fitness	Standard deviation	Best fitness	Mean fitness	Standard deviation
30	8	8.8333	0.7915	7	8.6333	0.6149	8	9.1000	0.7589
60	14	15.6000	1.0034	13	14.3667	0.8087	14	15.2667	0.6397
120	24	26.0333	1.0981	22	23.9000	0.9595	23	24.8333	1.0199
200	37	40.2333	1.6955	39	41.2333	1.3566	35	36.8333	1.2888

Table 3. Scheduling results on total processing time of four Microgrid networks.

	BABC			BPSO			BGA		
Sensors	Best fitness	Mean fitness	Standard deviation	Best fitness	Mean fitness	Standard deviation	Best fitness	Mean fitness	Standard deviation
30	104.6440	112.8563	5.4702	99.2851	109.3217	4.8430	108.8943	122.6962	8.7115
60	194.8804	218.4362	11.3580	184.7583	200.0373	9.3450	190.0892	231.4354	13.9566
120	367.6203	393.7597	16.9261	335.3734	359.0274	13.3140	367.3497	396.2536	16.2738
200	597.5693	637.1824	23.4425	614.2976	644.0646	18.9624	567.9033	604.5292	19.6896

For BABC, the Limit parameter was set to be 4. For BGA, single point crossover operation with the rate of 0.8 was employed and mutation rate was set to be 0.01. For BPSO, the learning rate parameters were set to the values $c1=c2=2$ and the inertia weight $w=1$.

Table 1 lists the processing times for this 30 Micro sources network case. Due to the limited space, the other Micro sources processing times for these four tested cases are omitted.

The experimental results, including the best, mean and standard deviation of the total processing time and the number of time steps found in 30 runs are proposed in Table 2–3. From the scheduling results, the BABC algorithm can constantly find an optimal schedule results (i.e., less time steps and total processing time). In fact, with an increasing in the number of the Micro sources (hence the degree of the graph), the problem of finding best solution becomes intractable. However, it can be see cleanly, our proposed method is able to find the optimal schedule results of the larger scale Microgrid networks robustly and consistently.

6 CONCLUSIONS

This paper is devoted to giving a new strategy for scheduling Micro sources networks in Space-based ubiquitous computing environment. That is, a novel optimal scheduling scheme for Microgrid networks using a binary version of artificial bee colony algorithm is presented in this work. Experiments on four test cases with four different scales that ranging from 10 to 200 readers, have been presented to demonstrate the performance of the proposed methodology. Simulation results show that the proposed algorithm can accommodate a considerable potential for scheduling large scale Microgrid networks.

ACKNOWLEDGMENTS

This work was supported in part by the China Postdoctoral fund Post-doctor No:172709, Grant No:2016M601332 and Liaoning province Doctoral Fund 201601217.

REFERENCES

Asztalos A., S. Sreenivasan, B.K. Szymanski, G. Korniss, Cascading failures in spatially-embedded random networks. PLOS One 9, e84563 (2014).

Asztalos A., S. Sreenivasan, B.K. Szymanski, G. Korniss, Distributed flow optimization and cascading effects in weighted complex networks. Eur. Phys. J. B 85, 288 (2012).

Bashan A., Y. Berezin, S.V. Buldyrev, S. Havlin, The extreme vulnerability of interdependent spatially embedded networks. Nat. Phys. 9, 667–672 (2013).

Brummitt C.D., R.M. D'Souza, E.A. Leicht, Suppressing cascades of load in interdependent networks. Proc. Natl. Acad. Sci. U.S.A. 109, E680–E689 (2012).

Kennedy J., R.C. Eberhart. A discrete binary version of the Particle Swarm algorithm. Proceeding of the Conference on Systems, Man and Cybernetics, (1997) 4104–4109.

Korkali M., J.G. Veneman, B.F. Tivnan, P.D.H. Hines, Reducing cascading failure risk by increasing infrastructure network interdependency. arXiv:1410.6836v3 Preprint (2014). D. Karaboga, B. Basturk. On the performance of artificial bee colony (ABC) algorithm. Applied Soft Computing 8(2008) 687–697.

Weiser M. The computer of the 21st century, Scientific American, (1991) 94–100.

Innovation and Application of Engineering Technology – Oscar, Chen & Wang (Eds)
 ISBN 978-1-138-05057-0

A study on permeability characteristics of coal samples for soft coal and its similar materials

S.Q. He, C. Zhang & T. Le
School of Civil and Environmental Engineering, University of Science and Technology Beijing, Beijing, China

L.Z. Jin
School of Civil and Environmental Engineering, University of Science and Technology Beijing, Beijing, China
Mine Emergency Technology Research Center, Beijing, China

ABSTRACT: Soft coal mine gas accidents happen frequently; the experiments in the mine were difficult and the success rate to obtain the coal core was low, and so we chose the soft coal sample and a similar material sample of briquette as the research object. The STY-6 permeability measuring system was applied to the comparison study on the permeability difference between the soft coal sample and the similar material coal sample with the method of changing the confining pressure and the gas pressure. The results show that the permeability of the soft coal sample and the similar material coal sample would have the same variation tendency when the confining pressure and the gas pressure changed. With the increase of confining pressure and gas pressure, the permeability presented a quadratic polynomial function. The study lays a foundation to build a physical model by using the developed material and carried out gas extraction, gas outburst, and other physical simulation experiments.

1 INTRODUCTION

Over the years, coal mine gas accidents frequently occur. Due to the low permeability of soft coal seams, it is difficult to control gas in the soft and high-gas coal seams. And therefore, it becomes a high incidence area of coal mines in China (Yuan, 2010). With the increase of coal demand, the mining depth is increasing and the accidents are increasing, and experiments are more difficult in the mine. It is the main research method in this field to configure the similar material and to construct the physical model to simulate the experiment (Cui *et al.*, 2011).

There have been studies based on the similarity criterion; they configured different types of similar materials to carry out physical simulation experiments. However, only the mechanical properties and percolation characteristics of various materials are studied (e.g., see Li *et al*, 2015, p. 80–86, Li *et al*, 2016, p. 204–206, Lin *et al*, 2016, p. 672–679, and Lv *et al*, 2016, p. 25–27). As important factors which affect coal and gas outburst and drainage efficiency, the confining pressure and gas pressure have not been used to comparatively study the permeability of raw coal and similar materials. The physical model constructed by this kind of similar materials is not reliable to perform the experiment of coal seam gas seepage flow. Based on this, this paper is a comparative study on the permeability of raw coal and similar materials by changing the confining pressure and gas pressure, developing materials similar to the law of coal seepage. This study lays a foundation to gas extraction, gas outburst, and other physical simulation experiments.

2 EXPERIMENT DESIGN

2.1 *Specimen preparation*

The test specimens were obtained from a soft coal mine in Huaibei. Processed raw coal and similar materials were configured by different concentrations of humic acid sodium solution and water into ø50 × 100 mm specimens, as shown in Figure 1. In order to avoid the internal and external structural differences and the machining errors, the molding pressure and pressing speed were controlled strictly and standard specimens which were smooth and without obvious cracks were selected.

2.2 *Experimental system*

The structure of the test system is shown in Figure 2, which is mainly composed of a gas cylinder, a pressure gauge, a sensor, a holder, an air valve, and a gas flowmeter.

2.3 *Experimental method*

Pure methane gas was used in this test. Under the condition of constant confining pressure of 1.4 MPa and 2 MPa, the test was carried out when the gas pressure was 0.2, 0.4, 0.6, 0.8, and 0.9 MPa, respectively. Under the condition of constant gas pressure of 0.4 MPa and 0.6 MPa, the test was carried out when the confining pressure was 1, 1.2, 1.4, 1.6, and 1.8 MPa, respectively. The effects of different conditions on the permeability of the raw coal specimen and similar material coal specimen with different humic acid sodium solutions were simulated. Each experiment was performed 3 times, and finally the permeability was calculated according to equation (1) (Zhou *et al.*, 2010).

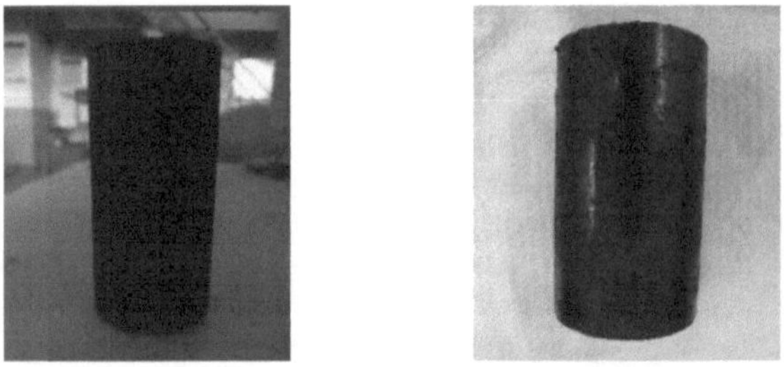

(a) Raw coal specimen (b) Similar material coal specimen

Figure 1. Pictures of raw coal and similar material coal specimens.

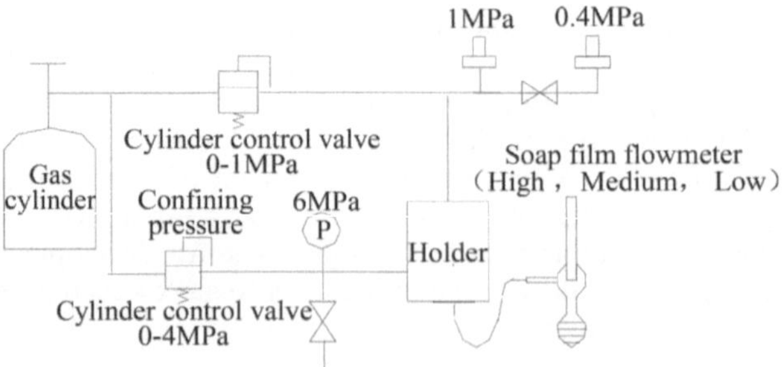

Figure 2. Schematic of the STY-6 permeability measuring system.

$$k = \frac{2QP_a\mu L}{A_s(p_1^2 - p_2^2)} \quad (1)$$

where P_a represents atmospheric pressure, 0.1 MPa; A_s represents the specimen cross-sectional area, cm^2; L represents the length of the test specimen, cm; μ represents air viscosity, 18.44×10^{-6} Pa·s; Q represents gas flow through the test specimen, cm^3/s; p_1 represents gas inlet pressure, MPa; and p_2 represents outlet gas pressure, MPa.

3 EXPERIMENTAL RESULTS

3.1 *Comparative analysis of the influence of confining pressure on the permeability of soft coal and similar material coal*

Based on the experimental data, the relationship between the confining pressure and the permeability of the specimen is shown in Figure 3. The permeability of the soft coal specimen and similar material coal specimen exhibits a similar change with the increase of confining pressure. It can be divided into two stages by a difference of 1.4 MPa. In the initial stage, when the gas pressure remains constant with the increase of confining pressure, the permeability decreased rapidly. It was mainly because the specimen is a porous medium composed of micro-pores and solid skeleton, with the increase of confining pressure, the side wall of the specimen was compressed, some micro-pores were compacted and the microstructure inside the specimen was changed, resulting in poor permeability of the specimen. Therefore, at the same gas pressure, the greater the confining pressure, the lower the permeability of the coal specimen.

When the confining pressure reached a certain value, the adjustment of the internal pore structure of the specimen was slow, which shows the trend of gas permeability change was not obvious. The permeability of raw coal was in the middle of a similar material with different cement concentrations. The permeability of the similar material decreased with the increase of the concentration of the cement. The permeability of raw coal changes faster than that of similar materials with the decrease of confining pressure. The main reason was that with the increase of the concentration of the cement, the coal powder with different particle size distributions is more compact and cohesive, reduces the pore channels of gas permeation inside the specimen and hinders the effective way of gas release. The specimen of raw coal was pressed without any additives. There was a lot of space in the middle of raw coal and the compressed space was large. After the confining pressure increased, the deformation of raw coal was greater than that of the similar material coal specimen. Therefore, the change trend of permeability with confining pressure is stronger than that of the similar material coal specimen.

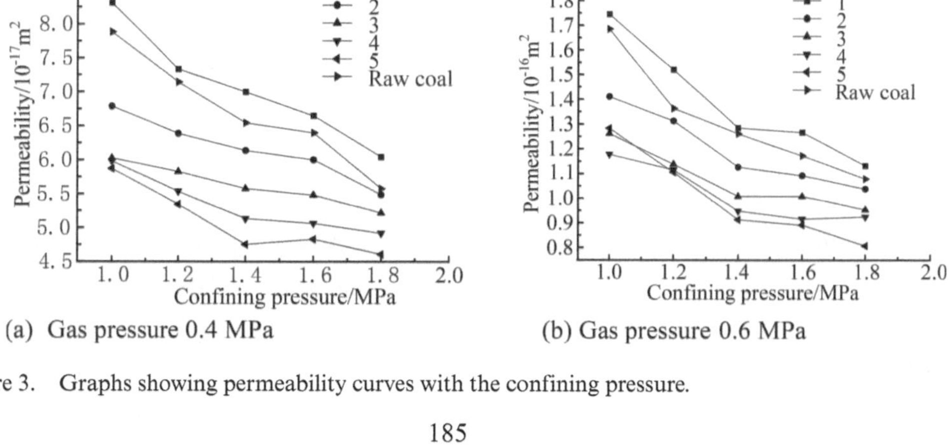

Figure 3. Graphs showing permeability curves with the confining pressure.

Fitting the experimental data, the results are given in Tables 1 and 2, which approximate to a polynomial function. The correlation coefficient R^2 is in the range of [0.8737, 0.9774] and the precision is high. The results show that the relationship between permeability and confining pressure can be expressed by using equation (2) when the gas pressure was fixed.

$$k = B_2 p^2 - B_1 p + C \tag{2}$$

where k represents permeability; p represents confining pressure; and B_1, B_2, and C represent fitted parameters.

3.2 *Comparative analysis of the influence of gas pressure on the permeability of soft coal and similar material coal specimens*

The relationship of soft coal permeability and gas pressure is shown in Figure 4.

As shown in Figure 4, the permeability of the soft coal raw material and similar material coal sample has similar change with the increase of gas pressure. When the confining pressure is constant, the permeability increases with the gas pressure. This is consistent with Cao et al.'s (2010) analysis conclusion. Mainly because, when confining pressure fixed, with the increase of gas pressure, the specimen inside the free state of the gas increased, the specimen's internal micro-pores and cracks expand in the presence of gas, and the number of gas seepage channels increased. Simultaneously, with the increase of the gas pressure, the higher the gas pressure difference between the coal specimens, the stronger is the driving force of the gas from the coal specimen, and the gas flow rate of the specimen is also increased gradually (Yin *et al.*, 2009). When the gas pressure increases to 0.8 MPa, the change rate of permeability of soft coal is accelerated and the change rate of permeability of similar materials is almost unchanged. This is mainly because the similar material specimen is composed of pulverized coal and humic acid sodium binder, and the particles are tightly bonded. While the raw coal

Table 1. Fitting relation between confining pressure and permeability at a gas pressure of 0.4 MPa.

Number	Gas pressure/MPa	Factor	Fit relation	R^2
1			$k = 1.3036p^2 - 6.258p + 13.1667$	0.9415
2			$k = -0.1946p^2 - 0.9445p + 7.8790$	0.9331
3	0.6	Confining pressure	$k = 0.0446p^2 - 1.1065p + 7.0830$	0.9716
4			$k = 1.6643p^2 - 5.964p + 10.2795$	0.9774
5			$k = 2.2786p^2 - 7.904p + 11.4979$	0.9142
Raw coal			$k = 0.5145p^2 - 4.1209p + 11.4280$	0.9354

Table 2. Fitting relation between confining pressure and permeability at a gas pressure of 0.6 MPa.

Number	Gas pressure/MPa	Factor	Fit relation	R^2
1			$k = 0.7123p^2 - 2.7346p + 3.7644$	0.9522
2			$k = 0.4323p^2 - 1.6939p + 2.6859$	0.9420
3	0.4	Confining pressure	$k = 0.4788p^2 - 1.7145p + 2.4975$	0.9456
4			$k = 0.4962p^2 - 1.7429p + 2.4444$	0.8737
5			$k = 0.6398p^2 - 2.3747p + 3.0193$	0.9630
Raw coal			$k = 0.8427p^2 - 3.0610p + 3.8785$	0.9501

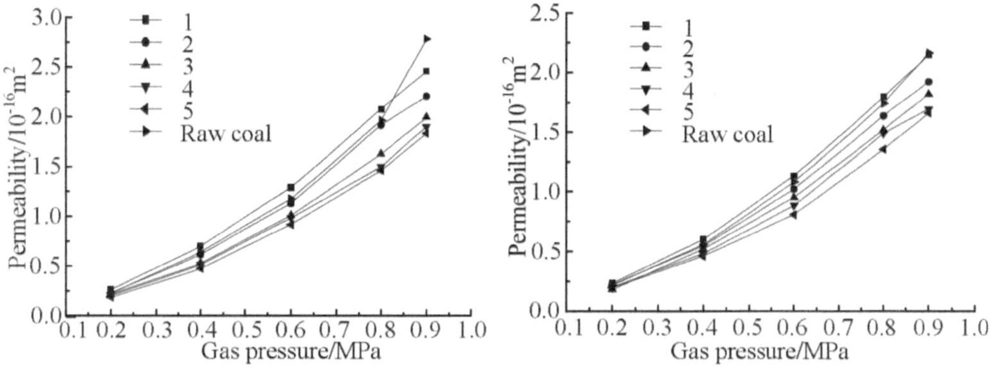

(a) Confining pressure – 1.4 MPa

(b) Confining pressure – 2 MPa

Figure 4. Relationship curve of permeability with gas pressure.

Table 3. Fitting relation of gas pressure and permeability at a confining pressure of 1.4 MPa.

Number	Gas pressure/MPa	Factor	Fit relation	R^2
1			$k = 1.8974p^2 + 1.0760p - 0.0308$	0.9990
2			$k = 1.8538p^2 + 0.8467p - 0.0176$	0.9947
3	1.4	Gas pressure	$k = 1.9264p^2 + 0.4248p + 0.0547$	0.9999
4			$k = 1.7310p^2 + 0.4841p + 0.0445$	0.9972
5			$k = 1.8214p^2 + 0.3305p + 0.0501$	0.9993
Raw coal			$k = 3.9935p^2 - 0.9291p + 0.2908$	0.9827

Table 4. Fitting relation of gas pressure and permeability at a confining pressure of 2 MPa.

Number	Gas pressure/MPa	Factor	Fit relation	R^2
1			$k = 1.6821p^2 + 0.9069p - 0.0189$	0.9997
2			$k = 1.4820p^2 + 0.8411p - 0.0132$	0.9986
3	2	Gas pressure	$k = 1.3333p^2 + 0.8822p - 0.0478$	0.9998
4			$k = 1.3855p^2 + 0.6843p - 0.0027$	0.9929
5			$k = 1.7684p^2 + 0.1456p + 0.1025$	0.9989
Raw coal			$k = 2.1340p^2 + 0.4235p + 0.0495$	0.9998

specimen is composed of pulverized coal and water, the adhesion between the particles is not good. When gas pressure is high, with the increase of gas pressure inside the specimen, the voids and cracks are easy to deform.

As shown in Tables 3 and 4, the correlation coefficient of permeability and gas pressure is in the range of [0.9827, 0.9999] and the accuracy is high. It is indicated that the relationship can be expressed by using equation (3) when confining pressure is fixed,which is:

$$k = \mathrm{A}_2 p^2 + \mathrm{A}_1 p + \mathrm{D} \tag{3}$$

In this equation, k is the permeability; p is the confining pressure; and A_1, A_2, and D are fitting parameters.

According to the studies, the variation tendency of the gas permeability of soft coal and the similar material is similar with changes in confining pressure and gas pressure and the permeability of raw coal is in the middle of that of the similar material. The relationship between gas permeability and gas pressure or confining pressure is a quadratic polynomial function. It is suggested that when the similar material meet similar criteria, the physical model can be constructed according to the permeability characteristics of different soft coal mines. And the physical model can be constructed to solve the problem in the deep mine, where the relevant experiment cannot be carried out. Physical simulation experiments can be carried out in a laboratory to solve the problems encountered under actual operating conditions.

4 SUMMARY

At fixed gas pressures, the variation of permeability of coal increasing with confining pressure had the same trend with the similar material. The gas permeability decreased with increase of the confining pressure. There was a quadratic polynomial function relationship, which is given as follows: $k = B_2p^2 - B_1p + C$. The permeability of raw coal is in the middle of that of the similar material with different binder concentrations. Similar material permeability decreased with increase of binder concentration.

Fixed confining pressure, the permeability of the soft coal and similar materials with the gas pressure changes had followed similar laws. The gas permeability increased with gas pressure, and there is a non-linear relationship which is given as follows: $k = A_2p^2 + A_1p + D$. When the gas pressure increased to a certain extent, the permeability of raw coal changed rapidly.

The laws of similar material permeability change with confining pressure or gas pressure were almost unanimous and the permeability is not much different in value. The permeability was in the middle of similar materials in different concentrations. Therefore, according to the soft coal mine penetration characteristics, scientists and engineers can adjust the similar material ratio and develop similar materials, which conform to the coal permeability and similarity criteria. The similar materials can be used to build a physical model, carry out permeability test, and study coal and gas outburst in the laboratory. The study puts forward a new scheme in order to solve the problem of difficulty and poor intuition in underground experiments.

ACKNOWLEDGMENTS

This study was supported by the National Natural Science Foundation of China (no. 51574017) and The National Key Research and Development Program of China (no. 2016YFC0801706).

REFERENCES

Cao, Shugang., Guo, Ping. & Li, Yong., et al. 2010 Effect of gas pressure on gas seepage of outburst coal. *Journal of China Coal Society* 35(4):595–599.

Cui, Ximin., Miao, Xiexing. & Su, Deguo., et al. 2002. Error analysis in similar material simulation test of the movement of rock strata and surface. *Chinese Journal of Rock Mechanics and Engineering* 21(12): 1827–1830.

Li, Li., Qu, Yulong. & Li, Shugang., et al. 2016. Research on Permeability Characteristics of Similar Material of Coalrock and Solid-gas Coupling. *Coal Technology* 35(6):204–206.

Li, Shugang., Zhao, Pengxiang. & Lin, Haifei., et al. 2015. Study on character of physical simulation similar. *Journal of China Coal Society* 40(1):80–86.

Lin, Haifei., Qu, Yulong. & Zhao, Pengxiang., et al. 2016 Research on mechanics and permeability characteristics of similar material of solid-gas coupling based on orthogonal design. *Journal of China Coal Society* 41(3):672–679.

Lv, Shuaifeng., Wu, Xiaomei. & Su, Dongmei., et al. 2016 Making method and seepage flow characteristics of artificial Coal Core. *Coal Technology* 35(1):25–27.

Yin, Guangzhi., Li, Xiaoshuang. & Zhao, Hongying., et al. 2009 Experimental study of effect of gas pressure on gas seepage of outburst coal. *Chinese Journal of Rock Mechanics and Engineering* 28(4):697–702.

Yuan, Liang. 2010. Concept of gas control and simultaneous extraction of coal and gas. *China Coal* (6):5–12.

Zhou, Junping., Xian, Xuefu. & Li, Xiaohong., et al. 2010 Effect of different adsorptional gases on Permeability of Coal. *Chinese Journal of Rock Mechanics and Engineering* 29(11):2256–2262.

Innovation and Application of Engineering Technology – Oscar, Chen & Wang (Eds)
© 2017 Taylor & Francis Group, London, ISBN 978-1-138-05057-0

A study on the relationship between security and investment in education and accidents by combining the Cobb–Douglas production function

Yan Zhang & Fuchuan Jiang
Kunming Polytechnic University, Kunming, China

ABSTRACT: There is a complex relationship between safety inputs and accidents. At present, many experts and scholars have carried out detailed analysis and research on the relations between the two, but the analysis of the safety input of the "people" in the composition of the safety input safety investment analysis of the influence of the whole enterprise is less. In this paper, large coal mining enterprises are taken as an example and combined with the Cobb–Douglas production function, the optimal value of the safety input of the enterprise is calculated, and the EXCEL table is used to express the relationship between the security inputs. Over the years, for the safety of investments, the enterprise has combined the composition of the safety management and safety training into a specific analysis, which reflects the current existence of China's security education.

Keywords: Security input; Safety management input; Safety training input; Douglas production function; optimal value; education

1 INTRODUCTION

To carry out the control and prevention of the accident, whether from Heinrich's theory of cause of the accident or otherwise from the theory of dangerous sources to analyze, one can take the appropriate measures from two aspects. On one hand is the "human" factor; on the other hand is the "material" factor. And the degree of freedom is much larger in the case of the "human" factor in relation to the "material" factors and therefore, the control is extremely difficult. Many of the existing data analysis and models have shown that the ratio of accident control and safety inputs is optimal as long as a reasonable investment is made for safety and related matters. At the same time, these data and analysis results also show that the reason why security incidents occur in China is mainly due to the fact that investments are not up to the standards; also to some extent, the essence of security is not the highest priority in the safety investment funds. And then, as long as the adoption of targeted measures (50%) has been controlled, and the remaining 50% is uncontrollable, the rate of occurrence of the accidents is not limited to the accident rate. It is at least 50%, and this is because there is a small accident probability; uncertainty that occurs can greatly reduce the incidence, This fact is not the case, since people's life and property are constantly under threat; security incidents never stopped indicating that there must be another important factor - "people".

In addition to taking control of the factors of "people", the focus should be the safety education and training; safety education in the most important attribute of education and the formation of attitude does not occur overnight; it requires a lot of manpower and resources and the surrounding. The impact of the environment can be formed. Heinrich's research and view show that the establishment and maintenance of workers' interest in accident prevention are the first principles of accident prevention, followed by continuous analysis to cope with the problem. Maintain the staff's interest in the accident that is through safety education to train workers on the safety of production and the attitude of accident prevention.

2 THE CONCEPT OF SECURITY INPUTS AND ACCIDENTS

Safety investment refers to the sum of all human, material and financial resources invested in security activities. People, technology, facilities and other inputs, safety education and training, labor protection and health care costs, accident relief and prevention, accident casualties, such as the cost of treatment are regarded as safe inputs. Security is classified into safety national basic input and safety input. In this paper, the enterprise safety investment is mainly analyzed. It mainly includes the following: investments, security projects for upgrading technology, safety management, safety training, and safety input (Figure 1). The five kinds of safety input, safety input, and safety management are put into effect and are revealed, intuitive, and also are very popular in the enterprise. But this is only a tactical means; from a long-term point of view, process upgrades and talent education are more strategic aspects of both; talent training can bring innovation, innovation can promote technological change, technology upgrades and ultimately, a significant change in the security situation in our country.

An accident refers to the production and business units in the production and business activities (including production and business-related activities) that are involved in causing sudden hurt of personal safety and health, or damage to facilities, or economic losses, resulting in the suspension or permanent termination of original production and business activities (including production and operation-related activities) Suspension or permanent termination of the accident. Accidental loss is commonly used to measure the severity of the accident. There is a complex relationship between direct economic loss and safe investments.

Taking a large state—owned coal mine as an example, in this paper, the data of safety input and accident loss in 2001–2010 are summarized and listed in tables.

It can be seen from Table 1 that the enterprise has been investing in safety management and production safety as its main investment direction and the safety education expenses for enterprise employees training are low. The average investment costs only 3.78605 million yuan/year. Figure 2 shows the relationship between the total safety input and the total loss of the accident. In addition to the increase in the total loss of total safety losses in 2003, the rest of the time is increased as the total safety input, but the total loss of the accident is not obvious; therefore, it fails to achieve the purpose of safe investment. The following is the analysis of the relationship between the input of the safety inputs and the total loss of the accident: the investment in the upgrading of the mining process, the investment in safety engineering and the investment in safety training increased from 2000 to 2006, and the three inputs are basically equal. Production safety has stabilized since the sharp decline in 2002 and has rebounded in 2006. It is reported that the reason for the emergence of such a situation is that business management that acquired safety materials, safety equipment and other new facilities considered that there is no need to continue a lot of investments; followed by

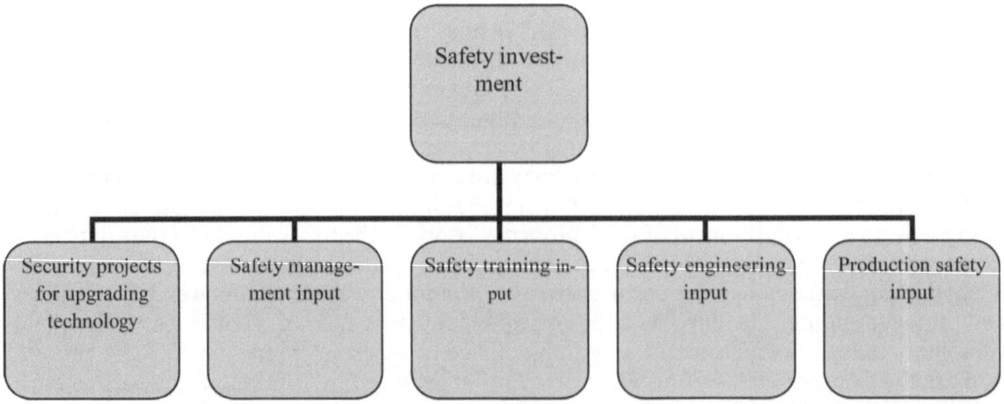

Figure 1. Flowing of method.

Table 1. Safety inputs and total loss of statistics.

Safety inputs and total loss statistics (Million yuan/a)							
Year	Security project	Safety engineer	Safety management	Safety training A	Production safety	Safety total	Accident total
2001	137.85	310.76	1347.04	203.12	1867.67	3866.44	885.88
2002	176.53	334.44	1400.71	232.58	1659.82	3804.08	962.44
2003	247.32	357.73	1270.13	301.16	1570.32	3746.66	1126.27
2004	321.53	465.95	1290.75	353.38	1045.97	3477.58	1375.27
2005	344.71	461.78	1645.78	385.57	1028.21	3866.05	1056.98
2006	348.95	496.55	1859.71	392.85	1009.93	4107.99	986.35
2007	473.16	460.58	1678.85	432.67	1072.52	4117.78	1026.98
2008	480.53	502.87	1730.91	486.03	1400.57	4600.91	980.27
2009	482.23	520.52	1859.95	496.31	1514.58	4873.59	913.81
2010	495.95	535.39	1859.63	502.38	2003.58	5396.93	882.39

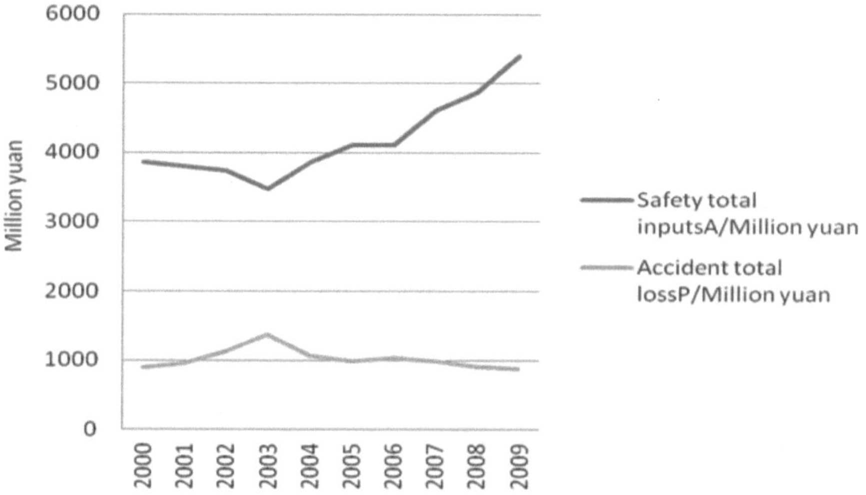

Figure 2. The relationship between the total safety input and the total loss of the accident.

the increase in investment, accidents occurred due to time changes caused by the production technology equipment behind or aging. From 2000 to 2003, total loss of the accident and did not decrease but increased and the safety input was lagging; after 2003, the reduction in accidents has not been obvious. On the whole, the main reason for the frequent accidents in the enterprises is that the safe investment is considered unreasonable and the operating personnel lack professional knowledge.

Since the relationship between the input of "human" and the accident in the safety input is focused in this paper, the following analysis is made on the safety management and safety training of the total safety input: investments in safety training is generally increasing year by year, indicating that the enterprise has been aware not only of the importance of training for operators, but also to take the appropriate measures; but the increase in the trend is not obvious; enterprise management funds into the sector because there is no scientific security investment allocation program, which is the reason behind the tentative increase in safety training investment; the importance given to safety training awareness is not enough, and there is no scientific and reasonable security investment allocation program. In contrast (Figure 3), safety

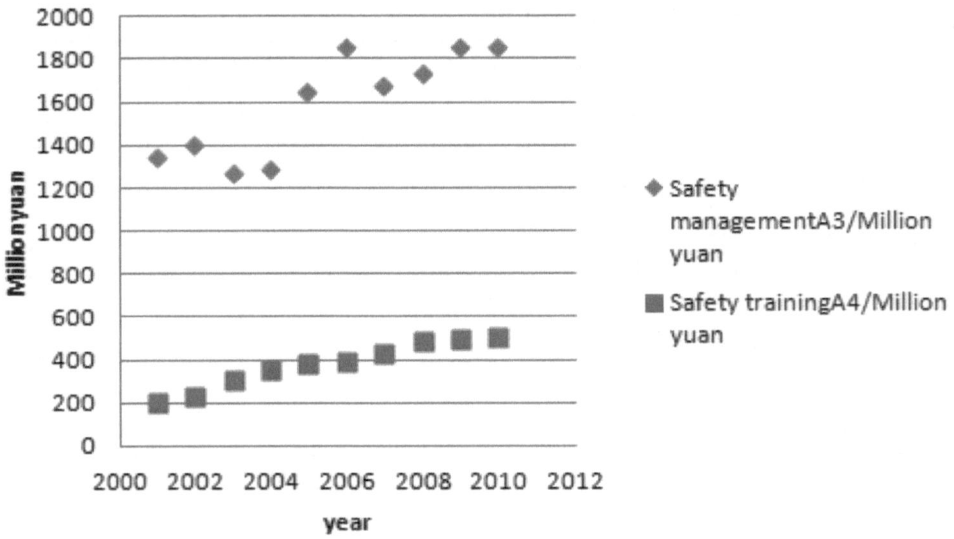

Figure 3. Relationship between safety management and safety training.

management investment is much higher than the safety training investment, and the overall trend is increasing. Ignoring the safety training investment and a substantial increase in safety management investment is bound to lead to the safety of management personnel to enhance the quality of the safety of laid off workers down, resulting in polarization. Management's efforts to enhance investments are also more prone to psychological stress and other reasons caused by irregularities or improper handling of emergency and so on. The light shed on the substantial increase in safety management investment is not enough. In order to further explore the relationship between safety management input and safety training input and accidents, we use the Cobb-Douglas production function to calculate the optimal allocation value of every safety input in 2000–2010 and use the optimal allocation value and safety management investment and safety training inputs to quantitatively analyze the relationship between safety inputs and accidents.

3 BRIEF ANALYSIS OF THE RELATIONSHIP BETWEEN SAFETY EDUCATION INVESTMENT AND ACCIDENT EDUCATION INVESTMENT AND ACCIDENT

3.1 *Cobb–Douglas production function*

The Cobb–Douglas production function was originally created by the American mathematician C. W. Cobb and the economist Paul Douglas to discuss the relationship between input and output; the name is based on American mathematician C. W. Cobb and economist Paul. H. Douglas's name that is made in the general form of the production function. An industrial system that is used to predict national and regional analysis and development or production of large enterprises producing pathway economic mathematical models is called the production function. It is one of the most widely used forms of production in economics and it has an important position in the research and application of mathematical economics and econometrics. The general formula for the production function is as follows:

$$Y = A\,(t)\,k \wedge aL \wedge b$$

where Y is the total industrial output value; at the level of integrated technology, L is the number of labor inputs (in person or people), K is the capital invested, which generally refers to the net fixed assets (in million or $ million, but labor must correspond to the number of units, such as the labor force by people as an unit, net fixed assets to use billion on the unit), a is the elasticity of labor output, and b is the capital-output elasticity.

The results of the application of the Cobb–Douglas production function in the allocation of quantitative and safe inputs of an enterprise are analyzed. The results of the analysis of the data of the calendar year show that there is a minimum value of the safety inputs in the period from 2001 to 2010. The total investment in safety engineering, safety training, and production safety is basically the same, but the total loss of accidents increased in 2007, and the investment in 2001–2005 was relatively reasonable. As a result of the above considerations, this study will select the 2001–2005 safety input data to test whether the five years of safe investments are reasonably allocated.

3.2 *Data selection and calculation*

According to the data in Table 1 on the accident loss and the safety of the natural logarithmic fit, the use of Matlab software programming, the source program:

```
A=[1, 2.13940677, 2.492425113, 3.129380492, 2.307538823, 3.271300143:
   1, 2.246818521, 2.524318215, 3.146348229, 2.366572366, 3.220060993:
   1, 2.393259238, 2.553555363, 3.103848174, 2.478797288, 3.195988162:
   1, 2.507221501, 2.668339316, 3.110842134, 2.548241966, 3.019519228:
   1, 2.537453882, 2.664435119, 3.21637178, 2.586103235, 3.012081823:
   0,1,1,1,1,1]:
P=[2.947374897, 2.983373664, 3.051642516, 3.138387969, 3.02406677,1]':
a=A\P:
```

The result is:

Variable	Value
A1	137.8500
A2	310.7600
A3	1270.130
A4	375.3497
A5	1009.930

As the main focus of this article is the safety management and safety training investment, in particular, the two safety inputs of the optimal value are acquired and analyzed:

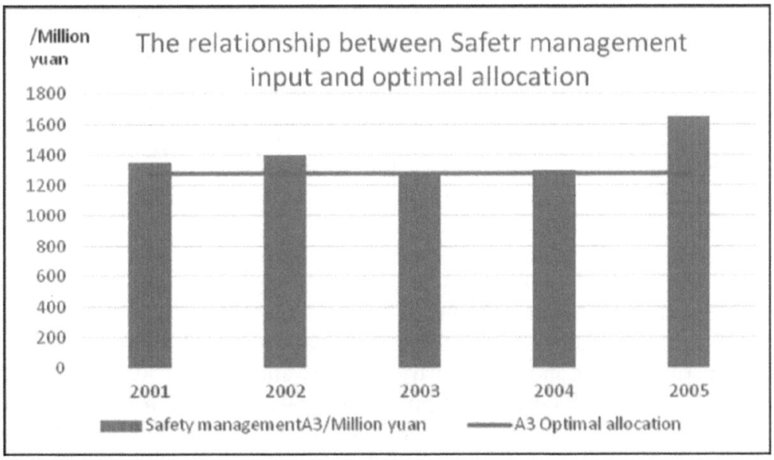

Figure 4. Shows that the investment in safety management was higher than the optimal value in the period from 2001 to 2002.

Figure 5. Relationship between safety training input and optimal allocation.

Figure 4 shows that the investment in safety management was consistent with the optimal value from 2003 to 2004, but from 2001–2003, accidents did not decrease but gradually increased. For the total loss of the accidents to reach the maximum, in 2005, a substantial increase in safety management costs was implemented. In the past five years, the total loss of the accidents has not changed much, and there is an unreasonable distribution of the overall safe investment.

It can be seen from Figure 5 that the enterprise safety training investment, although it has been on the rise, only in 2005 it could achieve the optimal allocation of the required value. Indicating that the safety training of the enterprise is not taken seriously, resulting in low quality of operating personnel, do not have well ability operating and the potential unsafe state cannot be identified and take appropriate measures, resulting in the total loss of the accident changes unreasonable.

4 CONCLUSION

The use of the Cobb—Douglas production function to solve the problem of coal mine safety into the optimal allocation provides a scientific basis. On the whole, the safety investment is not only unreasonable in distribution, but also there is a serious shortage of inputs. It is the main factor that causes the accident loss to change obviously. And lack of human education and training is the cause of the accidents caused due to serious reasons. Therefore, the enterprise should strengthen safety management work and constantly establish and improve the work of the accident prevention organization, improve the safety regulations, safety systems, safety regulations, establish the leadership of the business leaders, including safety management personnel, security and technical personnel, including the accident prevention system, and strictly monitor its effectiveness. And the safety of the perpetrators and operators of training, or to persuade people to educate, personnel adjustment and other acts should be carried out, to improve their safety awareness, thereby reducing the occurrence of unsafe behavior. In the event of accident prevention, safety technology and equipment are important. Safety management mechanisms, rules and regulations are also important, but the most important thing is the safety concept, security attitude, and security atmosphere.

In the event of accident prevention, the purpose of safe investment is to prevent accidents from occuring. Accidents may cause losses, including loss of life and loss of money in the event of an accident. Since ancient times, China has "life off days" argument, which is the main responsibility for production.

Therefore, to improve people's safety awareness and attitude is the eternal theme of safe work; the allocation of investments in safe inputs must be done by paying due attention to the "people".

A large part of the attitude of the social environment is caused by the impact and influence. According to statistics, since the promulgation and implementation of the safety production law, there has been an overall good effect; and the safe production situation improved year after year; the number of accidents from 1.073 million in 2002 fell to 28.2 million in 2015, the death toll from 13.9 million fell to 6.6 million people; It has taken 13 consecutive years to reduce the number of accidents and the death toll "double down." The newly revised safety production laws highlight the problem-oriented solution: emphasize the main responsibility of production and operation units and the responsibility of government supervision, strengthen the responsibility of the production and business units and the responsibility of government supervision, and strengthen the responsibility for safe production, and provide a more favorable way to promote stable production and legal protection. This aspect is the change of the environment, so that the concept of "safety first" enjoys popular support. On the other hand, it also shows the importance of security attitude; only if the attitude is correct, the policy can play a fundamental role.

The following recommendations are laid out to the current security education in China:

1. In the primary and secondary schools, safety education-related courses should be started, from an early age, to develop children's attitude towards safety, through more modern technologies such as VR, AR, MR interactive information technology simulation training, enhancing people's self-help emergency response capability.
2. Relevant policies are indeed implemented.
3. Rewards and punishments to encourage the people.
4. The use of accident simulation accountability, so that business people in leadership take responsibility.

Today, there are still a lot of people who will be safe and have an "only I will be safe" attitude rooted in the human brain; In the following circumstances such as making people feel dangerous and face the danger to set it right, and even if it is not done, the reaction of security behavior, and sometimes when they do not respond to the effect, the danger is much greater than the effect of the reaction. From post-treatment to pre-proactive change, enhance people's awareness of the crisis and increase the ability of national risk identification and self-help escape ability.

Based on the Cobb–Douglas production function, in this paper, the relationship between the safety education input and the accident of an enterprise is analyzed and the lack of safety education investment is reported. Safety awareness and attitude of safety management personnel and safety and technical personnel are the causes of the accidents. And also, the rational allocation of security investment and strengthening of safety education methods and recommendations are put forward. Only focus on safe investment and reasonable allocation of safety education investment; safety first and I want to be safe should be rooted in the concept of the social atmosphere, in order to achieve security for everyone, so that everything is safe.

ACKNOWLEDGMENT

This study was financially supported by the National Science Foundation of China (51674127).

REFERENCES

Li Xiang. Economic Analysis and Benefit Evaluation of Safety Investment [J]. *Journal of Chinese Society for Safety Sciences*, 2005, 3 (3); 26~29.

Luo Chunhong, Xie Xianping. Safety investment in the safety benefits of the forecast and correlation analysis [J]. *Chinese Journal of Safety Science*. 2007, 33 (8).
Mei Qiang. Study on the Direction of Security Investment Decision [J]. *Journal of the Chinese Society for Safety Science*, 1999, 9 (5): 42–46.
Niu Xiao Zheng, Lu Li-xing, Yin Xin. Application of Cobb-Douglas Production Function in Quantitative Safety Input Distribution of a Coal Mine Enterprise [A]. *Technology and Industry*, 2011.
Peng Hong-jun, Li Xin-chun, Li Yuan. Nonlinear Optimization Model of Safety Input [J]. *Journal of Chinese Society for Safety Sciences*.
Wan Mu Sheng, Chen Guohua, Zhang Hui. Safety Statistics Economics [M]. *Guangzhou: South China University of Technology Press*, 2008.
Wu Qiong, Xu Kaili. Safety Management [M]. *Beijing: Coal Industry Press*, 2002.

Innovation and Application of Engineering Technology – Oscar, Chen & Wang (Eds)
© 2017 Taylor & Francis Group, London, ISBN 978-1-138-05057-0

The size effect on the axial behavior of RC square columns wrapped with CFRPs

Meng Liang, Zuo-yun Mei, Xiao-guang Song, Shi-hong Chu, Shou-cai Li, Xing Bi & Ming-hai Li
Shandong Academy of Building Research, Jinan, Shandong Province, China

Wei Wang
School of Civil Engineering, Harbin Institute of Technology, Harbin, Heilongjiang Province, China

ABSTRACT: Investigations published on the behavior of Fiber-Reinforced Polymer (FRP)-confined concrete columns in axial compression have been generally based on specimens with small scales. In this paper, the axial behavior of carbon FRP (CFRP)-Confined Reinforced Concrete (RC) square columns with four sizes is focused, aiming to clarify the influence of size on the behavior of different scale specimens. The experimental results show that there are threshold values for the size effect on the axial compressive strength and stress-strain relationship of the columns, beyond which there are no size effects.

1 INTRODUCTION

In recent years, fiber-reinforced polymer (FRP) jackets have been increasingly adopted to the retrofit of existing reinforced concrete (RC) structures. One important application of FRP strengthening is to enhance the strength and ductility of RC columns in axial compression. A number of experimental and theoretical studies have been conducted on FRP-confined columns. However, experimental investigations mainly focus on small-sized specimens due to costs and test conditions in the laboratory (e.g., Mirmiran & Shahawy, 1997; Rochette & Labossière, 2000; Xiao & Wu, 2000; Shahawy *et al.*, 2000; Pessiki *et al.*, 2001; Lam & Teng, 2004; Berthet *et al.*, 2005; Cui & Sheikh, 2010). Generally, large-sized columns, especially RC square columns, have been employed in practical engineering applications, which inevitably result in an incomplete understanding between small and large-scale columns. For this reason, some research concerning the size effect on the axial behavior of confined concrete columns has been carried out; but most investigations are within the area of circular concrete columns so far (Thériault *et al.*, 2004; Carey & Harries, 2005; Rocca *et al.*, 2005; Rocca 2007; Youssef *et al.*, 2007; Yeh & Chang, 2007; Wang & Wu, 2011; Elsanadedy *et al.*, 2012; Liang *et al.*, 2012). With regard to confined circular specimens subjected to axial compression, the conclusions basically can be drawn that size has no influence on axial behavior, except a little research. It can be supposed that the uniform confinement provided by the FRP jacket to the circular cross-section is significantly efficient.

However, studies on the size effect on confined noncircular columns in axial compression are very limited (Masia *et al.*, 2004; Rocca 2007; Toutanji *et al.*, 2010; Wang & Wu, 2011; Luca *et al.*, 2011; Wang *et al.*, 2012). Masia *et al.* (2004) studied the axial behavior of CFRP-confined plain concrete (PC) square columns with three different sizes ($B \times H$: 100 × 300 mm, 125 × 375 mm and 150 × 450 mm). The test results demonstrated that the FRP confinement efficiency deteriorated with the increase of specimen size. Rocca (2007) tested CFRP-confined RC square columns with cross-sectional widths of 324 mm, 457 mm, 648 mm and 914 mm. The test results presented obvious differences in the axial stress-strain response between small-scale and large-scale specimens. Moreover, Rocca

(2007) indicated that the size effect on the axial behavior (e.g. ultimate strength) of confined square columns had still remained under unclear conditions due to inadequate and discrete test data. Wang and Wu (2011) examined how the dimension affected the compressive behavior of Aramid FRP (AFRP)-confined PC prisms with the sizes covering 70 × 210 mm, 100 × 300 mm and 150 × 450 mm. The experimental results showed that the specimen size affected its ultimate strength significantly, but influenced the ultimate strain, stress-strain relationship, and failure mode slightly. The investigation (Wang *et al.* 2012) on the stress-strain behavior of CFRP-confined RC square columns ($B \times H$: 204 × 612 mm and 305 × 915) was carried out, which mainly included test parameters of specimen size, CFRP thickness, hoop reinforcement ratio and load pattern. The test results revealed the specimen dimension obviously affected the peak stress, ultimate stress, ultimate strain as well as stress-strain curve. Toutanji *et al.* (2010) and Luca *et al.* (2011) specially studied the axial behavior of large-sized square columns in compression. They indicated that existing strength models developed from test data of small-sized columns generally overestimated the ultimate strength of large-sized columns. They also pointed out that the model of Lam and Teng (2003b) can give an acceptable prediction in the published strength models of confined prisms.

The purpose of this research is to clarify the size effect on the axial behavior of CFRP-confined RC square columns. An experiment of 26 confined and four unconfined prismatic specimens with different sizes, CFRP thicknesses and corner radii under various confinement levels were conducted in this study. It demonstrates that the size effect on the ultimate strength and stress-strain response weakens and even disappears as the lateral confinement reaches a relatively heavy level.

2 EXPERIMENTAL PROGRAM

2.1 *Specimens and materials*

A total of 30 RC square columns were produced to investigate the size effect on the axial behavior. All the specimens had four different sizes including 100 × 300 mm, 200 × 600 mm, 300 × 900 mm and 400 × 1200 mm (cross-sectional width B by height H), which were referred to S1, S2, S3, and S4, respectively. The same longitudinal steel reinforcement ratio (ρ_g = 1.5%) and the volumetric ratio of stirrup (ρ_v = 0.8%) were adopted for these RC prisms. Moreover, the plies of CFRP jackets, corner radius and concrete cover thickness were designed in proportion to the column sizes in order to reach the same lateral confinement at a certain level.

For the confining level, according to the study of Lam and Teng (2003a), the theoretical confinement pressure provided by the CFRP jacket can be expressed as follows:

$$\text{For the circular specimens: } f_{lf} = \frac{2E_f n t_f \varepsilon_{fe}}{D} \tag{1a}$$

$$\text{For the RC square specimens: } f_{lf} = \frac{2E_f n t_f \varepsilon_{fe}}{B} \cdot \kappa_a = \frac{2E_f n t_f \varepsilon_{fe}}{B} \cdot \left[1 - \frac{2(B-2r_c)^2}{3A_g(1-\rho_g)}\right] \tag{1b}$$

where E_f is the elastic modulus of the CFRP (MPa); n is the number of CFRP layers; t_f is the nominal thickness of one ply of the CFRP (mm); ε_{fe} is the effective rupture strain of the CFRP jacket; D is the diameter of circular specimens; B is the side length of square specimens; κ_a is the shape factor; r_c is the corner radius; ρ_g is the ratio of the longitudinal steel to the gross cross-section area; and the gross cross-section area is given by the following equation:

$$A_g = B^2 - (4-\pi) r_c^2 \tag{2}$$

The four confinement levels can be designed in accordance with Equations (1) and (2). Table 1 reveals the details of the experimental specimens, which includes confinement levels, sizes, corner radii, CFRP jacket thicknesses, concrete cover thicknesses and reinforcements.

The target strength grade of concrete was C30 (GB50010, 2010) and the 28-day average cubic compressive strength was 35.70 MPa. Crushed bluestone with a maximum diameter of 10 mm was used as coarse aggregates. Commercial concrete was cast into stainless steel moulds with chamfered corners and angle iron moulds welded with steel plates, as shown in Figure 1.

FRP sheets used in this test were a type of unidirectional carbon fabric. From the coupon test, the measured ultimate tensile strength, elastic modulus, and nominal thickness obtained were 4525 MPa, 250 GPa, and 0.17 mm, respectively. The steel bars applied in this test were of HRB400 grade. Figure 2 presents the steel cages of square specimens covering each size.

Table 1. Details of RC square columns at various confinement levels.

Confinement level	Sizes ($B \times H$)	Number of CFRP plies	Corner radius (mm)	Concrete cover (mm)	Longitudinal steel	Hoop reinforcement
Heavily confined	100 × 300	1	10	7	2Φ6+2Φ8	Φ4@90 mm
	200 × 600	2	20	14	4Φ14	Φ6@95 mm
	300 × 900	3	30	21	4Φ14+4Φ16	Φ8@110 mm
	400 × 1200	4	40	28	12Φ16	Φ10@130 mm
Medium-confined	100 × 300	1	5	7	2Φ6+2Φ8	Φ4@90 mm
	200 × 600	2	10	14	4Φ14	Φ6@95 mm
	300 × 900	3	15	21	4Φ14+4Φ16	Φ8@110 mm
	400 × 1200	4	20	28	12Φ16	Φ10@130 mm
Lightly confined	200 × 600	1	20	14	4Φ14	Φ6@95 mm
	400 × 1200	2	40	28	12Φ16	Φ10@130 mm
Minimum-confined	200 × 600	1	10	14	4Φ14	Φ6@95 mm
	400 × 1200	2	20	28	12Φ16	Φ10@130 mm

Note: The hoop reinforcement spacing in the above table signifies the spacing at the middle of the column height.

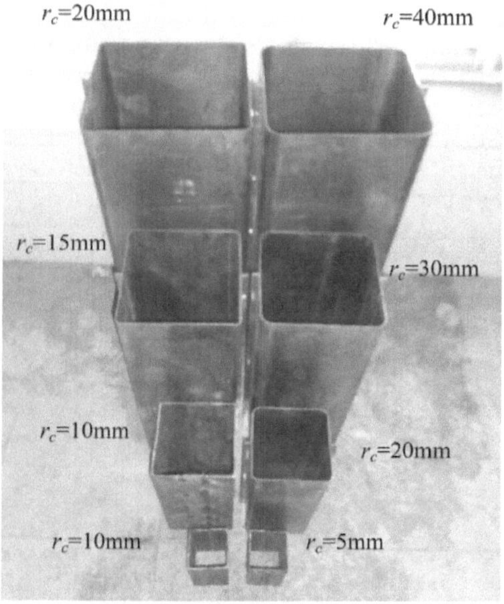

Figure 1. Stainless steel and angle iron molds.

Figure 2. Steel cages of prismatic specimens.

Figure 3. CFRP-confined specimens.

It is noteworthy that the total overlap of the CFRP jacket was taken as 50 mm, 200 mm, 450 mm and 800 mm for S1, S2, S3 and S4 specimens with continuous CFRP wrapping, respectively, to make the volume ratio between CFRP jacket and prismatic column identical. Figure 3 reveals each size of square columns wrapped with CFRP.

2.2 *Instrumentation and test procedure*

The test setup and instrumentation are shown in Fig. 4. All the specimens were tested in a compression machine with 10000 kN capacity. In the vertical direction, two linear variable differential transformers (LVDTs) mounted on steel frames were employed to measure the axial strain of specimens. The steel frames, consisting three scales due to various specimens sizes, were fixed at 1/4th and 3/4th the specimen height from top to the bottom, respectively. For the lateral strain measurement, two LVDTs were laid 180° apart to reflect the radial dilation of the square columns.

In this experiment, all the specimens were subjected to monotonic compressive loading until failure. A loading rate of 4 kN/s was adopted before reaching 80% peak compressive strength in accordance with theoretical prediction (Lam & Teng, 2003b) and then 2 kN/s remained up to complete failure.

3 TEST RESULTS AND DISCUSSION

3.1 *Failure modes*

All the CFRP confined-specimens failed suddenly in an explosive manner. The fracture of the CFRP wrap was a primary failure mode, which mainly appeared within the mid-height region. Once the jackets were removed, it became clear that RC square columns had severely crashed diagonally in the central zone. Moreover, the longitudinal steel bars had buckled

(a) S1 series specimen (b) S2 series specimen

(c) S3 series specimen (d) S4 series specimen

Figure 4. Test setup and instrumentation.

and the hoop reinforcement had bent outward, as shown in Figure 5. From the information mentioned above, it is obvious that the specimen size has little influence on the failure modes.

3.2 *Ultimate strength and strain*

The experimental results are summarized in Table 2. In Table 2, f'_{cc} and $\bar{f}'_{cc}$ are the individual and average compressive strengths of the CFRP-confined specimen, respectively. The term ε_{cc} is the ultimate axial strain, corresponding to f'_{cc}. The average value of ε_{cc} is denoted by $\bar{\varepsilon}_{cc}$. $f_{l,e}/f'_{co}$ is the average value of the effective confinement ratio, determined from the hoop strain within the mid-height region of the specimen. It should be noted that f'_{co} and ε_{co} are the ultimate compressive strength and corresponding strain of the unconfined RC square column in S2 series (200 × 600 mm). From Table 2, it is easy to obtain the average values of $f_{l,e}/f'_{co}$, which is equal to 0.149, 0.122, 0.069 and 0.046 and which corresponded to heavily, medium-, lightly, and minimum-confined level, respectively.

(a) S1 series specimen (b) S2 series specimen (c) S3 series specimen

(d) S4 series specimen (e) Typical post-yield pattern of longitudinal bars and stirrups

Figure 5. Pictures showing typical failure modes.

Figure 6 also displays the same relationship between specimen size and the ultimate strength. It can be seen that the specimen size has a significant effect on compressive strengths of unconfined RC square columns. For the CFRP-confined columns, the size effect on the ultimate strength of the specimen reduces with the increase of the lateral confinement level. Furthermore, it can be concluded, from Table 2 and Figure 6, that the strength of CFRP-confined RC square columns essentially exhibits no size effect while at the heavily confined level $\left(i.e.\ f_{l,e}/f'_{co} \geq 0.149\right)$.

Unlike the ultimate strength f'_{cc}, test data of ultimate strain ε_{cc} are relatively discrete, which hardly demonstrates a definite relationship to specimen size. Further research on this issue is necessary to be engaged in, which probably can produce some clear conclusions.

3.3 *Axial stress-strain response*

Axial stress-strain curves of all CFRP-confined specimens are categorized into four cases in accordance with various confinement levels. As shown in Figure 7, it can be recognized that the peak stress, ultimate strain, and trends of stress-strain curves exhibit significant differences between S2 and S4 specimens in the case of minimum-confined specimens. When the confinement reaches the lightly confinement level, the peak stress and ultimate strain still have some variance although the curve trend is similar for S2 and S4 specimens. Obviously,

Table 2. Compressive test results.

Specimen size ($B \times H$)	Confining level	$\frac{\bar{f}_{l,e}}{f'_{co}}$	f'_{cc} (MPa)	$\bar{f}'_{cc}$ (MPa)	$\frac{\bar{f}_{cc}}{f'_{co}}$	$\frac{\varepsilon_{cc}}{\varepsilon_{co}}$	$\frac{\bar{\varepsilon}_{cc}}{\varepsilon_{co}}$
S1C1R10-1	Heavily	0.135	40.38	41.68	1.49	4.89	3.78
S1C1R10-2			42.50			3.02	
S1C1R10-3*			42.16			3.42	
S2C2R20-1		0.153	42.84	42.88	1.53	6.62	5.86
S2C2R20-2			43.60			4.97	
S2C2R20-3*			42.20			5.99	
S3C3R30-1		0.176	40.98	41.58	1.49	5.22	5.00
S3C3R30-2			41.85			4.61	
S3C3R30-3*			41.91			5.17	
S4C4R40-1		0.131	41.82	40.93	1.47	5.94	4.96
S4C4R40-2			41.59			5.33	
S4C4R40-3*			39.37			3.60	
S1C1R5-1	Medium	0.118	37.51	37.98	1.36	2.52	4.29
S1C1R5-2			38.45			6.05	
S2C2R10-1		0.129	35.10	36.25	1.30	5.10	5.68
S2C2R10-2			37.39			6.26	
S3C3R15-1		0.129	37.10	35.50	1.27	6.08	5.41
S3C3R15-2			33.89			4.74	
S4C4R20-1		0.110	34.20	33.81	1.21	3.77	3.16
S4C4R20-2			33.41			2.56	
S2C1R20-1	Lightly	0.073	32.40	33.16	1.19	5.70	5.31
S2C1R20-2			33.92			4.91	
S4C2R40-1		0.065	29.11	29.23	1.05	2.45	2.42
S4C2R40-2			29.35			2.39	
S2C1R10	Minimum	0.045	32.47	32.47	1.16	2.15	2.15
S4C2R20		0.047	28.54	28.54	1.02	1.63	1.63

Note: (1) f'_{co} (28.01 MPa) and ε_{cc} (0.0046) indicate the compressive strength and corresponding strain of S2C0R0, respectively; (2) Designation marked with the asterisk * refers to the specimen wrapped with separate CFRP sheets.

the stress-strain relationships of all specimens are very similar except a little discrete data, which indicates that the size effect on the stress-strain response decreases and even essentially disappears while at the medium-confinement level or above.

3.4 *Confinement efficiency of different wrapping arrangements*

In this test, two CFRP wrapping arrangements, continuous wrapping and separate wrapping, were adopted for the case of heavily confined specimens, whose designation was marked with asterisks in Table 2. The use of a separate CFRP wrapping arrangement was to compare the confinement efficiency between different wrapping styles.

The confinement efficiency of the CFRP jacket can be determined by using Equation 3:

$$\rho_f = \frac{\left[4(B-2r_c)+2\pi r_c\right]\cdot t_f}{B^2-(4-\pi)\cdot r_c^2} \tag{3}$$

From Equation 3, the value of the confinement volume ratio for specimens with continuous wrapping is determined to be about $45t_f$, whereas $50t_f$ for ones with separate wrapping arrangements. It can be considered that the separate wrapping confinement has more efficiency theoretically. In fact, from Table 2, it can be understood that the ultimate strength and corresponding strain of specimens in a range from 100 × 300 mm to 300 × 900 mm ($B \times H$)

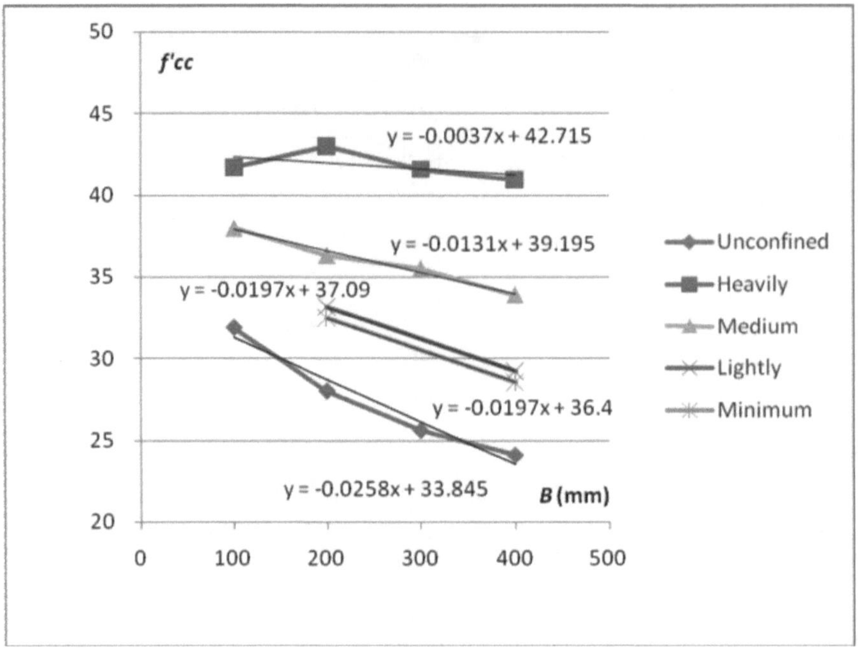

Figure 6. Relationship between specimen sizes and compressive strengths.

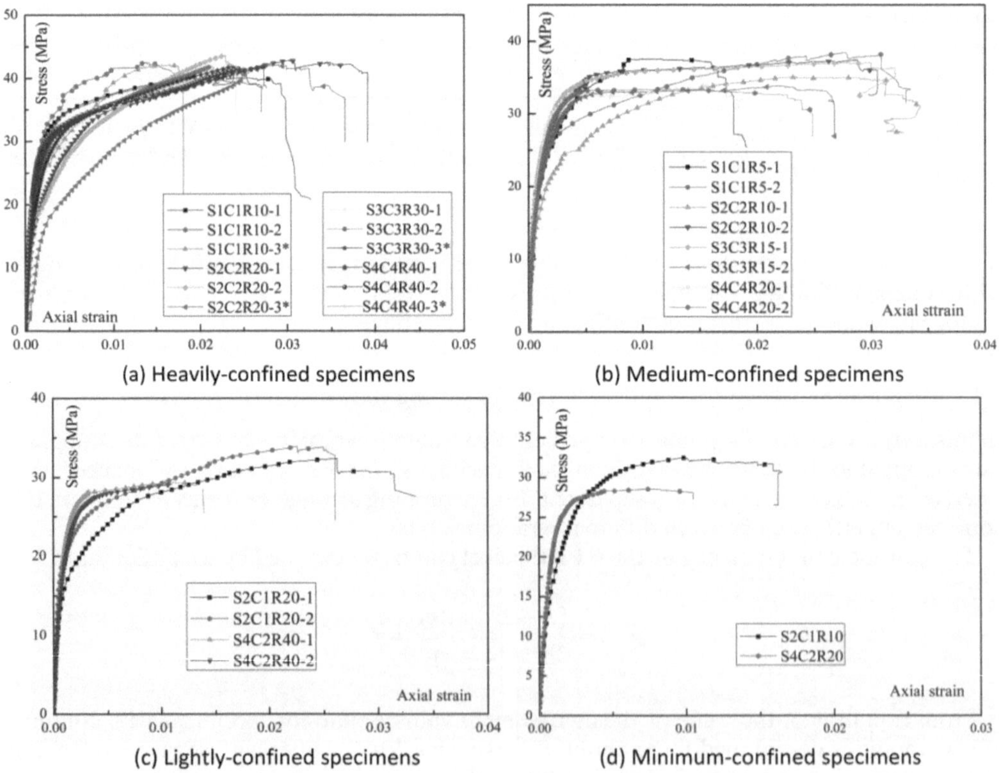

Figure 7. Axial stress-strain responses of specimens at various confinement levels.

are relatively close. But for the large-size columns (400 × 1200 mm), the ultimate strain of separate CFRP wrapping specimens is less about 40% when compared to the one of the continuous CFRP wrapping column, as well as the ultimate strength is lower about 5% for this case. As a result, it can be been realized that, under ultimate conditions, the axial performance of large size square columns with continuous CFRP wrapping is better than ones wrapped with separate CFRP sheets.

4 CONCLUDING REMARKS

The conclusions of this investigation are summarized in the following points.

1. The specimen size has no obvious effect on the failure modes for CFRP-confined RC square columns subjected to four different confinement levels.
2. When the confinement achieves the medium level or above $\left(f_{l,e}/f'_{co} \geq 0.122\right)$, the size effect exists on axial stress-strain relationships of confined specimens.
3. There is no size effect on the ultimate strength of confined RC square columns essentially while reaching the heavily confined level $\left(i.e. f_{l,e}/f'_{co} \geq 0.149\right)$.
4. The performance of RC square columns with continuous CFRP wrapping is better than the ones wrapped by separate CFRP sheets.

REFERENCES

Berthet, J.F., Ferrier, E, & Hamelin, P. (2005). Compressive behavior of concrete externally confined by composite jackets. Part A: experimental study. *Construction and Building Materials*, 19(3): 223–232.

Carey, S.A. & Harries, K.A. (2005). Axial behavior and modeling of confined small-, medium-, and large-scale circular sections with carbon fiber-reinforced polymer jackets. *ACI Structural Journal*, 102(4): 596–604.

Cui, C.Y. & Sheikh, S.A. (2010). Experimental study of normal- and high-strength concrete confined with fiber-reinforced polymers. *Journal of Composites for Construction*, 14(5): 553–561.

Elsanadedy, H.M., Al-Salloum, Y.A., Alsayed, S.H. & Iqbal, R.A. (2012). Experimental and numerical investigation of size effects in FRP-wrapped concrete columns. *Construction and Building Materials*, 29(4): 56–72.

Lam, L. & Teng, J.G. (2003a). Design-oriented stress-strain model for FRP-confined concrete. *Construction and Building Materials*, 17(6–7): 471–489.

Lam, L. & Teng, J.G. (2003b). Design-oriented stress-strain model for FRP-confined concrete in rectangular columns. *Journal of Reinforced Plastics and Composites*, 22(13): 1149–1185.

Lam, L. & Teng, J.G. (2004). Ultimate condition of fiber reinforced polymer-confined concrete. *Journal of Composites for Construction*, 8(6): 539–548.

Liang, M., Wu, Z.M., Ueda, T., Zheng, J.J. & Akogbe, R. (2012). Experiment and modeling on axial behavior of carbon fiber reinforced polymer confined concrete cylinders with different sizes. *Journal of Reinforced Plastics and Composites*, 31(6): 389–403.

Luca, A.D., Nardone, F., Matta, F., Nanni, A., Lignola, G.P. & Prota, A. (2011). Structural evaluation of full-scale FRP-confined reinforced concrete columns. *Journal of Composites for Construction*, 15(1): 112–123.

Masia, M.J., Gale, T.N., & Shrive, N.G. (2004). Size effects in axially loaded square-section concrete prisms strengthened using carbon fiber reinforced polymer wrapping. *Canadian Journal of Civil Engineering*, 31(1): 1–13.

Ministry of Housing and Urban-Rural Development of the People's Republic of China (2010) GB 50010–2010. Code for design of concrete structures. Beijing, MOHURD.

Mirmiran, A. & Shahawy, M. (1997). Behavior of concrete columns confined by fiber composite. *Journal of Structural Engineering*, 123(5): 583–590.

Pessiki S., Harries, K.A., Kestner, J.T., Sause, R. & Ricles, J.M. (2001). Axial behavior of reinforced concrete columns confined with FRP jackets. *Journal of Composites for Construction*, 5(4): 237–245.

Rocca, S. (2007). *Experimental and analytical evaluation of FRP-confined large size reinforced concrete columns*. PhD Dissertation, University of Missouri, Rolla, USA.

Rochette, P. & Labossière, P. (2000). Axial testing of rectangular column models confined with composites[J]. *Journal of Composites for Construction*, 4(3): 129–136.
Shahawy, M., Mirmiran, A. & Beitelman, T (2000). Tests and modeling of carbon-wrapped concrete columns. *Composites, Part B: Engineering*, 31(6–7): 471–480.
Thériault. M., Neale, K.W., & Claude, S. (2004). Fiber-reinforced polymer-confined circular concrete columns: Investigation of size and slenderness effects. *Journal of Composites for Construction*, 8(4): 323–331.
Toutanji, H., Han, M., Gilbert, J., & Matthys, S. (2010). Behavior of large-scale rectangular columns confined with FRP composites. *Journal of Composites for Construction*, 14(1): 62–71.
Wang, Y.F. & Wu, H.L. (2011). Size effect of concrete short columns confined with aramid FRP jackets. *Journal of Composites for Construction*, 15(4), 535–544.
Wang Z.Y., Wang D.Y., Smith S.T., & Lu, D. (2012). CFRP-confined square RC columns. I: Experimental investigation. *Journal of Composites for Construction*, 16(2): 150–160.
Xiao, Y. & Wu, H. (2000). Compressive behavior of concrete confined by carbon fiber composite jackets. *Journal of Materials in Civil Engineering*, 12(2): 139–146.
Yeh, F.Y. & Chang, K.C. (2007). Confinement efficiency and size effect of FRP confined circular concrete columns. *Structural Engineering and Mechanics*, 26(2): 127–150.
Youssef, M.N., Feng, Q., & Mosallam, A.S. (2007). Stress-strain model for concrete confined by FRP composites. *Composites Part B: Engineering*, 38(5–6): 614–628.

Innovation and Application of Engineering Technology – Oscar, Chen & Wang (Eds)
© 2017 Taylor & Francis Group, London, ISBN 978-1-138-05057-0

A study on new air supply terminals in mine fires

K. Li
China Academy of Safety Science and Technology, Beijing, China

Y. Shu & C. Liu
China Academy of Safety Science and Technology, Beijing, China
School of Civil and Resource Engineering, University of Science and Technology Beijing, Beijing, China

X.Y. Xie & G.D. Mei
China Academy of Safety Science and Technology, Beijing, China

ABSTRACT: Based on the two-dimensional model, the influence of different air supply terminals including cylindrical, conical, arc and new ones on the air and gas distribution was investigated. The results showed that the air jet from all types of air supply terminals had the flushing and entrainment effects on the airflow in the tunnel. Only the new air supply terminal had the eddy effect and the main area influenced can be characterized by three connected regions: entrainment region, flushing region and eddy region. The smoke (CO) concentration was quite low around the air jet and increased as the distance (less than 2 m) to the cylindrical, conical and arc terminals increased. A stable region with quite low smoke concentrations, which is less than 10 ppm, was created by the new air supply terminal due to the eddy effect. The safety zone with CO concentration less than 24 ppm was the largest when using the new air supply terminal and it had the best effect on the dilution and dispersion of the smoke gas.

1 INTRODUCTION

Mine fire is one of the major disasters in non-coal mines, which seriously threatens the safety of non-coal mine workers and hinders the normal production of the mine. As a result of the high temperatures and poisoning choke effect of fire, the single accident death rate of fire accident is much higher than any other mine disasters. And so, the industry standards, the regulations for the construction of a compressed-air self-help system in metal and nonmetal underground mines, were officially released by the State Administration of Work Safety in July, 2011. Up to now, the compressed-air self-help system is constructed in almost 40,000 non-coal underground mines. The compressed-air self-help system is an air supply system for the underground mine workers in fires and provided a new method for escaping during mine fires. However, the effect of the system on the dilution and dispersion of the smoke gas was the critical factor for saving the underground mine workers.

At present, there are many researchers, who are studying the control and prevention of non-coal mine fires, but only a few studies focus on the effect of the compressed-air self-help system in mine fires. With regard to mine fires studies, some researchers have mainly focused on the flow pattern of the smoke gas in fires (Duan 2014, Singh 2007). Some had also studied the division of the safety zone (Chen 2016, Deng 2015). A few had explored the secondary disasters caused by mine fires (Li 2010). With respect to the studies of the compressed-air self-help system, these have the parameters specification and system design as their main areas of focus (Li 2012, Lu 2010 & Zhang 2013). There were only a few researchers carrying out the numerical simulation of the compressed-air self-help system during mine fires (Zhang 2013). However, in those relative studies, the ordinary cylindrical air supply terminals were used in the system and few studies discussed the design of the air supply terminal and the influence on the smoke gas in mine fires. The air supply terminal was an important and essential part of the compressed-air self-help system, which had a significant influence on the dilution and dispersion of the smoke gas. Therefore, in this paper, it is aimed to analyze the influence of four different air supply terminals in horizontal tunnels during mine fires and determine the optimal form of the air supply terminal.

2 PHYSICAL AND NUMERICAL MODELS

The two-dimensional physical model of the horizontal tunnel was established, as shown in Figure 1. The horizontal tunnel was 10 m long and 4 m wide. The air supply terminal was located on the left-hand side of the horizontal tunnel, and the centerline of the air supply terminal is 3 m away from the inlet of the horizontal tunnel. Under the premise of the same inlet diameter and inlet velocity of the air supply terminal, the cylindrical, conical, arc and new air supply terminals were chosen to carry on the research. The sizes of four air supply terminals are shown in Figure 1.

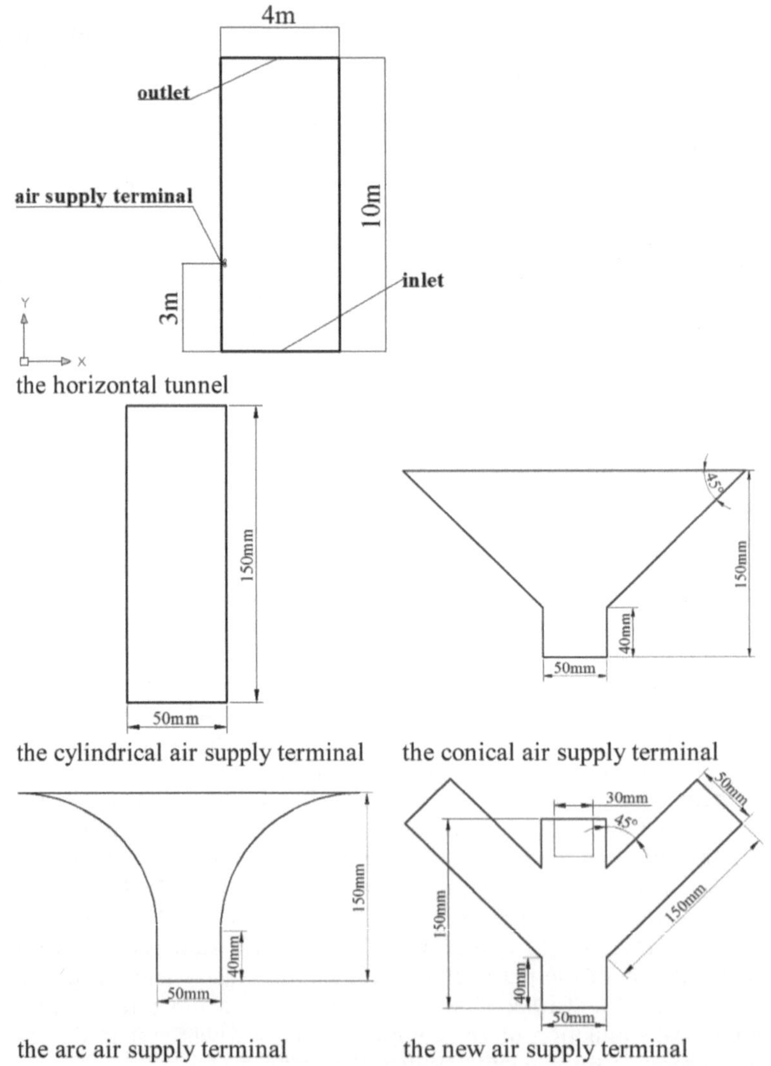

Figure 1. Models of the horizontal tunnel and four air supply terminals.

In this paper, the standard k-ε equation and the gas transport equation were used to calculate the turbulent flow and diffusion of the fluid. When fire broke out, the density of the heated air flow would change, and so the influence of gravity on the flow process was considered in the simulation.

When developing the mathematical model, the following assumptions were made: the wall of the horizontal tunnel was supposed to be adiabatic. There were no obstacles in the tunnel and the influence of dust is ignored; the specific combustion process and heat radiation of the fire source will not be considered; and the fire smoke was assumed to be CO.

The boundary conditions were set as follows: the air inlet of the horizontal tunnel was set as a velocity inlet. The air supply terminal was set as a velocity inlet. The air outlet of the horizontal tunnel was set as a pressure outlet.

According to the ventilation in real mines, the initial inlet velocity of the horizontal tunnel was set to be 0.7 m/s. The initial velocity of air jet from the air supply terminal was set to be 50 m/s. The initial CO concentration in the whole horizontal tunnel was set to be 50 ppm. Based on the model described above, the influence of different air supply terminals on the airflow, the smoke gas distribution and the safety zone were investigated.

3 RESULTS AND ANALYSIS

3.1 *Influence of the air supply terminal on the airflow*

When the air supply terminal was in operation, the air jet from the terminal would eject into the horizontal tunnel and change the air flow distribution. When the air jet velocities of the four air supply terminals were all 50 m/s, the airflow in the horizontal tunnel was as shown in Figure 2.

As shown in Figure 2, it can be found that the air jet from all the four different air supply terminals had an entrainment and flushing effects on the air flow in the horizontal tunnel. In addition, the air flow distribution being influenced by the cylindrical, conical and arc air supply terminals was similar. The flow field could be characterized by two regions: the entrainment region and the flushing region. However, the air flow distribution of the new air supply terminal was quite different from the other three types. As shown in Figure 3, the flow field influenced by the new air supply terminal could be characterized by three regions: the entrainment region, flushing region and eddy region. Moreover, the air flow status in the

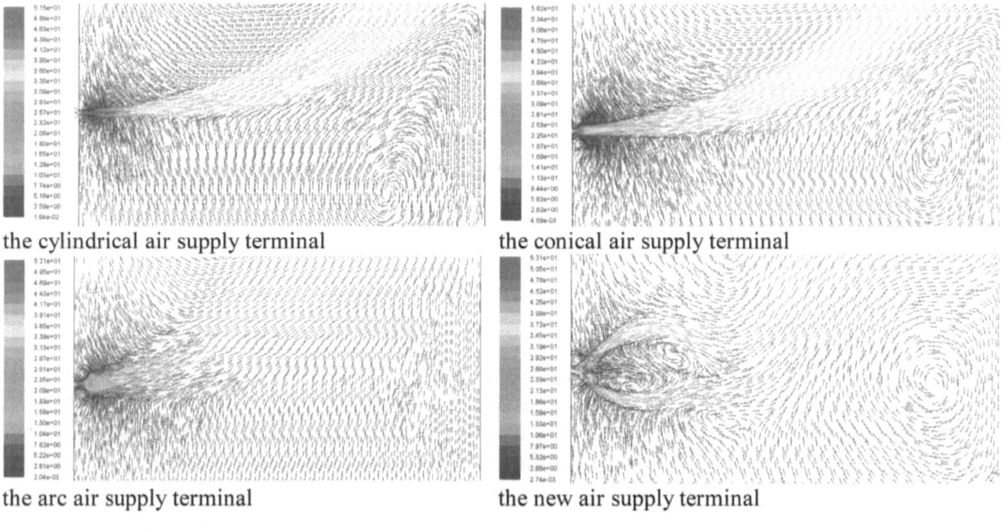

Figure 2. The air flow distribution in the horizontal tunnel.

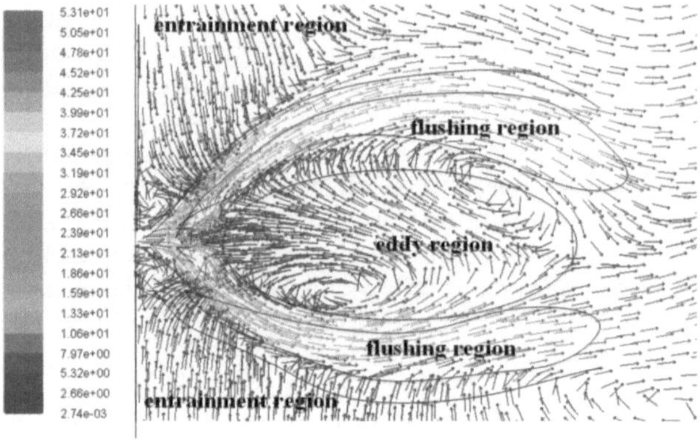

Figure 3. The air flow distribution influenced by the new air supply terminal in the horizontal tunnel.

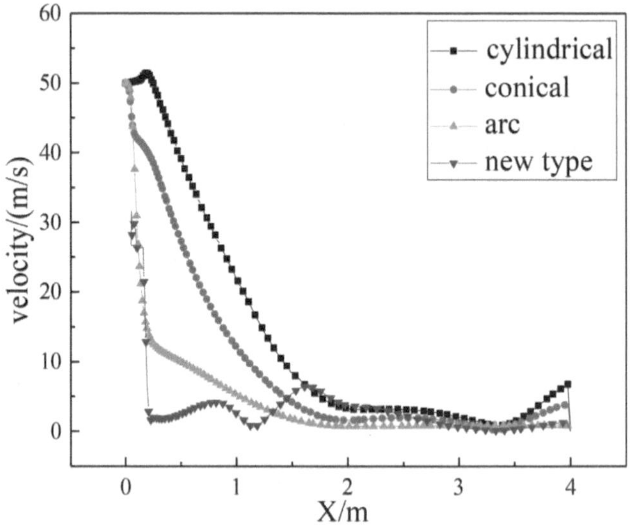

Figure 4. The variation of the jet velocity at y = 3 m.

eddy region is stable and the air flow velocity was very low, which is suitable for mine workers' breathing. The main reason for inducing this phenomenon is that there were two other air outlets on the side of the new air supply terminal and the middle air outlet had an obstacle in the center. Due to the interaction between the air jet in the middle and on both sides, two small eddies were formed and created the eddy region.

In order to further analyze the effect of four air supply terminals on the air flow, the variation of the air jet velocity is calculated at y = 3 m, as shown in Figure 4.

It can be seen from Figure 4 that as the distance to the air supply terminal increased, the air jet velocity decreased and tended to be stable when the distance was larger than 2 m. This meant that the flushing region was 2 m long. However, the decreasing rate of the air jet velocity of the four air supply terminals was not the same. The decreasing rate of the new air supply terminal is the highest and the air jet velocity tended to be stable and quite low around 3 m/s, when the air jet ejected into the horizontal tunnel. The decreasing rates of the other three air supply terminals were low and the air jet velocity was higher than 10 m/s. The main reason of this phenomenon was the formation of the eddy region by the new air supply terminal.

3.2 *Influence of the air supply terminal on gas distribution*

As discussed above, when the air supply terminal was in operation, the air flow in the horizontal tunnel changed significantly, and so the distribution of the smoke gas would be affected. Figure 5 showed the distribution of the smoke gas in the horizontal tunnel when four air supply terminals was in operation, respectively.

It can be seen from Figure 5 that, due to the flushing effect of the air jet from the cylindrical, conical and arc air supply terminals, the gas concentration in the front area decreased to a low level. However, due to the strong entrainment effect, the area influenced was small and narrow, basically right ahead of these three air supply terminals. As for the new air supply terminals, because of the low air jet velocity, the.entrainment effect was weak and the eddy effect was the main factor influencing the gas distribution; it showed a totally different pattern. The gas concentration in the eddy region was quite low and suitable for saving mine workers.

In order to further analyze the effect of four air supply terminals on the distribution of the smoke gas, the variation of the concentration of CO gas is calculated at y = 3 m, as shown in Figure 6.

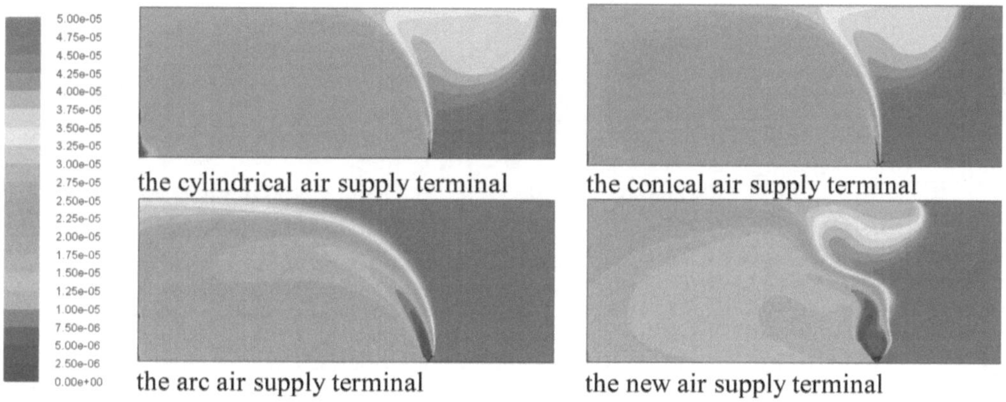

Figure 5. The distribution of the smoke gas in the horizontal tunnels.

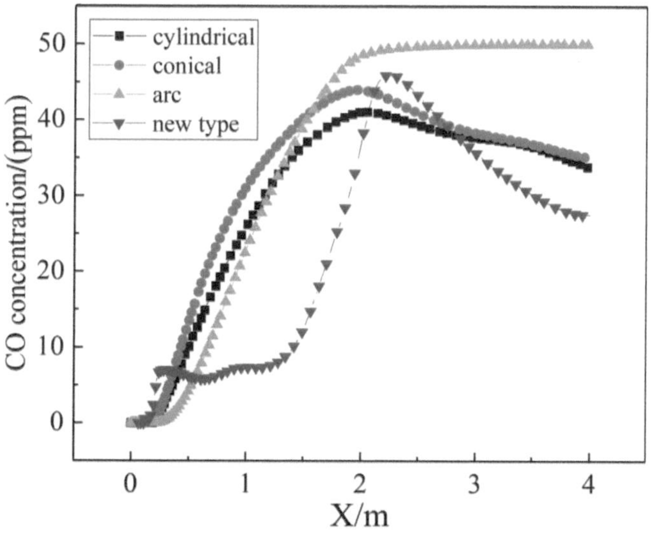

Figure 6. The variation of the gas concentration at y = 3 m.

It can be found from Figure 6 that, in the flushing region for the cylindrical, conical and arc air supply terminals, the gas concentration increased as the distance to the terminals increased. The reason is that, as the distance increased, the flushing effect became weaker and had less and less influence on the smoke gas distribution away from the terminal. With regard to the new air supply terminal, the gas distribution was quite different. The gas concentration increases slightly and then maintained at a constant value in the eddy region. In this eddy region, the gas concentration was lower than 10 ppm, which was safe for mine workers' breathing.

3.3 *Analysis of the safety zone*

In this paper, the safety zone was defined as the area, where the gas concentration in the horizontal tunnel was less than 24 ppm. The safety zone in the horizontal tunnel, when the air supply terminals were in operation, is shown in Figures 7 and 8.

It can be seen from Figure 7 that, due to the strong entrainment effect of the cylindrical and conical air supply terminals, the safety zone was small and narrow, basically right

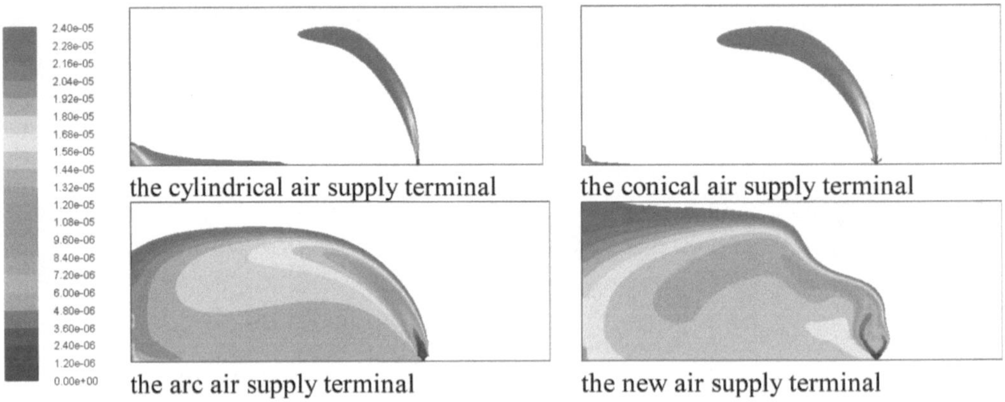

Figure 7. The safety zone distribution of four air supply terminals.

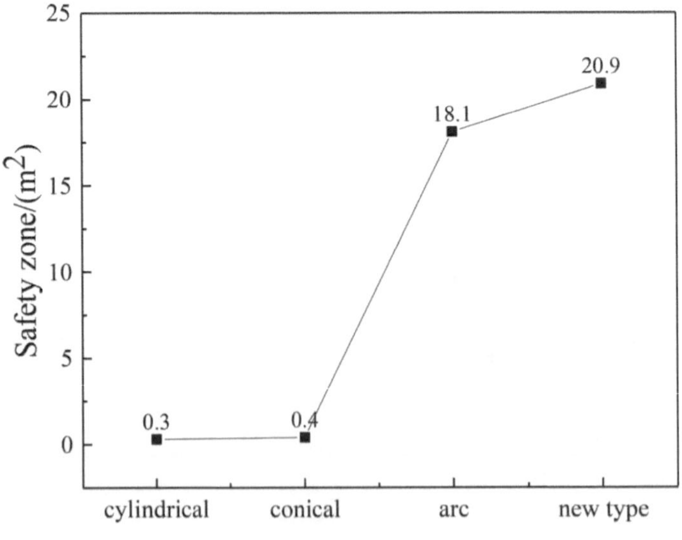

Figure 8. The area of the safety zone of four air supply terminals.

ahead of these two terminals. With respect to the other two terminals, especially the new one, the safety zone was large and the gas concentration was low. In the zone just right ahead of the new terminal, the safety zone was larger and the gas concentration was much lower than those of the arc terminal. To further analyze the safety zone of four air supply terminals, the area of the safety zone in the horizontal tunnel was calculated, as shown in Figure 8. The safety zone area formed by the cylindrical and conical air supply terminals were 0.3 m^2 and 0.4 m^2, while it was 18.1 m^2 and 20.9 m^2 by the arc and new terminals. Based on the analysis made above, it can be concluded that the cylindrical and conical air supply terminals widely used in the non-coal mines had poor ability to dilute the smoke gas and were actually not suitable for rescuing mine workers. The new air supply terminal was the best for diluting the smoke gas and much more suitable for the compressed-air self-help system.

4 CONCLUSIONS

The influence of different air supply terminals including cylindrical, conical, arc and new ones on the air and gas distribution and safety zone was investigated in this paper. Some conclusions can be obtained as follows:

1. The air jet from all types of air supply terminals has the flushing and entrainment effects on the airflow. The main area influenced by the new air supply terminal can be characterized by three connected regions: entrainment region, flushing region and eddy region.
2. Due to the eddy effect of the new air supply terminal, a stable region with quite low smoke concentration, which is less than 10 ppm, was created.
3. The cylindrical and conical air supply terminals widely used in the non-coal mines were actually not suitable for rescuing mine workers. The new air supply terminal was the best for the compressed-air self-help system in fire disasters.

ACKNOWLEDGMENTS

This work was supported by the Fundamental Research Funds for the China Academy of Safety Science and Technology (no. 2016 JBKY06) and the 2016 Scientific and Technological Projects for Solving Safety Problems (no. 2016PJNT05).

REFERENCES

Chen, L. & Wu, C.F. 2016. Numerical Simulation Study on the Regional Division of Mine Roadways Fire Safety. Fire Science and Technology 05: 633–636.

Deng, Q.L. 2015. The Numerical Simulation of the Smoke Flow during Fire and the Safety Zone Divide in Mine Roadways. Jiangxi University of Science and Technology.

Duan, J. & Lin, J.S. 2014. Numerical Simulation of Fire Smoke Flow in Tunnel Changes Based on Fluent. Coal Technology 09: 310–312.

Li, K. & Quan, J. 2010. Influence of Fire Location on Ventilation and Secondary Catastrophes in Tunnels. Journal of Tsinghua University (Science and Technology) 02: 270–273.

Li, L. & Huang, Y.Y. 2012. Design of Parameter Compressed-air Self-help System in Metal Mine. Metal Mine 08: 150–152.

Lu, J.Z. 2010. Compressed Air Self Rescue System in Underground Mine. Coal Science and Technology 38(12): 4–6, 32.

Singh, A.K. & Singh, R.V.K. 2007. Mine Fire Gas Indices and their Application to Indian Underground Coal Mine Fires. International Journal of the Coal Geology 69: 192–204.

Zhang, Y.H. & Huang, M.Y. 2013. Research on advanced compressed air self-rescue system in underground mine. Journal of Mining & Safety Engineering 01: 136–142.

Zhang, J.K. & Mu, S.B. 2013. Effects of compressed air self-help system on mine fires in excavation tunnels. Gold 11: 30–34.

Innovation and Application of Engineering Technology – Oscar, Chen & Wang (Eds)
© 2017 Taylor & Francis Group, London, ISBN 978-1-138-05057-0

Analysis of new air supply terminals on the fire smoke distribution in tunnels

K. Li
China Academy of Safety Science and Technology, Beijing, China

C. Liu & Y. Shu
China Academy of Safety Science and Technology, Beijing, China
School of Civil and Resource Engineering, University of Science and Technology Beijing, Beijing, China

X.Y. Xie & G.D. Mei
China Academy of Safety Science and Technology, Beijing, China

ABSTRACT: Based on the two-dimensional horizontal tunnel model and the new air supply terminal model, the influence of different air jet velocities from the terminal on the fire smoke gas distribution and safety zone was investigated. The results showed that, in the area right ahead of the terminal, the gas distribution can be characterized by three regions including the entrainment region, flushing region and eddy region at different air jet velocities. The gas concentrations in the flushing region and eddy region were lower than 10 ppm. The area of the eddy region with low gas concentration increased linearly as the air jet velocity increased. When the air jet velocity was 50 m/s, the average velocity of the airflow in the safety zone was lower than 2 m/s and the safety zone, especially the eddy region, was much more suitable for rescuing mine workers.

1 INTRODUCTION

Mine fires, one of the major mine disasters, have been seriously threatening the lives of mining workers and restraining the mining production. As the fire smoke contains toxic gases with high temperature, the single accidental death rate of mine fires is much higher than the other disasters'. At present, most studies on fire prevention are focused on numerical simulation of smoke gas flow and distribution (Li 2013, Yang 1994) and the critical velocity (Feng 2002, Hwang 2005, Kunsch 2000 & Wu 2000). The compressed-air self-help system is an air supply system for the underground mine workers in fires and provided a new method for escaping during mine fires. Up to now, the compressed-air self-help system has been constructed in almost 40,000 non-coal underground mines. However, there were only a few researchers studying the compressed-air self-help system during mine fires. These researchers are mainly focused on the design of the system parameters and functions (Li 2012, Xue 2012 & Zhang 2013) and few studied on the effects of the compressed-air self-help system (Zhang 2013). The ordinary air supply terminals cannot meet the safety requirements; and so, it is quite significant to study how the new air supply terminal affects the gas distribution and safety zone in horizontal tunnels.

In this article, the numerical simulation method was used for the study on the influence of the different air jet velocities on the gas distribution and safety zone.

2 PHYSICAL AND NUMERICAL MODELS

Based on the real non-coal mine tunnels, the two-dimensional horizontal tunnel and air supply terminal models were developed, as shown in Figure 1. The horizontal tunnel was 10 m

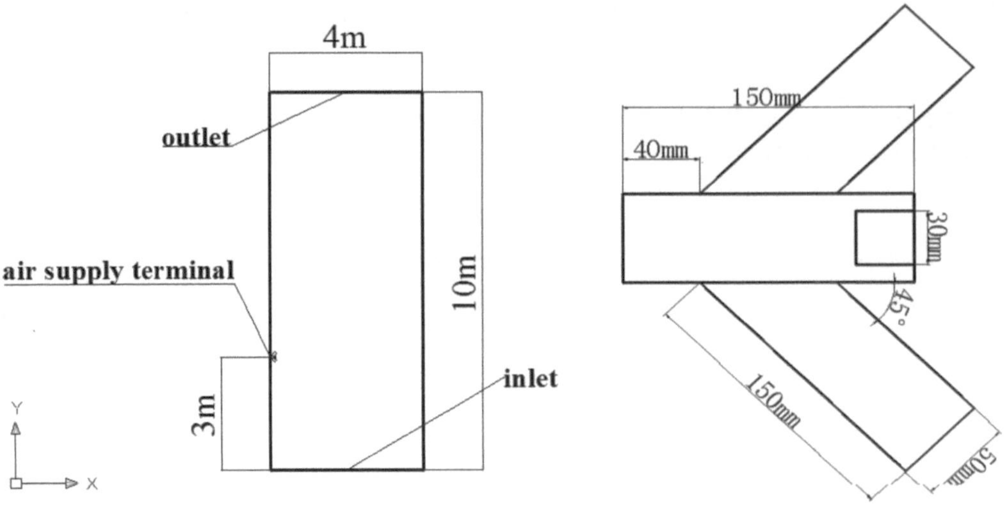

the horizontal tunnel

the air supply terminal

Figure 1. The tunnel and air supply terminal models.

long and 4 m wide. The air supply terminal was fixed on the left-hand side of the tunnel, and the centerline of the air supply terminal is 3 m away from the inlet of the tunnel. The diameter of the terminal inlet and outlet were both 50 mm. The angles between the outlet on the sides and central outlet were both 45°. The baffles in the central outlet was 30 mm wide.

To facilitate calculating, the following assumptions were made: the wall of the horizontal tunnel was supposed to be adiabatic. There were no obstacles in the tunnel, and the influence of dust is ignored; the specific combustion process and heat radiation of the fire source will not be considered; and the fire smoke was assumed to be CO.

The standard k-ε equation and the gas transport equation were used to calculate the turbulent flow and diffusion of the fluid. When fire broke out, the density of the heated air flow would change and therefore, the influence of gravity on the flow process was considered in the simulation.

The boundary conditions were set as follows: the air inlet of the horizontal tunnel was set as a velocity inlet. The air supply terminal was set as a velocity inlet. The air outlet of the horizontal tunnel was set as a pressure outlet.

According to the ventilation in real mines, the initial inlet velocity of the horizontal tunnel was set to be 0.7 m/s. The initial CO concentration in the whole horizontal tunnel was set to be 50 ppm. Based on models described above, the influence of the new air supply terminal on the gas and safety zone distribution under different compressed-air velocities (10 m/s, 20 m/s, 30 m/s, 40 m/s, 50 m/s) was analyzed.

3 RESULTS AND DISCUSSION

3.1 *Influence of the air jet on the gas flow and distribution*

Figure 2 shows the gas flow pattern around the terminal under different air jet velocities. It can be found that under different air jet velocities, the gas flow pattern changed a little. The air jet from the central outlet and the outlet on both sides ejected into the tunnel and formed the flushing region along the ejection direction. And the gas concentration in this region was quite low, but the velocity was high. When the air jet was ejecting into the tunnel, the gas around the air jet was attracted and forced into the jet due to the entrainment effect and formed the entrainment region on the outside of the flushing region. In the entrainment region, the velocity

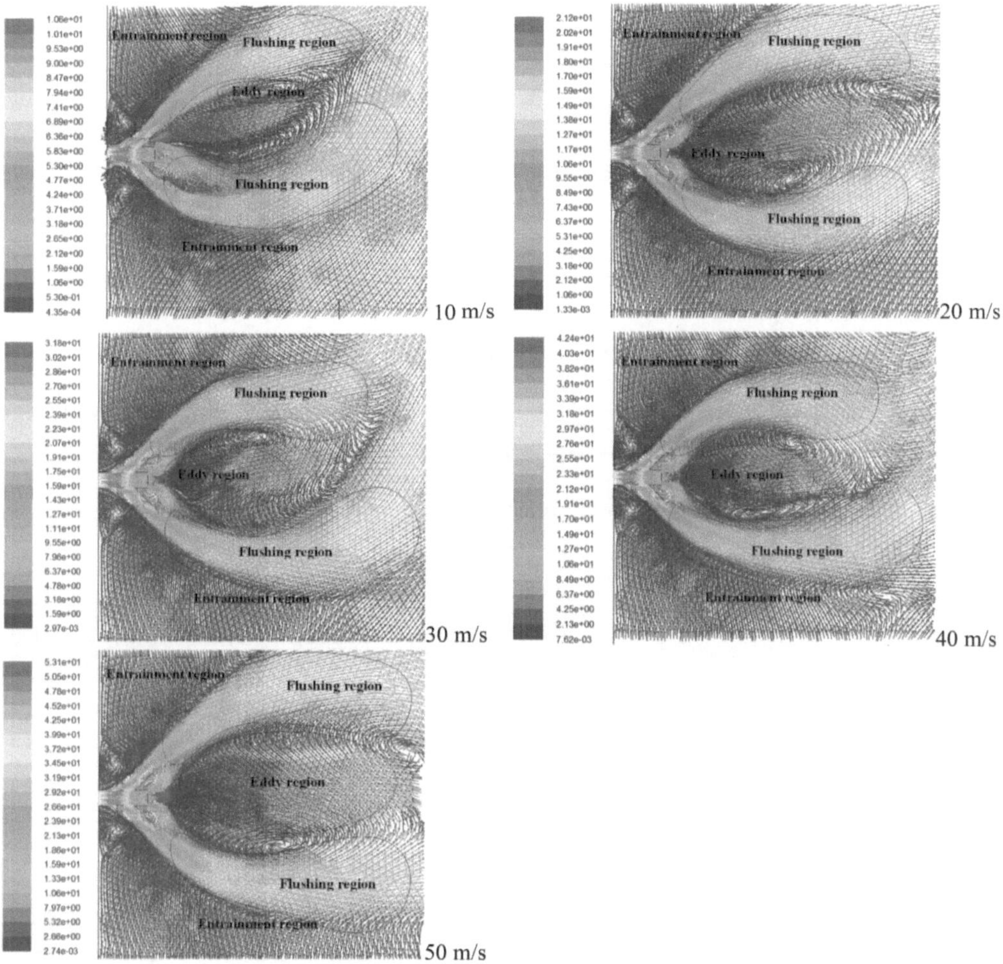

Figure 2. The gas flow under different air jet velocities.

was quite low, but the gas concentration was very high. As a result of the terminal design, there were symmetrical flushing regions and two eddies were created between the flushing regions and formed the eddy region. In this region, the velocity and the gas concentration were both low and it was the critical and main area for mine workers' breathing. Based on the analysis detailed above, the air jet velocity was not a critical factor influencing the gas flow pattern. Under different air jet velocities, the gas distribution right ahead of the terminal can be divided into three parts: entrainment region, flushing region, and eddy region.

Figure 3 showed the relationship between air jet velocity and the area of the eddy region. It can be found that, as the air jet velocity increased, the area of the eddy region increased and increased sharply when the air jet velocity was 50 m/s. The reason resulting in this phenomenon was that, as the air jet velocity increased, the flushing effect enhanced and the eddies between the flushing region became larger. The bigger eddy region was much safer for mine workers.

The air jet had the entrainment and flushing effects on the gas flow and distribution. Figure 4 showed the gas distribution in the tunnel under different air jet velocities.

As shown in Figure 4, the effect of the air jet on the dilution of the gas in the tunnel was positive and remarkable. The gas distribution near the tunnel inlet was not affected and it had changed a lot in the region right ahead of the terminal. The gas concentration was the

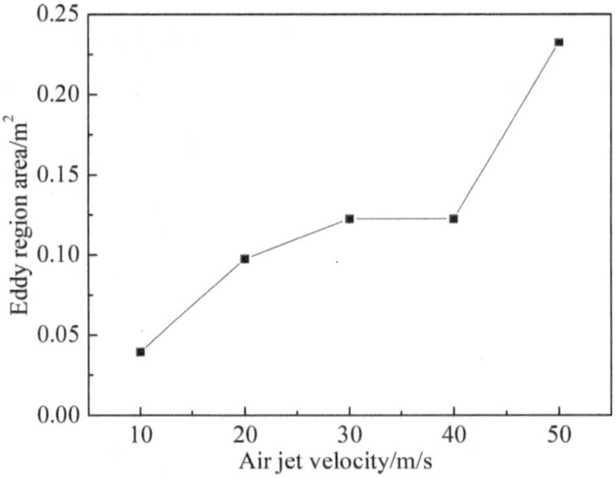

Figure 3. The area of the eddy region under different air jet velocities.

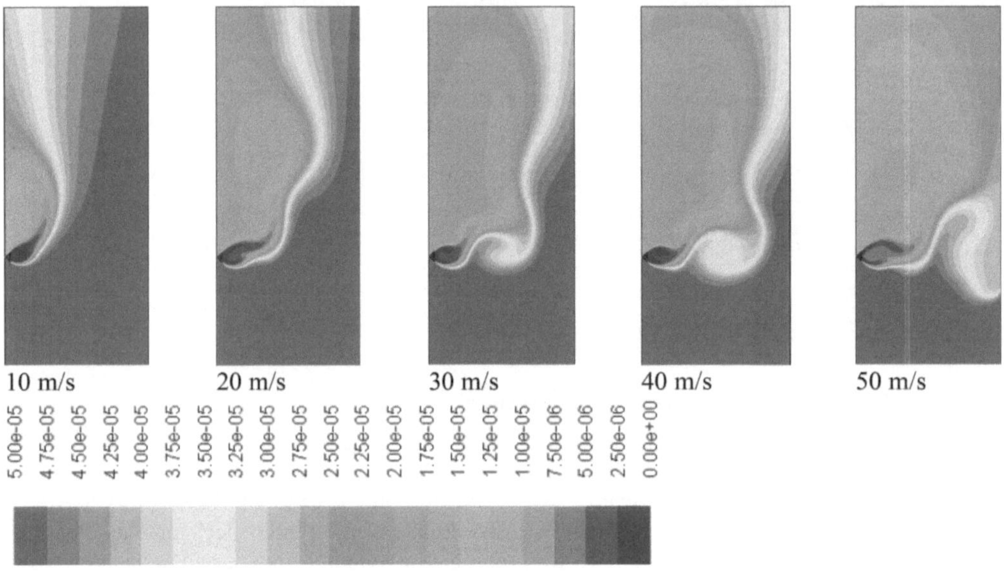

Figure 4. Gas distribution under different air jet velocities.

lowest in the flushing region and it was also quite low in the eddy region. However, as the air jet velocity increased the gas concentration in the eddy region increased to a certain extent due to the entrainment effect.

To further study the gas distribution in the tunnel, the gas distribution under different air jet velocities at x = 2 m and y = 3 m was analyzed, as shown in Figures 5 and 6. The gas concentration remained at a high level around the tunnel inlet (y = 0–2 m), and started to decline sharply around the terminal (Y = 2–4 m). The concentration in the domain (Y = 4–10 m) basically had no variation. The interval between the highest and lowest gas concentration increased as the air jet velocity increased and was maintained at a certain value when the air jet velocity was higher than 30 m/s. The reason was that the air jet flow has both entrainment and flushing effects on the gas around. When the air jet flow increased, the flushing effect

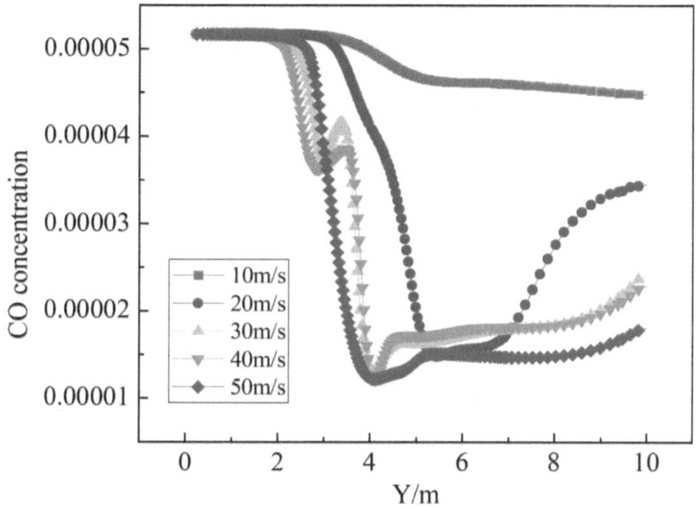

Figure 5. The variation of gas concentration at x = 2 m.

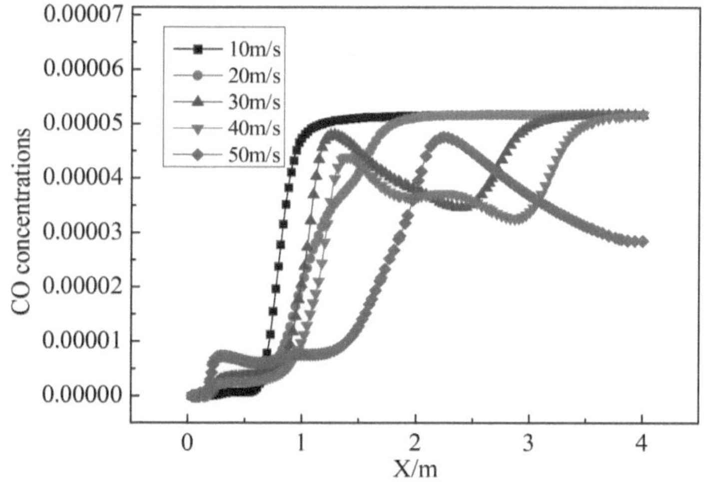

Figure 6. The variation of gas concentration at y = 3 m.

became stronger than the entrainment effect and achieved a balance when the air jet velocity was higher than 30 m/s.

It can be seen from Figure 6 that the gas concentration increased rapidly around the terminal inlet, and then remained at a certain value and increased as the distance to the terminal increased. Along with the ejection direction, the scope influenced increased linearly while the air jet velocity increased. In addition, the scope, with the gas concentration being maintained at a certain value (less than 10 ppm), which was known as the eddy region, increased linearly as the air velocity increased. Based on the analysis above, the area of the eddy region had a positive correlation with the air jet velocity. And from Figure 6, it can be seen that the gas concentration in the eddy region was lower than 10 ppm, which is appropriate for mine workers' breathing.

3.2 *Influence of the air jet on the safety zone*

The safety zone was defined as the zone where the gas concentration was lower than 24 ppm. Figure 7 showed the distribution of safety zones under different air jet velocities.

As shown in Figure 7, when the terminal was in operation, there was a safety zone in the horizontal tunnel and the safety zone spread right ahead and to the downstream region of the terminal. As the air jet velocity increased, the scope of the safety zone had a positive correlation with the air jet velocity. When the air jet velocity was 50 m/s, the scope of the safety zone increased sharply and reached to 1.7 m, which was large enough for rescuing several mine workers, as shown in Figure 8. However, whether the safety zone was suitable for saving mine workers or not was subjected to two factors including the area and gas velocity. Figure 9 showed the relationship between the average velocity in the safety zone and the air jet velocity. It can be found that, as the air jet velocity increased, the average velocity in the safety zone increased first and then decreased. When the air jet velocity was 50 m/s, the average velocity was lower than 2 m/s, which was comfortable for mine workers. Therefore, simultaneously considering the two factors influencing the safety zone, when the air jet velocity was 50 m/s, the area of the safety zone and the average velocity in the safety zone got optimal and these were the suitable operation parameters of the terminal.

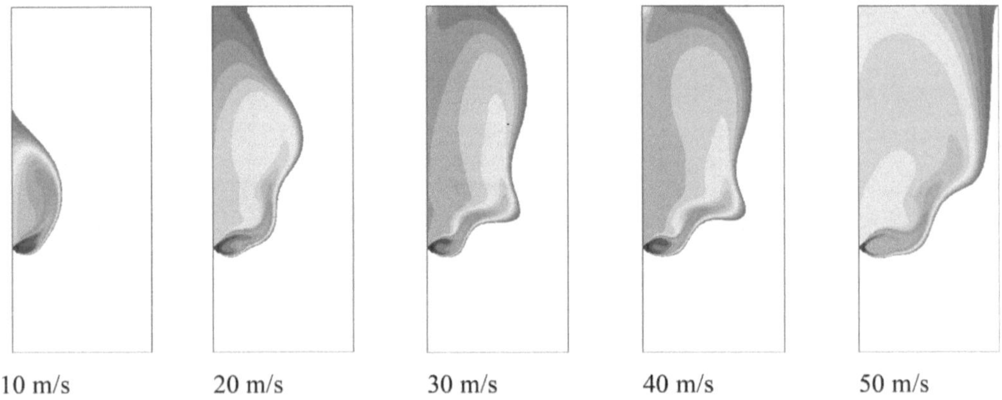

Figure 7. Distribution of the safety zone under different air jet velocities.

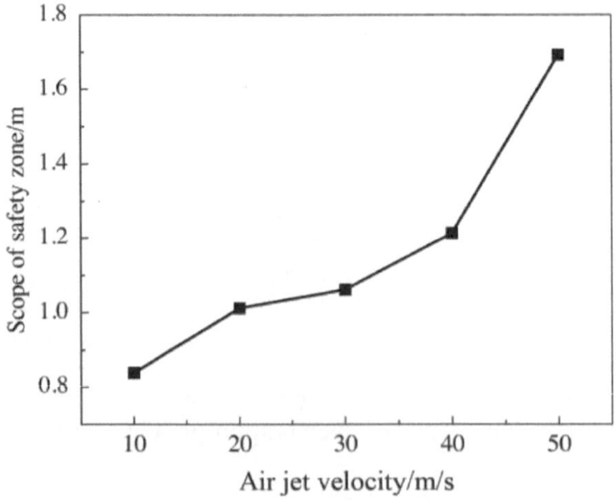

Figure 8. Graph showing the relationship between the safety zone and air jet velocities.

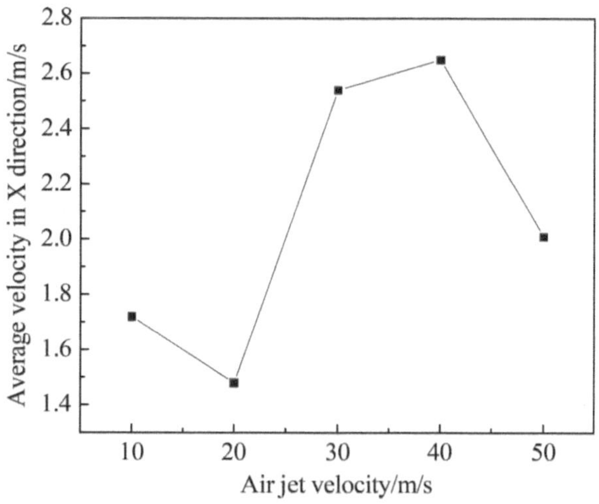

Figure 9. Graph showing the safety zone's average velocity in the X direction.

4 CONCLUSIONS

Numerical models for the new air supply terminal and horizontal tunnel has been established and the influence of the air jet on the fire gas distribution and safety zone was investigated. The results are as follows: in the area right ahead of the terminal, the gas distribution can be characterized by three regions including the entrainment region, flushing region, and eddy region under different air jet velocities. The gas concentrations in the flushing region and eddy region were lower than 10 ppm. The area of the eddy region with low gas concentration increased linearly as the air jet velocity increased. When the air jet velocity was 50 m/s, the average velocity of the airflow in the safety zone was lower than 2 m/s and the safety zone, especially the eddy region, was much more suitable for rescuing mine workers.

1. The air jet velocity was not a critical factor influencing the gas flow pattern. Under different air jet velocities, the gas distribution right ahead of the terminal can be divided into three parts: entrainment region, flushing region, and eddy region.
2. The gas concentrations in the flushing region and eddy region were lower than 10 ppm. The area of the eddy region with low gas concentration increased linearly as the air jet velocity increased.
3. The safety zone can be evaluated by two factors including the area and gas velocity. When the air jet velocity was 50 m/s, the area of the safety zone, especially the eddy region, and the average velocity in the safety zone became optimal and it was the suitable operation parameters of the terminal and was much more suitable for rescuing mine workers.

ACKNOWLEDGMENTS

This work was supported by the Fundamental Research Funds for the China Academy of Safety Science and Technology (no. 2016 JBKY06) and the 2016 Scientific and Technological Projects for Solving Safety Problems (no. 2016PJNT05).

REFERENCES

Feng, L. & Liu, Y.Q. 2002. Numerical simulation of subway fire smoke control. Underground Space 01: 61–64.

Hwang, C.C & Edwards, J.C. 2005. The critical ventilation velocity in tunnel fires—a computer simulation. *Fire Safety Journal* 40(3): 213–244.
Kunsch, J.P. 2000. Simple model for control of fire gases in a ventilated tunnel. *Fire Safety Journal.*
Li, L. & Hung, Y.Y. 2012. Design of parameter compressed-air self-help system in Metal Mine. *Metal Mine* 08: 150–152.
Li, C.P. & Cao, Z.G. 2013. 3-D simulation modeling techniques of fire fume spread process for underground mines. *Journal of China Coal Society* 02: 257–263.
Wu, Y. & Bakar, M.Z.A. 2000. Control of smoke flow in tunnel fires using longitudinal ventilation systems—a study of the critical velocity. *Fire Safety Journal* 35(4): 363–390.
Xue, Z.J. & Shen, J.J. 2012. Discussion of mine compressed-air self-help system construction and management. *Shaanxi Coal* 31(5): 29–30.
Yang, C.X. & Yuan, X.P. 1994. Numerical simulation of fire phenomenon and application. *China Safety Science Journal* 01: 17–22.
Zhang, H.L. 2013. Study on the technology of underground mining compressed-air intelligect system. *Nanjing University.*
Zhang, J.K. & Mu, S.B. 2013. Effects of compressed air self-help system on mine fires in excavation tunnels. *Gold* 11: 30–34.

Innovation and Application of Engineering Technology – Oscar, Chen & Wang (Eds)
© 2017 Taylor & Francis Group, London, ISBN 978-1-138-05057-0

Author index